Root-Galling Disease of Vegetable Plants

Faheem Ahmad • Gloria Nombela
Editors

Root-Galling Disease of Vegetable Plants

Springer

Editors
Faheem Ahmad
Department of Botany
Aligarh Muslim University
Aligarh, Uttar Pradesh, India

Gloria Nombela
Consejo Superior de Investigaciones Científicas
Instituto de Ciencias Agrarias
Madrid, Spain

ISBN 978-981-99-3891-9 ISBN 978-981-99-3892-6 (eBook)
https://doi.org/10.1007/978-981-99-3892-6

This Springer imprint is published by the registered company Springer Nature Singapore Pte Ltd.
The registered company address is: 152 Beach Road, #21-01/04 Gateway East, Singapore 189721, Singapore

Paper in this product is recyclable.

This volume is dedicated to the memory of Late Mrs. Sairun Nisha (سائرن نشا), Grandmother of the Editor, Dr Faheem Ahmad.

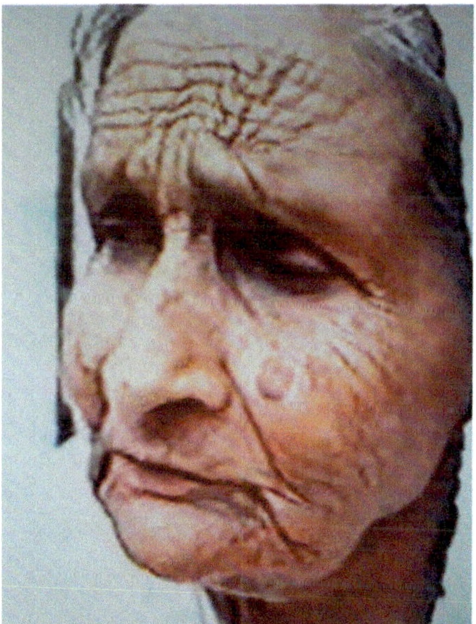

Foreword

Root-knot nematodes (*Meloidogyne* spp.) exhibit varying levels of specialization in terms of their preferred hosts. Plants are usually good hosts than weeds. The discussion of *Meloidogyne* spp. frequently focuses on the four major species: three tropical species, *M. javanica, M. incognita,* and *M. arenaria*, and the temperate species, *M. hapla*. Each has an extensive host range and is globally distributed, further contributing to their recognized importance. *Meloidogyne* spp. forms galls after infection, and the egg masses are often lodged within the galls and on their surfaces. The galls have a unique knot-like appearance and can be quite large, small, or barely noticeable on hosts. Damage and yield losses caused by plant pathogens, including *Meloidogyne* spp., are, on average, greater in tropical than in temperate regions because of great pathogen diversity, favourable environmental conditions for colonization, development, reproduction and dispersal, and lack of technical and financial resources to combat root infection in plants. Therefore, it is crucial to assess the root-galling illness and the recent management strategies that have been described. This book has contributed to many aspects of plant root-galling disease in the form of chapters written by academicians and scientists from different countries like Brazil, Nepal, and Morocco. I am confident that readers working in horticultural sciences, plant pathology, crop protection, gardening, and related fields will find the information offered by the contributors to be of great use. Additionally, people working commercially with vegetable plants to recognize and enhance the diagnosis of *Meloidogyne* spp. from an agricultural perspective will also benefit from this work.

The commitment and the efforts made by the editors Dr. Faheem Ahmad, A/Prof. in the Department of Botany at Aligarh Muslim University, Aligarh, and Dr. Gloria Nombela, senior scientist at the CSIC and Head of the Research Group "Interactions of Plants with Insects and Plant-parasites in Agroecosystems" in the Institute of Agricultural Sciences (ICA-CSIC) at Madrid (Spain), in designing this volume are appreciated and welcomed. This book's information is well-written and notable. I must applaud the editors and authors for compiling this book on the root-galling disease of vegetable plants.

(Rakesh Pandey)

CSIR-Central Institute of Medicinal and Aromatic Plants
Lucknow, India

Emeritus AcSIR & CSIR Emeritus Scientist
Lucknow, India

Indian Phytopathological Society, IARI
New Delhi, India

Rakesh Pandey

Preface

The yield losses of vegetable plants due to nematodes depend on the nematode genus, population level, plant species, and cultivars. The most important *Meloidogyne* species (root-knot nematodes) are the tropical *Meloidogyne incognita, M. arenaria, M. javanica, and the temperate M. hapla*. Typical symptoms include stunted growth, wilting, leaf discolouration, and deformation of the roots. The plant cells surrounding the nematode and its feeding site become hypertrophic and hyperplastic and result in root galls. These extreme modifications of root architecture result in devastating effects of RKNs on the quality and yield of vegetable plants.

Understanding the devastating impact of root-galling diseases on the yield of vegetable plants and from the agricultural point of view, the complete knowledge on better diagnosis and detection of root-gall disease is necessary for developing effective control methods to reduce the yield loss. Unfortunately, detailed and latest information on the root-galling biology of infected vegetable plants caused by RKNs is very scattered. Therefore, the current subject has recently attracted us to gather updated information in a comprehensive book, *Root-Galling Disease of Vegetable Plants*, covering the *Meloidogyne* species topics appropriate to vegetable plants. This book incorporates critical reviews on important root-galling diseases of different vegetable plants and their suitable management strategies. This volume contains 13 chapters, which cover comprehensive information on: (1) Root-knot nematodes (*Meloidogyne* spp.), (2) *Meloidogyne* species: Threat to Vegetable Produce, (3) Chemotaxis in Root-Knot nematode, (4) Phytohormone-Mediated Feeding Site Development, (5) Current and Future Studies on the Genes for Parasitism in *Meloidogyne*, (6) Natural Product Repertoire for Suppressing the Immune Response of *Meloidogyne* Species, (7) Epigenetic Mechanisms and their Role in Root Gall Formation, (8) Mass Spectrometry Imaging (MSI) and Root Gall Elucidation, (9) Root-Knot Disease Complexes: An Interactive Perspective with Microorganisms, (10) Breeding for Resistance in Vegetables against *Meloidogyne* spp. causing Root Gall Disease, (11) An Overview of Predacious Fungi for the Management of Root-knot Disease in Vegetables, (12) Biofertilizer of Organic Origin for Management of Root Galling Disease of Vegetables, (13) Prospects for the Use of Metabolomics Engineering in Exploring and Harnessing Chemical Signalling in Root Galls. The literature on root-galling disease is a global necessity because of the alarming nematode problem on agricrops.

As a professional reference, this comprehensive book will be beneficial for a broad readership, including university professors, researchers, development department officials, extension workers, as well as a wider community of readers (educators, scholars, policymakers, science writers, and students). In addition, those working commercially with vegetable plants to identify and improve the diagnosis of *Meloidogyne* species from an agricultural point of view will also be benefited.

We are indebted to the contributors who made the book possible. Finally, we acknowledge our publisher Springer Nature and in particular Ms. Aakanksha Tyagi (Senior Editor—Books, Life Sciences) for agreeing to publish the book, and Ms. Muthuneela Muthukumar (Project Coordinator—Books) for their assistance in this endeavour.

Aligarh, India Faheem Ahmad
Madrid, Spain Gloria Nombela

Acknowledgements

Our sincere thanks are extended to academicians, scientists, and researchers who happily agreed to contribute chapters for this book on various aspects of root-galling disease commonly caused by the root-knot nematode. This volume emphasizes in-depth explanations of root-galling issues, which will be helpful to readers actively engaged in agricultural subjects, including plant pathology, crop protection, gardening, and horticultural sciences.

Dr. Ahmad also acknowledges to the University Grants Commission (UGC) for financially supporting a start-up grant to establish the lab, which was quite useful during the compilation of this volume. I further express my gratitude to Aakanksha Tyagi (Senior Editor, Books, Life Sciences, Springer Nature), especially for agreeing to send the book proposal to reviewers for their input and acceptance. Moreover, I extend my deepest gratitude to my wife, Sana Ahmad, who provided complete collaboration throughout the preparation of this book in several unnoticed ways.

Contents

Editors and Contributors

About the Editors

Faheem Ahmad graduated from the Aligarh Muslim University, Aligarh. Earlier, he worked as a postdoctoral fellow at the Ehime University, Japan; North-West University, South Africa, and National Sun Yat-Sen University, Taiwan. He is the Senior Assistant Professor and Principal Investigator of UGC-BSR Start-up Grant at the Department of Botany, Aligarh Muslim University. His research interests include plant-microbe interaction, biological control of plant-parasitic nematodes, and plant disease studies of economic crops. He has more than 50 peer-reviewed research publications in prestigious national and international journals and has delivered several scientific talks as oral and poster presentations in many scientific meetings. He is also an active life member of several professional societies, including the Nematological Society of India, the Indian Phytopathological Society, the Indian Science Congress Association, and Bose Science Society.

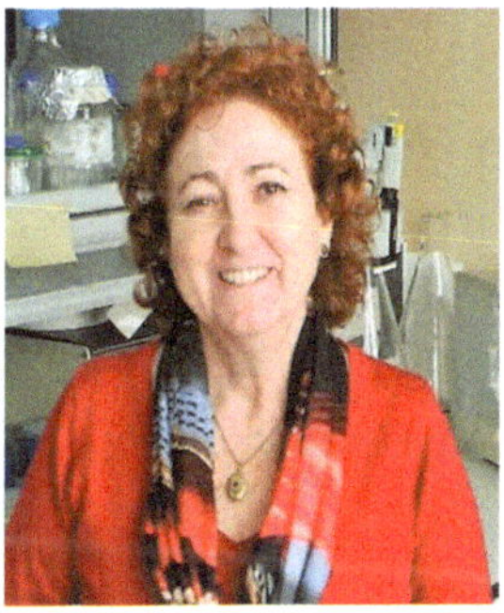

Gloria Nombela is a Senior Scientist at the CSIC and Head of the Research Group "Interactions of Plants with Insects and Plant-parasites in Agroecosystems" in the Institute of Agricultural Sciences (ICA-CSIC) at Madrid, Spain. She has also collaborated as an Expert Advisor of CSIC for the Delegated Intervention of the Spanish Ministry of Finance in the Ministry of Agriculture, for the reception of scientific assignments on Plant Protection issues. Her experience is contrasted after a long career in the fields of Entomology and Nematology, leading research projects and publishing a number of scientific contributions. Among her research lines are the study of plant resistance and susceptibility to plant nematodes and insect pests, analysis of gene expression in compatible and incompatible host/parasite interactions, and the biotic and abiotic factors that affect the impact of pests and plant diseases. Dr. Nombela develops a wide range of activities through the tutoring of PhD theses, national and foreign students hosted in her laboratory for End-of-degree Projects, Master's Thesis, Erasmus grants, and undergraduate internships. Also outstanding are her regular contributions to research management as an evaluator in tribunals for scientific and technical staff positions, research projects, and scientific articles for prestigious national and international journals. She was recently a co-editor of the topic "Integrated Pest Management of Crops", which includes five open-access journals of the MDPI editorial group.

Contributors

Faheem Ahmad Department of Botany, Aligarh Muslim University, Aligarh, India

Said Amiri Phytopathology Unit, Department of Plant Protection, Ecole Nationale d'Agriculture de Meknès, Menkes, Morocco

Shahid Anwar Ansari Department of Plant Protection, Faculty of Agricultural Sciences, Aligarh Muslim University, Aligarh, India

T. S. Archana Department of Plant Pathology, School of Agriculture, Lovely Professional University (LPU), Phagwara, Punjab, India

Zineb Belabess Plant Protection Laboratory, Regional Center of Agricultural Research of Oujda, National Institute of Agricultural Research, Oujda, Morocco

Daniele De Brum Department of Phytopathology, Federal University of Lavras, University Campus, Lavras, Minas Gerais, Brazil

Vicente Paulo Campos Department of Phytopathology, Federal University of Lavras, University Campus, Lavras, Minas Gerais, Brazil

Monalisa Chakraborty Department of Biotechnology, Mount Carmel College (Autonomous), Karnataka, Bengaluru, India

Anwesha Chatterjee Department of Microbiology, St. Xavier's College (Autonomous), Kolkata, West Bengal, India

Saba Fatima Department of Botany, Aligarh Muslim University, Aligarh, India

Pritha Ghosh Department of Entomology, School of Agriculture, Lovely Professional University (LPU), Phagwara, Punjab, India

Khadija Goura Phytopathology Unit, Department of Plant Protection, Ecole Nationale d'Agriculture de Meknès, Menkes, Morocco

Rishil Gupta Department of Botany, Aligarh Muslim University, Aligarh, India

Hajar El Hamss Phytopathology Unit, Department of Plant Protection, Ecole Nationale d'Agriculture de Meknès, Menkes, Morocco

Mohammad Haris Plant Pathology and Nematology Section, Department of Botany, Aligarh Muslim University, Aligarh, India

Touseef Hussain Division of Plant Pathology, ICAR-Indian Agricultural Research Institute, New Delhi, India

Touseef Hussain Plant Pathology and Nematology Section, Department of Botany, Aligarh Muslim University, Aligarh, India

Jihane Kenfaoui Phytopathology Unit, Department of Plant Protection, Ecole Nationale d'Agriculture de Meknès, Menkes, Morocco

Najwa Seddiqi Khalil Phytopathology Unit, Department of Plant Protection, Ecole Nationale d'Agriculture de Meknès, Menkes, Morocco

Abrar Ahmad Khan Plant Pathology and Nematology Section, Department of Botany, Aligarh Muslim University, Aligarh, India

Amir Khan Plant Pathology and Nematology Section, Department of Botany, Aligarh Muslim University, Aligarh, India

Arshad Khan Department of Botany, Aligarh Muslim University, Aligarh, India

Faryad Khan Department of Botany, Aligarh Muslim University, Aligarh, India

Faryad Khan Krishna Devi Balika P.G. College, Farrukhabad, India

Matiyar Rahaman Khan Division of Nematology, ICAR-Indian Agricultural Research Institute, New Delhi, India

Adesh Kumar Department of Plant Pathology, School of Agriculture, Lovely Professional University (LPU), Phagwara, Punjab, India

Devendra Kumar Department of Plant Pathology, School of Agriculture, Lovely Professional University (LPU), Phagwara, Punjab, India

Vipul Kumar Department of Plant Pathology, School of Agriculture, Lovely Professional University (LPU), Phagwara, Punjab, India

Rachid Lahlali Phytopathology Unit, Department of Plant Protection, Ecole Nationale d'Agriculture de Meknès, Menkes, Morocco

Ikram Legrifi Phytopathology Unit, Department of Plant Protection, Ecole Nationale d'Agriculture de Meknès, Menkes, Morocco

Oishi Mitra Department of Biotechnology, School of Biosciences & Technology, Vellore Institute of Technology (VIT), Vellore, Tamil Nadu, India

Fouad Mokrini Biotechnology Unit, Regional Center of Agricultural Research, INRA-Rabat Morocco, Rabat, Morocco

A. Mounika Department of Plant Pathology, Indira Gandhi Krishi Viswavidyalaya, Raipur, India

Hera Nadeem Department of Botany, Aligarh Muslim University, Aligarh, India

Sagnik Nag Department of Biotechnology, School of Biosciences and Technology, Vellore Institute of Technology (VIT), Vellore, Tamil Nadu, India

Denilson Ferreira De Oliveira Department of Chemistry, Federal University of Lavras, University Campus, Lavras, Minas Gerais, Brazil

Ekta Pandey Department of Botany, Aligarh Muslim University, Aligarh, India

Lalith Pandey Department of Soil Science and Agricultural Chemistry, Banaras Hindu University, Varanasi, India

Prachi Pandey Drs Kiran & Pallavi Patel Global University, Vadodara, Gujarat, India

Ashwani Kumar Patel Department of Mycology and Plant Pathology, Institute of Agricultural Science, Banaras Hindu University, Varanasi, Uttar Pradesh, India

Shiv Shankar Patel Department of Mycology and Plant Pathology, Institute of Agricultural Science, Banaras Hindu University, Varanasi, Uttar Pradesh, India

Letícia Lopes De Paula Department of Phytopathology, Federal University of Lavras, University Campus, Lavras, Minas Gerais, Brazil

P. V. Phanindra Department of Mycology and Plant Pathology, Institute of Agricultural Sciences, Banaras Hindu University, Varanasi, India

Mandla Rajashekar Krishi Vigyan Kendra, Professor Jayashankar Telangana State Agricultural University, Hyderabad, India

Vandana Sahu Department of Mycology and Plant Pathology, Institute of Agricultural Science, Banaras Hindu University, Varanasi, Uttar Pradesh, India

Souvik Samanta Department of Biotechnology, School of Biosciences & Technology, Vellore Institute of Technology (VIT), Vellore, Tamil Nadu, India

Mohammad Shariq Department of Botany, Aligarh Muslim University, Aligarh, India

Sushmita Sharma Department of Entomology, Agriculture and Forestry University, Rampur, Chitwan, Nepal

Bitaisha Nakishuka Shukuru Department of Plant Pathology, School of Agriculture, Lovely Professional University (LPU), Phagwara, Punjab, India

Mansoor Ahmad Siddiqui Department of Botany, Aligarh Muslim University, Aligarh, India

Jorge Teodoro De Souza Department of Phytopathology, Federal University of Lavras, University Campus, Lavras, Minas Gerais, Brazil

William César Terra Department of Phytopathology, Federal University of Lavras, University Campus, Lavras, Minas Gerais, Brazil

Uday Kumar Thera Department of Plant Sciences and Landscape Architecture, University of Maryland, College Park, MD, USA

Ashmita Timsina Department of Mycology and Plant Pathology, Institute of Agricultural Sciences, Banaras Hindu University, Varanasi, India

Sparsh Tiwari Department of Mycology and Plant Pathology, Institute of Agricultural Sciences, Banaras Hindu University, Varanasi, India

Sachin Upadhayaya Department of Microbiology and Parasitology, Agriculture and Forestry University, Rampur, Chitwan, Nepal

Raman Kumar Walia Division of Nematology, ICAR-Indian Agricultural Research Institute, New Delhi, India

Saeeda Zaima Zeb Department of Botany, Aligarh Muslim University, Aligarh, India

Root-knot Nematodes (*Meloidogyne* spp.)

1

Raman Kumar Walia and Matiyar Rahaman Khan

Abstract

Root-knot nematodes (*Meloidogyne* spp.) are the most widespread, have a vast host range, vascular feeder endoparasites, and, therefore, are considered the most damaging among the plant-parasitic nematodes globally. This chapter describes the systematics of the major species of *Meloidogyne* based on morphological, morphometrical, enzyme phenotypes, and molecular parameters. The existence of host races and cytological races in general and the occurrence of economically important *Meloidogyne* species in India are tabulated along with estimations on recent crop losses. A brief account of the general biology, life cycle, and host-parasite relationship of *Meloidogyne* sp. is included. The damage symptoms of *Meloidogyne* spp. on different vegetable crops is depicted through images. The management of nematode vegetable cropping systems has been dealt in detail. This includes cultural/agronomic practices, biological control through fungal and bacterial bioagents, host plant resistance, newer chemical nematicides, and their integration. A dedicated section is included on managing root-knot nematodes in protected cultivation systems. Root-knot nematode dissemination through horticultural nurseries has been highlighted, along with practical methods to check it. Lastly, some emerging problems of root-knot nematodes have been reported.

Keywords

Vegetable crops · Root-knot nematodes · Economic losses · Disease complexes · Management · Dissemination · Protected cultivation

R. K. Walia (✉) · M. R. Khan
Division of Nematology, ICAR-Indian Agricultural Research Institute, New Delhi, India

F. Ahmad, G. Nombela (eds.), *Root-Galling Disease of Vegetable Plants*,
https://doi.org/10.1007/978-981-99-3892-6_1

1.1 Introduction

Among all the plant-parasitic nematodes, root-knot nematodes, *Meloidogyne* spp. were and remained the number one problem globally, including in India. Many reviews/monographs have been published, including a comprehensive treatise on root-knot nematodes. Notable among these are Taylor and Sasser (1978), Lamberti and Taylor (1979), Sasser and Kirby (1979), Barker et al. (1985), Sasser and Carter (1985), and Karssen (2002). Some of these reviews emanated from International *Meloidogyne* Project during the 1970s and 1980s at North Carolina State University, Raleigh, USA, under Dr. J. N. Sasser's stewardship, with collaborating centers worldwide, including India. The same project continued under the title "Crop Nematode Research and Control Project" during the 1990s.

In India, the All India Coordinated Research Project (AICRP) on Nematodes, operative since 1979, has been instrumental in generating significant information on the occurrence, losses, and management of major nematode pests of crops, including root-knot nematodes throughout the country. A comprehensive chapter on root-knot nematodes in India was contributed by Dasgupta and Gaur (1986). Khan et al. (2014) published a monograph on root-knot nematodes that included basic and applied aspects relevant to Indian conditions. Some important nematological events/problems have recently drawn our attention (e.g., nematode problems in protected cultivation systems and nematode dissemination through horticultural nurseries) for possible solutions. Gowda et al. (2019) have given an overview of root-knot nematode problems and their management in vegetable crops in India.

Vegetable crops are most vulnerable to the nematode. They are widely distributed (more than 146 countries) and responsible for global crop losses. Root-knot nematodes are sedentary endoparasites of many crops; more than 3000 host plants are affected by the nematode species. The root-knot nematode species that infect vegetable crops most often in India are *Meloidogyne incognita*, *M. javanica*, *M. enterolobii*, *M. arenaria*, and *M. hapla*. *M. hapla* is limited to high altitudes and temperate areas, while the first four species are distributed in tropical and subtropical areas.

1.1.1 Historical

The first-ever record on root-knot nematodes dates back to 1855 when Berkeley reported damage to glasshouse-grown cucumbers in England due to "vibrios." Greef (1872) and Cornu (1879) independently designated root gall-forming nematodes as *Anguillula radicicola* and *A. marioni*, respectively. During 1879–1948, the root-knot nematodes were placed along with cyst nematodes in the genus *Heterodera*. Goodey (1932) named it *Heterodera marioni*, but it was Chitwood (1949) who separated all root-knot nematodes from *Heterodera* and placed them in the genus *Meloidogyne* (Greek: *melon* = apple or gourd; *eidos* = resembling; *gyne* = female), a name originally coined by Göldi (1892) for coffee root-knot nematode, *Meloidogyne exigua*. Chitwood (1949) provided the diagnostic characters for the genus

Meloidogyne and recognized four existing species, namely, *Meloidogyne exigua* (Göldi 1892), *M. javanica* (Treub 1885), *M. incognita* (Kofoid & White 1919), and *M. arenaria* (Neal 1889) besides adding a new species *M. hapla* and a new variety *M. incognita acrita*.

In India, Barber (1901) recorded this nematode for the first time on tea (as *Heterodera radicicola*) from the Devala estate of Kerala. Subsequently, many other reports of its occurrence poured in from vegetables, other crops, and areas (Ayyar 1926, 1933, 1934).

1.2 Systematics

1.2.1 Systematic Position (as per Siddiqi 2000)

- Phylum: Nematoda
- Class: Secernentea
- Order: Tylenchida
- Suborder: Tylenchina
- Superfamily: Hoplolaimoidea
- Family: Meloidogynidae
- Subfamily: Meloidogyninae
- Genus: *Meloidogyne*

1.2.2 Diagnostic Characters of Genus *Meloidogyne* (Hunt and Handoo 2009, Modified After Siddiqi 2000)

- *Mature Female*: Round to pear-shaped with short projecting neck, white, sedentary. No cyst stage. Vulva and anus located close together, terminal; perineum with a fingerprint-like cuticular pattern, usually flattened, rarely elevated. Phasmids dot-like, slightly anterior to, and on either side of the anus. Cuticle striated. Stylet slender, generally 12–15 μm long, with small basal knobs. Excretory pore anterior to median bulb, often just posterior to the base of the stylet. Genital tracts paired, prodelphic, convoluted. Six large rectal glands secrete gelatinous material in which eggs are deposited; eggs not retained in body.
- *Male*: Vermiform, up to 2 mm long, tail end twisted, developing by metamorphosis within a swollen juvenile. Cuticle strongly annulated; lateral field with four incisures. Labial region not sharply offset, with distinct labial disc and few (1–3) annules; lateral sectors wider than submedian sectors, appearing as "cheeks." Stylet robust (18–25 μm), with large basal knobs. Pharyngeal glands lie mostly ventral to the intestine. Spicules slender, generally 25–33 μm long, gubernaculum 7–11 μm long. Testis single, but paired when sex reversal occurs. Tail rounded. Phasmids dot-like, located near cloacal aperture, which is subterminal. Bursa absent.

- *Juveniles*: First stage with a blunt tail tip, moulting within egg; second and third moults occurring within cuticle of the second stage. Second stage vermiform, migratory, infective, straight to arcuate upon death. Labial region with coarse annules (1–4), a distinct labial disc, framework lightly sclerotized, lateral sectors wider than submedian sectors, stylet slender, under 20 μm, excretory pore posterior to hemizonid. Median bulb with large oval refractive thickenings. Tail with conspicuous hyaline region, tip narrow, irregular in outline. Third stage sedentary, swollen, sausage-shaped with a short blunt tail. Stylet absent. Fourth stage sedentary, swollen, with terminal anus. Stylet absent.

Type species: *Meloidogyne exigua* (Göldi 1887)

1.3 Major Identification Tools for *Meloidogyne* Species

The morphological and morphometrical characteristics of females (body shape, perineal pattern, the position of the excretory pore, head region, stylet, and stylet knobs, dorsal oesophageal gland orifice, etc.), males (length, excretory pore, lateral field, head region, stylet, dorsal oesophageal gland orifice, spicules, and gubernaculum), and second-stage juveniles (body length, lateral field, head region, stylet, dorsal oesophageal gland orifice, tail, hyaline tail terminus, etc.) are used for identification of *Meloidogyne* species. In addition, biochemical parameters, especially the esterase isozyme profiles, are very useful in distinguishing species of *Meloidogyne*. Molecular approaches are nowadays routinely used for the characterization of species. Differences in host tests are also helpful in identifying species/ races of common *Meloidogyne* species.

1.3.1 Major Morphological Characteristics

Adult females of root-knot nematode are swollen, saccate bodies (Fig. 1.1a–f) that measure about 0.44–1.30 mm in median length and 0.33–0.70 mm in median width.

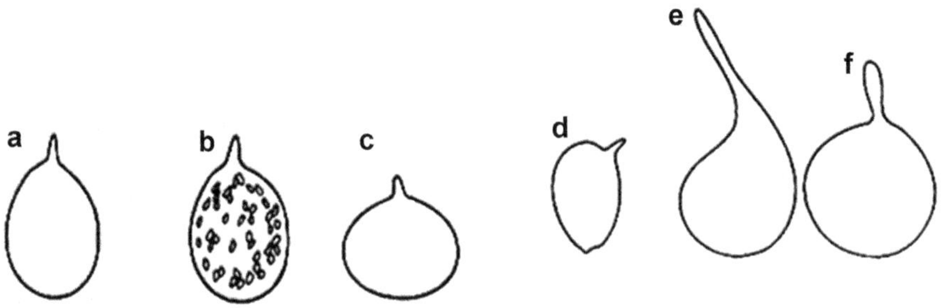

Fig. 1.1 Body shape of the genus *Meloidogyne* species (Source: Nickle 1991)

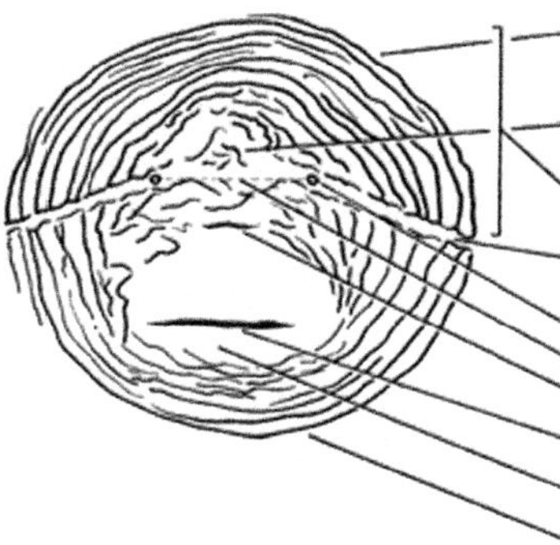

Fig. 1.2 Basic structure of perineal pattern in *Meloidogyne* species (Source: MACTODE)

They have short- to medium-size neck protruding anteriorly; the vulva and anus are located terminally, the posterior end mostly smooth and round, often with a slightly raised protuberance (Fig. 1.1d).

The perineal pattern remains the most important tool for the preliminary identification of root-knot nematode species. The basic structure of the perineal pattern is shown in Fig. 1.2. The variations in the perineal patterns of economically important species in India are depicted in Fig. 1.3 and Table 1.1.

The important and contrasting characteristics of female stylets of the most common species of *Meloidogyne* are included in Table 1.2 and Fig. 1.4.

The distinctive chacharacteristics head region and stylets of males are provided in Table 1.3 and Figs. 1.5 and 1.6.

Important characters on head shape and stylets in respect of second-stage juveniles of the common species are presented in Table 1.4 and Figs. 1.7 and 1.8.

1.3.2 Isozyme Phenotypes for Identification of *Meloidogyne* Species

As the number of *Meloidogyne* species increases, species identification purely based on morphological observations becomes increasingly challenging. To overcome the constraints of morphological characterization and differential host testing, taxonomists for this genus rely on novel taxonomic approaches. Dickson et al. (1970) found the stability of the protein profile of the root-knot nematode and demonstrated its use in species identification. Gel electrophoresis revealed that protein and enzyme composition patterns were beneficial for distinguishing *Meloidogyne* species (Hussey 1985; Esbenshade and Triantaphyllou 1985). The four enzyme patterns (non-specific esterase, malate dehydrogenase, superoxide dismutase, and glutamate-oxaloacetate transaminase) have been used widely to differentiate the root-knot and other nematode species. However, beta esterase is the most useful for identifying the common *Meloidogyne* species (Esbenshade and Triantaphyllou 1985; Cofcewicz et al. 2004). Esbenshade and Triantaphyllou (1985) used isozyme phenotypes to differentiate *Meloidogyne* spp. and reported esterase patterns from 16 *Meloidogyne* species, with the most prevalent phenotypes are A2

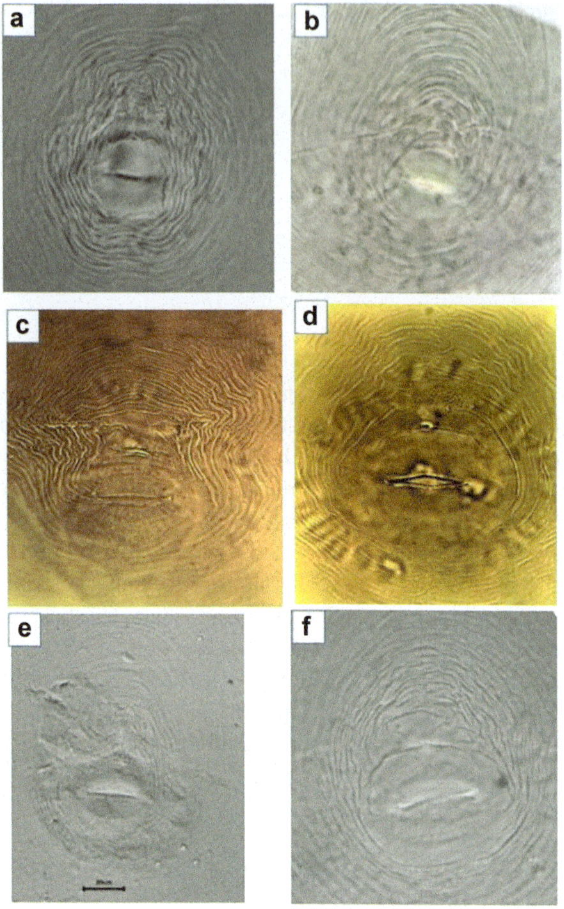

Fig. 1.3 Perineal patterns of most common species of *Meloidogyne* in India: (**a**) *Meloidogyne incognita,* (**b**) *M. javanica*, (**c**) *M. arenaria*, (**d**) *M. hapla*, (**e**) *M. enterolobii*, (**f**) *M. graminicola* (Source: MACTODE C–D)

and A3 in *M. arenaria*, H1 in *M. hapla*, I1 in *M. incognita*, and J3 in *M. javanica* (Fig. 1.9). Later, Esbenshade and Triantaphyllou (1990) utilized isozymes for around 300 populations of *Meloidogyne* representing 65 different countries and continents. The isozyme patterns from various surveys and works of the International *Meloidogyne* Project have been compiled for *Meloidogyne* species (Berge and Dalmasso 1975; Dalmasso and Berge 1978; Fargette 1987; Janati et al. 1982; Esbenshade and Triantaphyllou 1985, 1990; Carneiro et al. 2000; Hernandez et al. 2004). Enzyme phenotypes are identified primarily by the number of bands found; phenotypes with the same number of bands are distinguished by small letters (Esbenshade and Triantaphyllou 1985, 1990). The enzyme patterns are often compared to *M. javanica* as standard; this species is included in the electrophoresis to measure migration distances. Miniaturization and automation of electrophoresis equipment, along with precasting polyacrylamide gels, have made isozyme phenotyping more affordable and attractive (Esbenshade and Triantaphyllou 1985; Karssen et al. 1995; Chen et al. 1998; Molinari 2001). Malate dehydrogenase (Mdh)

Table 1.1 Important diagnostic characters of perineal patterns of the most common *Meloidogyne* species

Species	Dorsal arch	Lateral field	Striae	Tail terminus
M. incognita	High, squarish	Lateral ridges absent, marked by breaks and forks in striae	Coarse, smooth to wavy, sometimes zigzaggy	Often with distinct whorl
M. javanica	Low, rounded	Distinct lateral ridges	Coarse, smooth to slightly wavy	Often with distinct whorl
M. arenaria	Low, rounded, indented near lateral fields	Lateral ridges absent, marked by short, irregular, and forked striae	Coarse, smooth to slightly wavy	Usually without distinct whorl
M. hapla	Low, rounded	Lateral ridges absent	Fine, smooth to slightly wavy	Whorl absent, marked by subcuticular punctations
M. enterolobii	Moderately high to very high and square-rounded	Lateral ridges not distinct	Coarse and smooth	Whorl absent
M. graminicola	Dorsal arch moderately high and rounded, dorsoventrally ovoid	Lateral ridges absent	Widely spaced and broken striae in the dorsal part, an anal fold and often a fold in perivulval area	Clear but two dorsolateral striae forming a V shape leading from anus to adjacent phasmid

Source: Eisenback (1985), Karssen et al. (2012), Yang and Eisenback (1983)

has been used to separate *M. hapla* from *M. incognita*, *M. arenaria*, and *M. javanica*. In contrast, glutamate dehydrogenase has been utilized to separate *M. incognita* from *M. javanica*, *M. arenaria*, and *M. hapla* (Esbenshade and Triantaphyllou 1985). To confirm the identity of existing *Meloidogyne* species and help identify and describe new species, biochemical and molecular methods have been considered useful diagnostic tools in recent years (Blok and Powers 2009).

1.3.3 Molecular Characterization of *Meloidogyne* Species

The development of molecular techniques, mainly polymerase chain reaction (PCR), species-specific molecular markers, and DNA sequencing (Harris et al. 1990;

Table 1.2 Important diagnostic characters of stylets of females of the most common *Meloidogyne* species

Species	Stylet cone	Stylet shaft	Stylet knobs	Stylet length
M. incognita	Anterior half cylindrical, dorsally curved	Slightly wider posteriorly	Set off, rounded to elongate transversely, sometimes indented anteriorly	16 μm mean 15–17 μm range
M. javanica	Slightly curved dorsally	Cylindrical	Set off, transversely elongate	16 μm mean 14–18 μm range
M. arenaria	Straight, broad, and robust	Wider posteriorly	Not set off, backward sloping, merging with shaft	15.5 μm mean 13–17 μm range
M. hapla	Slightly curved dorsally, narrow, and delicate	Slightly wider posteriorly	Set off, small, and round	15.5 μm mean 13–17 μm range
M. enterolobii	Slight dorsal curvature	Broadens posteriorly	Set off, each knob transversely ovoid with a deep, median longitudinal indentation	15.1 μm mean 13.2–18 μm range
M. graminicola	Slight dorsal curvature, tapers gradually to the apex	Broadens in the posterior half and narrows a little just anterior to the junction with knobs	Smooth, pear-shaped and backwardly slopping	13.5 μm mean 12–15 μm range

Source: Eisenback (1985), Karssen et al. (2012), Yang and Eisenback (1983)

Powers and Fleming 1998), has expedited and simplified the nematode identification process. All PCR-based techniques are relatively quick, highly dependable, and irrespective of nematode life stage (Zijlstra 2000). DNA-based techniques commonly employ mitochondrial DNA (Harris et al. 1990; Powers and Harris 1993), satellite DNA (Castagnone-Sereno et al. 1995), ribosomal DNA (Zijlstra et al. 1995; Peterson and Vrain 1996; Petersen et al. 1997; Zijlstra 1997), and randomly amplified polymorphic DNA fragments (RAPDs). The conserved regions (ETS, ITS I&II, IGS I&II, and the D2–D3 expansion segment of 28S ribosomal rDNA) have been employed to diagnose many plant-parasitic nematodes (Fig. 1.10). Available DNA-based approaches for detecting genetic differences are being utilized or developed for diagnostic and taxonomic purposes (Curran 1991; Curran and Robinson 1993, De Giorgi et al. 1994; Hyman and Whipple 1996; Powers and Fleming 1998). DNA sequence analysis has been widely utilized in nematode biosystematics (Powers et al. 2005) and for identifying nematodes (Williamson

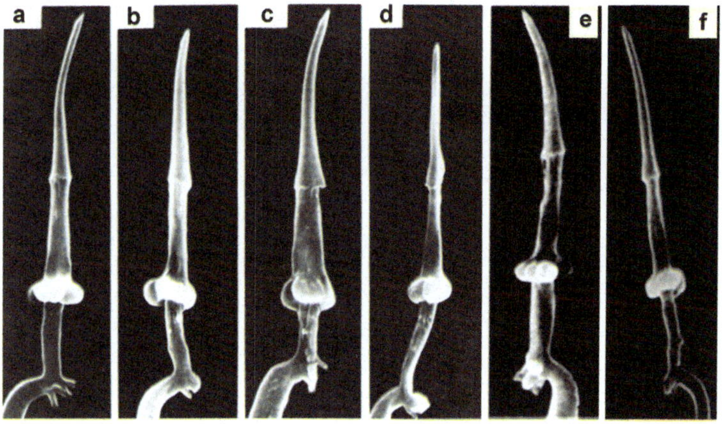

Fig. 1.4 Scanning electron micrographs of excised stylets of root-knot nematode females: (**a**) *Meloidogyne incognita*, (**b**) *M. javanica*, (**c**) *M. arenaria*, (**d**) *M. hapla* (Source: Eisenback et al. 1981), (**e**) *M. enterolobii* (Yang and Eisenback 1983), (**f**) *M. graminicola* (Nickle 1991)

et al. 1997; Randig et al. 2002a, b; Zijlstra et al. 2004). Adam et al. (2007) suggested a systematic diagnostic key for identifying seven of the most prevalent and economically important *Meloidogyne* spp. and provided a logical procedure for the molecular identification of individual nematodes. Various DNA marker-based methods (RAPD, RFLP, SCAR, Multiplex PCR, and AFLP) and PCR are now used to identify *Meloidogyne* species. Pokharel et al. (2007) performed a phylogenetic analysis based on the rRNA genes' partial internal transcribed spacer (ITS) sequences. They found that all Nepalese isolates formed a separate clade within the known isolates of *M. graminicola*. Based on species-specific RAPD fragments, several species-specific primers for identifying *Meloidogyne* species have been designed (Williamson et al. 1997; Zijlstra 2000; Dong et al. 2001a). Meng et al. (2004) developed Sequence Characterized Amplified Regions (SCAR) primers for the identification of *M. incognita*, *M. javanica*, and *M. arenaria*. Tesarova et al. (2003) designed PCR primers based on the gene sequence, which were used to detect and distinguish *M. incognita* from other *Meloidogyne* species. Dong et al. (2001b) isolated DNA from 26 different single eggmass nematodes, including seven *M. arenaria*, three *M. hapla*, eleven *M. incognita*, and five *M. javanica*, identified species-specific sequence tagged sites, and found variations among isolates of each species, particularly within *M. arenaria* and *M. hapla*. SCAR and species-specific markers (sat DNA, ITS, D2–D3, IGS, and mtDNA) have been used successfully to identify *M. arabicida*, M. *arenaria*, *M. chitwoodi*, *M. enterolobii*, *M. ethiopica*, *M. exigua*, *M. fallax*, *M. graminis*, *M. hapla*, *M. incognita*, *M. izalcoensis*, etc.

Tanaka et al. (2012) developed a simple and rapid DNA preparation method for nematodes. ITS region sequences were utilized to confirm the identification of *M. mayaguensis* (= *M. enterolobii*) populations (Blok et al. 2002; Brito et al. 2004). Jeyaprakash et al. (2006) investigated the mitochondrial AT-rich area of

Table 1.3 Important diagnostic characters of male head shape and stylet morphology of the most common *Meloidogyne* species

Species	Head cap	Head region	Stylet cone	Stylet shaft	Stylet knobs	Stylet length	DEGO[a]
M. incognita	Flat to concave, labial disc raised above the medial lips	Not set off, usually marked by 2–3 incomplete annulations	Tip blunt, blade-like	Usually cylindrical, often narrows near knobs	Set off, rounded to elongate transversely, sometimes indented anteriorly	24 μm mean, 23–25 μm range	Short, 3 μm mean, 2–4 μm range
M. javanica	High rounded, set off from head region	Not set off, smooth, or marked by 2–3 incomplete annulations	Tip pointed, cone straight	Usually, cylindrical	Set off, low and very wide	20 μm mean, 18–22 μm range	Short, 3 μm mean, 2–4 μm range
M. arenaria	Low sloping posteriorly	Not set off, smooth or marked by 2–3 incomplete annulations	Tip pointed, cone broad and robust	Usually cylindrical, often broadens near knobs	Not set off, backward sloping, merging with shaft	22 μm mean, 20–25 μm range	Long, 5.5 μm mean, 4–7 μm range
M. hapla	High and narrow	Set off, smooth, larger diameter than first body annule	Tip pointed, cone narrow and delicate	Cylindrical, often wider, or narrow at its base	Set off, small and round	20 μm mean, 18–22 μm range	Moderately long, 5 μm mean, 4–6 μm range
M. enterolobii	High and rounded	Slightly set off, not annulated	Tip gradually pointed, straight	Cylindrical, unusually uneven, narrow distinctly at base	Knobs large, ovoid to rounded, slightly slopping backward	23.4 μm mean, 21–25 μm range	4.7 μm mean, 3.7–5.3 μm range
M. graminicola	High, rounded, set off from head annule	Not set off, smooth	Tip pointed, often narrows near base	Cylindrical to distinctly angular	Set off, pear shaped knobs, backward slopping	17.4 μm mean, 15.3–19.4 μm range	3.3 μm mean, 2.3–4.5 μm range

Source: Eisenback (1985), Karssen et al. (2012), Yang and Eisenback (1983)
[a] DEGO = Dorsal oesophageal gland orifice

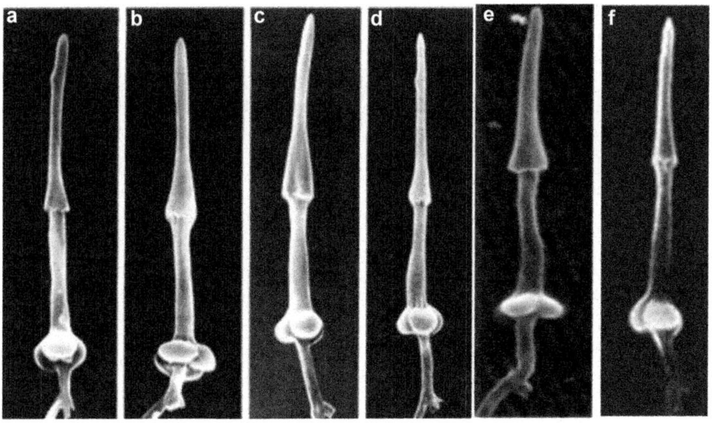

Fig. 1.5 Scanning electron micrographs of the stylets of males of the most common species of *Meloidogyne*: (**a**) *M. incognita*, (**b**) *M. javanica*, (**c**) *M. arenaria*, (**d**) *M. hapla* (Source: Eisenback et al. 1981), (**e**) *M. enterolobii* (Yang and Eisenback 1983), (**f**) *M. graminicola* (Nickle 1991)

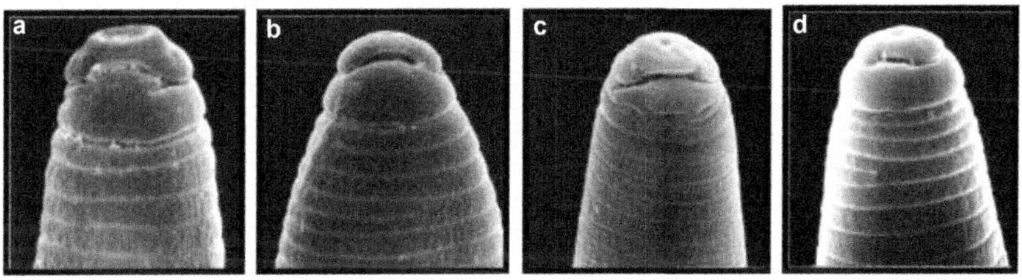

Fig. 1.6 Scanning electron micrographs of the anterior portions of males of the most common species of *Meloidogyne*: (**a**) *M. incognita*, (**b**) *M. javanica*, (**c**) *M. arenaria*, (**d**) *M. hapla* (Source: Eisenback et al. 1981)

M. floridensis, *M. arenaria*, *M. mayaguensis*, *M. incognita*, and *M. javanica*. They developed a molecular approach for differentiating *M. floridensis* from other species.

Subbotin and Burbridge (2021) standardized recombinase polymerase amplification (RPA) tests targeting the IGS rRNA gene of *M. hapla* and viewed this technique as a tool for rapid diagnosis and a sensitive tool for *M. hapla*. There needs to be more publications on the molecular characterization of root-knot nematode populations in India. Gaur et al. (1996) distinguished two groups of *Meloidogyne* species by PCR-RFLP of rDNA and discovered variation in ITS regions. Using RAPD analysis, Swain et al. (1999) recognized four subspecies of *M. incognita*. Umarao et al. (2003) identified races of *M. incognita* based on ITS rDNA sequences; their phylogenetic relationship revealed that race 3 was distantly related to other races (1, 2, and 4). Meher et al. (2003) recorded genetic polymorphism in *M. incognita* populations from the northern (aubergine-Delhi), southern (tomato-Bangalore, Karnataka), eastern (okra-Bhubaneswar, Odisha), and western (cowpea-Anand, Gujarat) regions of India using RAPD. Hinge et al. (2010) distinguished four *Meloidogyne* species using

Table 1.4 Important characters of second-stage juvenile head shape and stylet morphology of the common *Meloidogyne* species

Species	Head cap	Head region	Stylet width	Stylet knobs	DEGO[a]
M. incognita	Anteriorly flattened, elongate	Usually marked by 1–3 incomplete annulations	Moderately sized cone and shaft	Set off, posteriorly rounded sloping backward	Short, 3 μm mean, 2–3 μm range
M. javanica	Anteriorly flattened, elongate	Smooth or marked by 1–3 incomplete annulations	Moderately sized cone and shaft	Set off, posteriorly, rounded, sloping backward, transversely elongate	Moderately long, 3.5 μm mean, 3–4 μm range
M. arenaria	Anteriorly flattened, elongated	Smooth or marked by 1–3 incomplete annulations	Broad cone and shaft	Not set off, posteriorly rounded, merging with shaft	Moderately long, 3.5 μm mean, 3–4 μm range
M. hapla	Rounded and narrow	Rounded, usually smooth	Narrow cone and shaft	Set off small and rounded	long, 4.5 μm mean, 4–5 μm range
M. enterolobii	Anterior end truncate, rounded	Slightly set off, not annulated	Straight narrow cone, slightly pointed	Large rounded, separate from each other, set off from shaft	3.42 μm mean, 2.8–4.3 μm range
M. graminicola	Anteriorly flattened, elongate	Usually smooth, rounded	Narrow cone and shaft	Set off, small and rounded	2.8–3.4 μm range

Source: Eisenback (1985), Karssen et al. (2012), Yang and Eisenback (1983)
[a] DEGO = Dorsal oesophageal gland orifice

RAPD-DNA. Furthermore, phylogenetic analysis revealed variation among *M. indica*, *M. incognita*, and *M. javanica* populations in Gujarat and clustered these groups. Genomic DNA extracted from *Meloidogyne* spp. from various Indian states, including Manipur (mulberry), Kerala (rice), Delhi (rice), Andaman (okra), Assam (rice), West Bengal (Hooghly, brinjal), Odisha (Basella), Gujarat (tomato), West Bengal (Bankura, brinjal), Kerala (rice), and West Bengal (Hooghly, brinjal) (Pongalam, okra) and the amplified PCR products 800bp contains 18S rDNA partial sequence, the internal transcribed spacer 1, 5.8S rDNA gene, the internal transcribed spacer 2 complete sequence, and 28S rDNA gene partial sequence (Fig. 1.11). These amplified PCR products were successfully sequenced and compared with the sequence database using BLAST. Five sequences of *M. incognita*, two of *M. javanica*, and three of *M. graminicola* were analyzed.

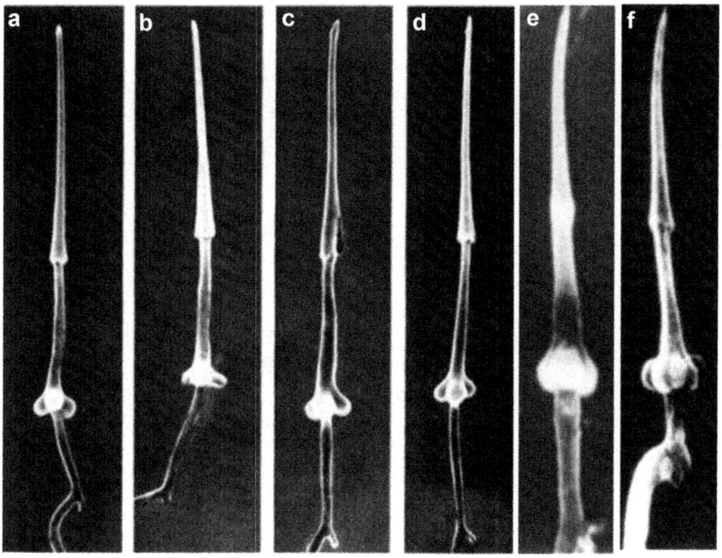

Fig. 1.7 Scanning electron micrographs of the excised stylets of second-stage juveniles of the most common species of *Meloidogyne*: (**a**) *M. incognita*, (**b**) *M. javanica*, (**c**) *M. arenaria*, (**d**) *M. hapla* (Source: Eisenback 1982), (**e**) *M. graminicola*, (**f**) *M. enterolobii* (Source: Jepson 1987)

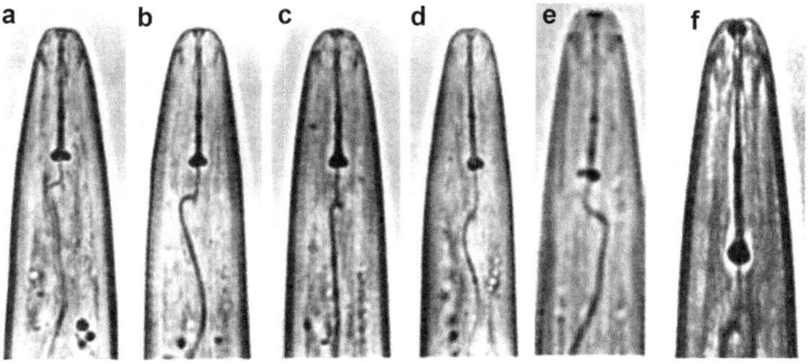

Fig. 1.8 Light micrographs of the head and stylets of second-stage juveniles of *Meloidogyne* species: (**a**) *M. incognita*, (**b**) *M. javanica*, (**c**) *M. arenaria*, (**d**) *M. hapla* (Eisenback 1982), (**e**) *M. enterolobii* (Yang and Eisenback 1983), (**f**) *M. graminicola* (Nickle 1991)

Some sequences exhibited considerable nucleotide polymorphism among *M. graminicola* isolates, particularly the one from Kerala (rice). Like other Indian isolates of *M. incognita*, the Odisha (Basella) isolate demonstrated substantial polymorphism.

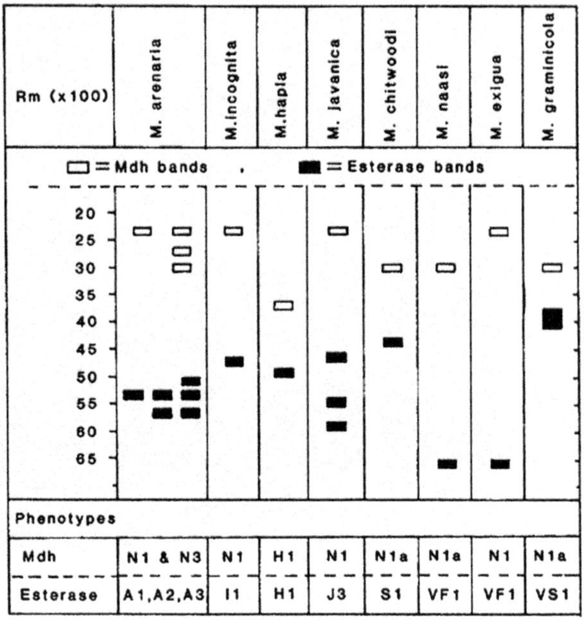

Fig. 1.9 Isozyme phenotypes of some important *Meloidogyne* species (Source: Esbenshade and Triantaphyllou 1990)

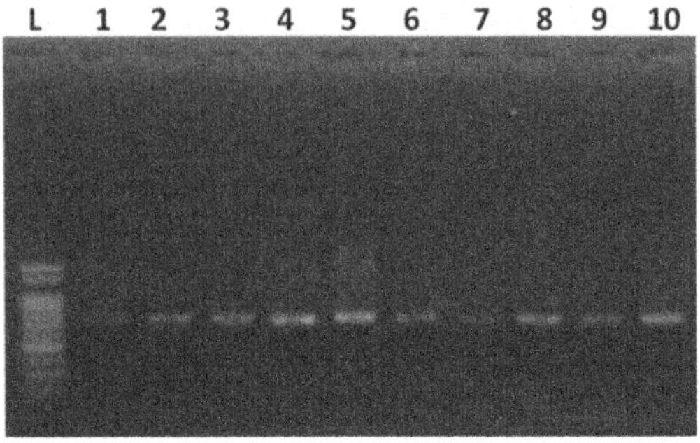

Fig. 1.10 Amplification of PCR product (800bp) of root-knot nematode populations from different states of IndiaL-100bp ladder. 1 Manipur (mulberry), 2 Kerala (rice), 3 Delhi (rice), 4 Andaman (okra), 5 Assam (rice), 6 West Bengal (Hooghly, brinjal), 7 Odisha (*Basella*), 8 Gujarat (tomato), 9 West Bengal (Bankura, brinjal), 10 Kerala (Pongalam, okra) (Source: AICRP-Nematodes)

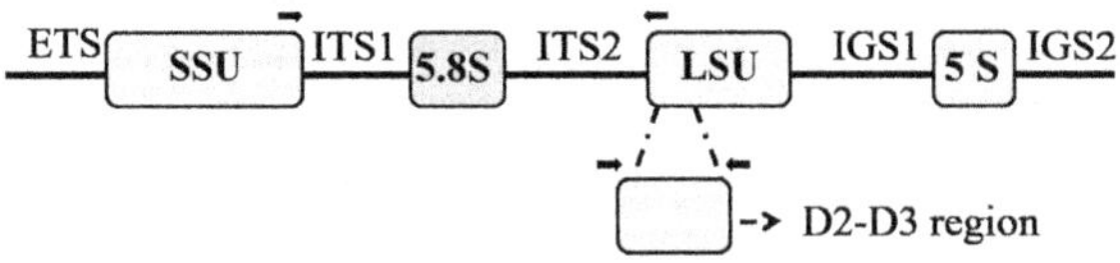

Fig. 1.11 Nuclear rRNA gene of Eukaryotic cells: ETS – external transcribed spacer, SSU – 18 small subunit, ITS1&2 – internal transcribes spacers, LSU – 28 large subunit, IGS 1&2 – intrageneric spacer regions, D2–D3 expansion region of 28S LSU (arrowhead indicating possible primer amplification)

1.3.4 Existence of Host Races/Cytological Races Within *Meloidogyne* spp.

Sasser (1952) observed the variations in the host range within the four common species of *Meloidogyne*. Increased inconsistencies in host responses across the globe led to the development of the North Carolina (NC) differential host test (Table 1.5) and the discovery of host races (Hartman and Sasser 1985). The NC host differentials are Cotton cv. Deltapine 16, Tobacco cv. NC 95, Pepper cv. California Wonder, Watermelon cv. Charleston Gray, Peanut cv. Florunner, and Tomato cv. Rutgers. Using host differential, races in *M. incognita, M. arenaria, M. javanica,* and *M. hapla* have been identified based on host susceptibility or resistance (Taylor and Sasser 1978). There are six races of *M. incognita* and *M. javanica*; two cytological races (race A and B) of *M. hapla* and two races (1 and 2) of *M. arenaria* (Table 1.6).

Four races of *M. incognita* (races 1, 2, 3, and 4) have been reported worldwide, along with two of *M. javanica*, two of *M. arenaria*, and two of *M. chitwoodi*. Most abundant among the four races of *M. incognita* are races 1, 2, and 3 (Khan and Haider 1991). In the West Mediterranean region of Turkey, races 2 and 6 of *M. incognita*, race 1 of *M. javanica*, and races 2 and 3 of *M. arenaria* were found (Zubeyir and Sogut 2011; Devran and Sogut 2011). Races 5 and 6 of *M. incognita*, races 1 and 5 of *M. javanica*, and race 3 of *M. arenaria* were identified from Spain (Robertson et al. 2009), and two possible isolates of *M. graminicola* from South-East Asia (Pokharel et al. 2010). Recently, Uysal et al. (2017) have identified 2, 4, and 6 races of *M. incognita* and 1 and 3 races of *M. javanica* infecting vegetables in Turkey.

Before developing nematode-resistant cultivars against the target population of any *Meloidogyne* species, it is essential to identify the races. The root-knot

Table 1.5 North Carolina differential host test reaction chart

	Differential host plant					
***Meloidogyne* Species & Race**	**Cotton Deltapine16**	**Tobacco NC95**	**Pepper California Wonder**	**Watermelon Charleston Gray**	**Peanut Florunner**	**Tomato Rutgers**
M. incognita						
Race 1	–	–	+	+	–	+
Race 2	–	+	+	+	–	+
Race 3	+	–	+	+	–	+
Race 4	+	+	+	+	–	+
M. javanica	–	+	–	+	–	+
M. arenaria						
Race 1	–	+	+	+	+	+
Race 2	–	+	–	+	–	+
M. hapla	–	+	+	–	+	+

Box indicates key differential host plant; + = reproduced; – No reproduction

Box indicates key differential host plant; + = reproduced; – No reproduction
Source: Sasser and Carter (1985)

Table 1.6 Chromosome numbers and modes of reproduction of the four major *Meloidogyne* species

Meloidogyne species and cytological race	Range of chromosome numbers	Mode of reproduction
M. incognita		
Race A	40–46	Mitotic parthenogenesis
Race B	32–36	Mitotic parthenogenesis
M. javanica	42–48	Mitotic parthenogenesis
M. arenaria		
Race A	54(50–56)	Mitotic parthenogenesis
Race B	34–37	Mitotic parthenogenesis
M. hapla		
Race A	14–17	Facultative meiotic parthenogenesis
Race B	30–32, 43, 45, 48	Mitotic parthenogenesis

Source: Sasser and Carter (1985)

nematodes from India infest crops in various Indian regions. The occurrence of races 5 and 6 in *M. incognita* and races 4, 5, and 6 in *M. javanica* were reported in India (Khan et al. 2014). An overview of the economically important *Meloidogyne* species and the occurrence of races in India is given in Fig. 1.12.

Morphological and morphometric comparison of 14 populations of *M. graminicola* from various agro-ecological zones in India divided the population into two groups: Anand, Bhubaneswar, Hyderabad, Jammu, Jorhat, Kalyani, Kanpur, Ludhiana, Mandya, Palampur, Vellayani clustered with *M. graminicola*, *M. triticoryzae*, and *M. salasi*; whereas, Hisar, New Delhi, Samastipur clustered with *M. oryzae* and *M. graminis*. Despite morphological variations, ITS-based molecular phylogenetic analyses revealed that these populations belonged to *M. graminicola* (Salalia et al. 2017). Host differential studies undertaken at various AICRP (Nematodes) facilities with local populations of *M. graminicola* offer support to the possibility of host races within *M. graminicola*, if not a grouping of closely related species (Walia et al. 2018a).

1.4 Economically Important *Meloidogyne* Species in India

The genus is currently comprised of more than 100 species. Out of 1000 root-knot nematode populations procured from 75 countries under the aegis of the International *Meloidogyne* Project, *M. incognita* represented 52%, *M. javanica* 30%, *M. arenaria* 8%, *M. hapla* 8%, and other species comprised only 2% of the remaining populations (Taylor and Sasser 1978). Thus, these four most common species mentioned above comprise 98% of the total root-knot nematodes encountered worldwide.

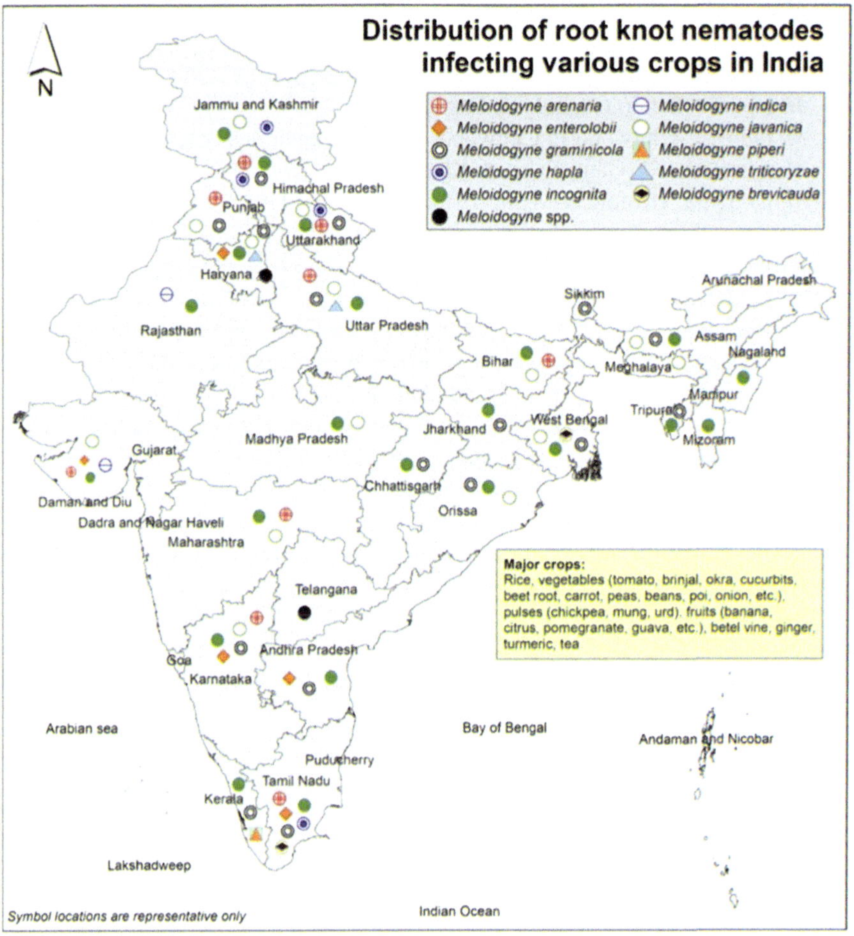

Fig. 1.12 Distribution of economically important *Meloidogyne* spp. in India

1.4.1 List of *Meloidogyne* Species Recorded in India (see also Fig. 1.12)

1. *Meloidogyne africana* (Whitehead 1968)
2. *M. arenaria* (Neal 1889; Chitwood 1949)
3. *M. brevicauda* (Loos 1953)
4. *M. enterolobii* (Yang and Eisenback 1983)
5. *M. exigua* (Göldi 1887)
6. *M. graminicola* (Golden & Birchfield 1965)
7. *M. graminis* (Sledge & Golden 1964; Whitehead 1968)
8. *M. hapla* (Chitwood 1949)
9. *M. indica** (Whitehead 1968)

10. *M. javanica* (Treub 1885; Chitwood 1949)
11. *M. lucknowica** (Singh 1969)
12. *M. naasi* (Franklin 1965)**
13. *M. piperi** (Sahoo Ganguly & Eapen 2000)
14. *M. thamesi* (Chitwood in Chitwood, Secht, & Havis 1952; Goodey 1963)
15. *M. triticoryzae** (Gaur, Saha, & Khan 1993)
16. *M. incognita* (Kofoid & White 1919; Chitwood 1949)

* Described from India; **Recently recorded by Suresh et al. (2017)

Out of the above mentioned species, *M. incognita* and *M. javanica* are most widespread across the country and attack vegetables, fruits, pulses, oilseeds, ornamentals, spices, and fibre crops. *M. arenaria* is also quite widespread, but it is considered a major problem on groundnut in Gujarat. *M. hapla* is confined to temperate areas of Jammu and Kashmir, Uttarakhand, Himachal Pradesh, and the Nilgiris in Tamil Nadu where it attacks potatoes and carrots. *M. enterolobii*, the guava root-knot nematode, is a recent interception from Tamil Nadu (Poornima et al. 2016), with subsequent reports from other states associated with guava, on which it appears to be particularly virulent. Recently the occurrence of *M. enetrolobii* has been reported from Andhra Pradesh, Gujarat, Haryana, Karnataka, Kerala, Madhya Pradesh, Rajasthan, Tamil Nadu, Telangana, Uttar Pradesh, Uttarakhand, and West Bengal (Khan et al. 2022). Citrus root-knot nematode (*M. indica*) appears to be restricted in certain parts of Gujarat, where it damages citrus (acid lime) severely. It also attacks Bt cotton, castor (Patel et al. 1999; Khan et al. 2018), and neem (Phani et al. 2018). Three species, viz., *M. incognita acrita*, *M. thamesi*, and *M. lucknowica*, which were earlier reported from India, have been redefined as synonyms of *M. incognita*, *M. arenaria*, and *M. javanica*, respectively (Hunt and Handoo 2009). *M. graminicola* is currently regarded as a national issue, as it has been discovered in nearly every region of the nation, primarily parasitizing upland rice. In India, the other species are not considered economically important.

1.5 Recent Estimations on Crop Losses

Loss estimations due to nematodes in different crops are a continuous programme operating at different centers of the AICRP on Nematodes across the country. Expression of crop losses in monetary terms is essential for project formulations and helps in showcasing, convincing, and seeking research support from policy planners. Besides, the private sector engaged in producing chemical and biological products for nematode management may find it helpful in planning appropriate inputs in different crops and areas.

Field trials on loss estimations employing t-tests are conducted regularly in various crops in different seasons using the nematicidal product (carbofuran) at varying initial nematode populations (>Economic Threshold Level).

Data on area, production, the yield of principal crops, and minimum support price (MSP) of various agricultural commodities were obtained from the most reliable sources (Anonymous 2014, 2015, 2016). Barring a few crops such as ginger, turmeric (Ray et al. 1995), sunflower (Devappa et al. 1998), and papaya (Jonathan et al. 2001), the data generated by AICRP on Nematodes have been used for the assessment of yield losses. Loss estimations varied for each crop according to locations, seasons, and years; therefore, data obtained from different centers were averaged for calculations. Out of the total area for a particular crop, only 10% has been considered nematode-infested, and the same has been used for calculating yield losses. The major nematode pest in most of the crops is a root-knot nematode, *Meloidogyne* spp., except for a few crops, for example, citrus. To keep the information comprehensive, the complete data is retained, although it includes some other nematodes as well (Kumar et al. 2020).

Root-knot nematodes (*Meloidogyne* spp.) alone are responsible for Rs. 77,373.87 million losses in different crops that constitute about 75.83% of the total estimated losses (Rs. 102,039.79 million); thus, proving to be the economically most important of all the plant-parasitic nematodes.

The losses (Fig. 1.13a, b) in 19 horticultural crops were assessed at Rs. 50,224.98 million, while for 11 field crops, it was estimated at Rs. 51,814.81 million. The mean per cent losses were higher in horticultural crops, that is, 23.03% (fruits 25.5%, vegetables 19.6%, and spices 29.5%) than in field crops, that is, 18.23% (cereals 18.8%, pulses 23%, oilseeds 11.8%, and fibre crops 19.75%). The economic losses in rice due to rice root-knot nematode, *M. graminicola* alone, were maximum (Rs. 23,272.32 million) among all the crops and nematodes considered. Citrus (Rs. 9828.22 million) and banana (Rs. 9710.46 million) among fruit crops, and tomato (Rs. 6035.2 million), brinjal (Rs. 3499.12 million), and okra (2480.86 million) among the vegetable crops suffered comparatively more losses that is partly attributable to areas of production in respect of these crops (Kumar et al. 2020).

Some more interesting facts can be deduced when this data. First, an overall 4.84 times increase in economic losses since 2007 is attributable to an escalation in MSP, an increase in area under cultivation, additional crops included in the study, etc. Second, the emergence of *M. graminicola* as the most important and national problem of rice relegating *Aphelenchoides besseyi*, *Ditylenchus angustus*, and *Hirschmanniella* spp. Third, relatively more shift toward horticulture, and therefore, nematode problems in these cropping systems.

It is emphasized here that the figures presented here pertain to quantitative losses only. The qualitative losses such as forking in carrots, infection in underground edible plant parts like tubers in potato, rhizomes in turmeric, ginger, etc. often result in the non-acceptability of produce at the level of consumers; such parameters have not been taken into account while assessing losses due to nematodes. Furthermore, the information on nutritional quality parameters in crop production due to nematodes is totally lacking.

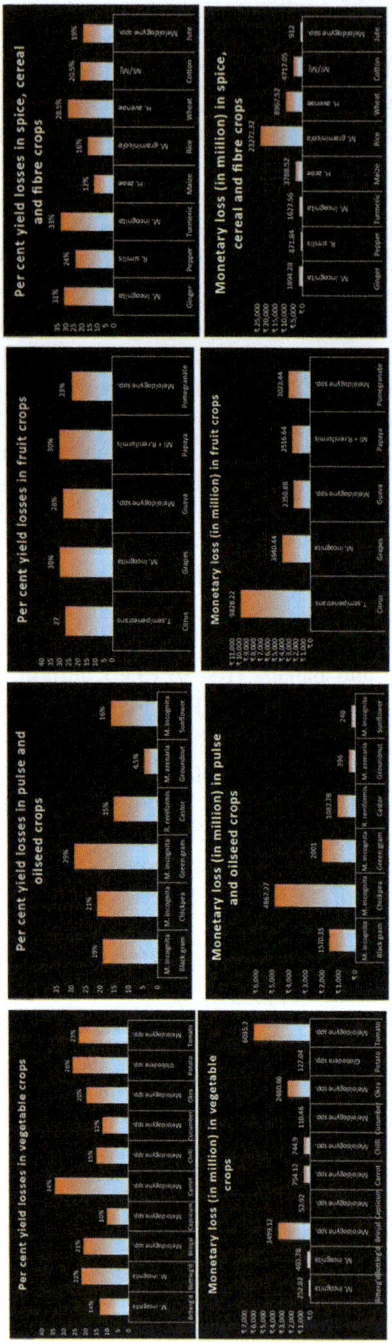

Fig. 1.13 Per cent and monetary losses in different crops due to plant-parasitic nematodes in India

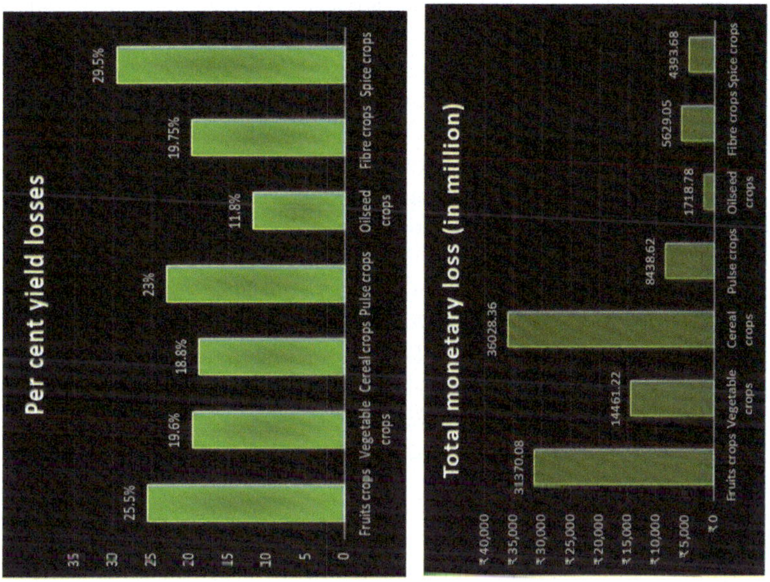

Fig. 1.13 (continued)

1.6 Biology and Life Cycle

Root-knot nematodes are sedentary endoparasites. Saccate females are completely embedded inside the root galls, with heads located near the vascular tissues and terminal portions near the root epidermis. Reproduction is generally parthenogenetic. Males are vermiform, wander in soil, and are non-parasitic. The rectal glands of the females secrete a gelatinous substance on the root surface where eggs are deposited. Oviposition continues for 10–12 days; each female lays about 200–400 eggs held together in an eggmass or eggsac (Fig. 1.14a). Occasionally, eggmasses may be formed inside the roots, particularly in compound galls.

Embryogenesis takes about 1–2 weeks and each eggmass contains eggs at different stages of development. First-stage juveniles (J1) moult while still within the eggshell and become second-stage juveniles (J2). Root-knot nematodes generally do not require any specific stimulus from the host and hatch freely in the water. Depending upon the availability of suitable temperature and moisture, J2 hatches out, moves freely in the soil in search of new roots of the same plant or some other plant, and is the only infective stage. This stage is also known as preparasitic J2. Under normal conditions, J2 can remain in the soil for several days, deriving energy from reserve food material. However, under adverse environmental conditions, J2 can survive in the soil for several months in an inactive phase (to cut down metabolism) when the body shrinks and coils.

Initially, the J2 move in soil randomly, but once in the vicinity of host roots, they are attracted to them due to the presence of exudates emanating from the roots. The J2 penetrates the roots just behind the root tip (meristematic zone). Penetration is facilitated by repeated stylet thrusts and/or enzymes secreted by the nematode oesophageal glands. The J2 moves through the root cells and positions itself with the head near the vascular tissues, while the rest of the body is completely inside the cortex (Fig. 1.14b). At this stage, J2 becomes sessile and initiates the development of feeding sites (giant cells).

As the feeding process begins, the J2 starts assuming swollen shape, now called parasitic J2. Sex differentiation occurs at this stage; the juveniles destined to become females acquire V-shaped genital primordium, while in males it is I-shaped (Fig. 1.14a). Under optimum conditions, second moult occurs in about a week and J3 is formed. The third moult follows quickly and the juvenile changes to J4. J3 and J4 retain the old cuticles, the pointed tail of J2 still visible, and hence are also called "spike-tailed stages" (Fig. 1.14a). The body grows in width, genital primordia develop further, but these stages are non-feeding as they lack stylet. At the last moult, the adult female becomes sac-like, stylet reappears, and the reproductive system gets fully developed with a vulval opening making its appearance. The adult males are, however, vermiform, coiled inside the J4 cuticle, emerge out and leave the roots to come out into the soil. They are short-lived. Adverse environmental conditions after penetration may induce maleness in the developing juveniles.

The whole life cycle is completed in about 25 days at 25–30 °C, which is optimum for most species. During the winter season under North Indian conditions,

a

b

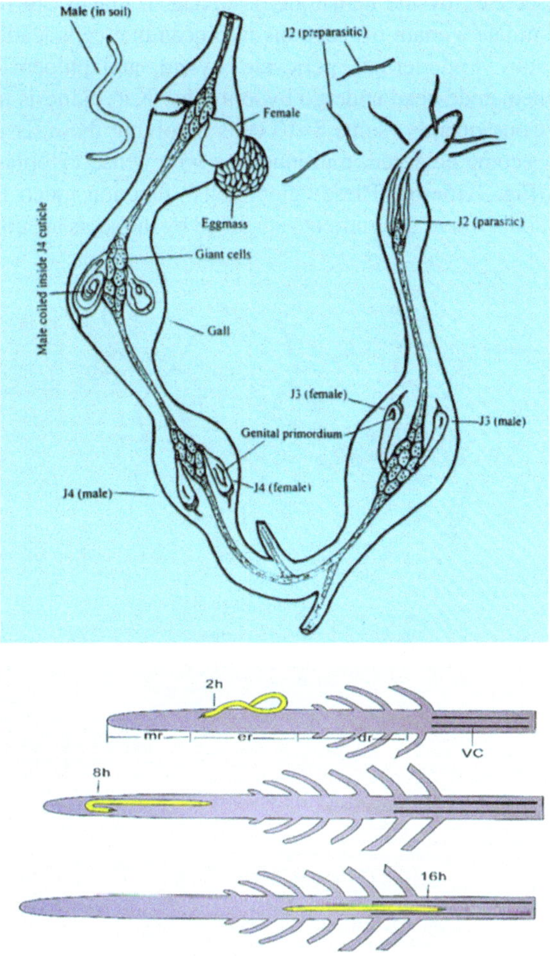

Fig. 1.14 (**a**) Typical life cycle of *Meloidogyne* sp. (Source: Walia and Bajaj 2014) (**b**) Process of root invasion by the second-stage juvenile (Source: MACTODE)

the life cycle duration may be prolonged to 60–80 days depending upon prevailing temperature. Thus 7–8 overlapping generations are completed in a year.

1.7 Parasitic Relationships with Host Plants

The preparasitic infective J2 are attracted to the zone of elongation, where they penetrate the root and then migrate intercellularly, separating cells at the middle lamella in the cortical tissue. The juveniles usually migrate down to the root tip and then turn around in the region of the root apical meristem (Fig. 1.14b). They then migrate up the center of the root to the zone of differentiation. This process appears to include mechanical force and enzymatic secretions from the nematode. Enzymes secreted by the nematode oesophageal glands are released into the host cells and initiate a chain of reactions in procambial cells leading to the formation of feeding sites. Endodermal, pericycle, xylem, and phloem tissues in the vicinity of the nematode head undergo hypertrophy. Karyokinesis takes place without cytokinesis. Consequently, some 8–10 cells involving these tissues around the nematode head become enlarged, multinucleate with dense cytoplasm, showing hypermetabolism (Fig. 1.15a-b). These "giant cells" function much like transfer cells or metabolic sinks where the nutrients absorbed by the roots are continuously pooled and diverted

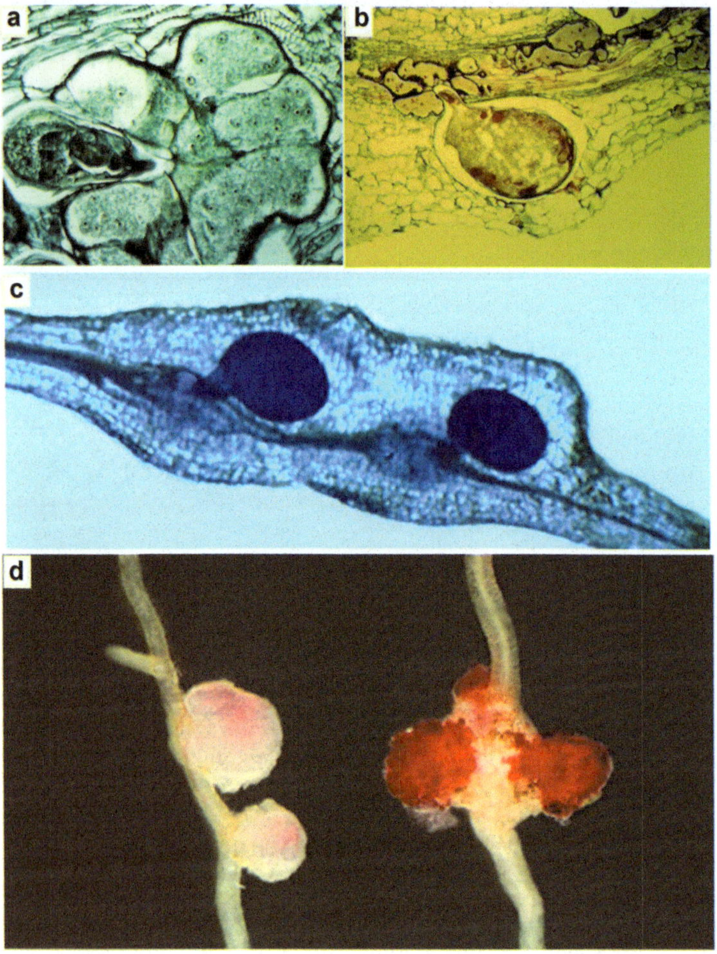

Fig. 1.15 (**a**) Multinucleate giant cells around the head of female are clearly visible in the transverse section and (**b**) longitudinal section of infected root; (**c**) Fully formed females with their necks in vascular tissues; the swollen areas within vascular tissues are sites of giant cell formation; (**d**) Rhizobium nodule (left) and nematode gall with eggmasses (right) (Source: MACTODE)

to nematode for its growth and development. The disruption in the continuity of conducting vessels hampers the flow of nutrients and water to the shoots, leading to reduced plant growth and yield. The formation of giant cells is essential for a successful host-parasite relationship, and if a nematode fails to induce the formation of these feeding sites, it does not develop further and dies. Such a situation arises in incompatible hosts.

Simultaneously, the protease enzymes released by nematodes act on host proteins, breaking them into amino acids. The concentration of amino acids, particularly tryptophan, a precursor of IAA (indole acetic acid), leads to the accumulation of auxins or hormonal imbalance at the site of infection. Thus, instead of growing longitudinally, the roots grow axially due to hyperplasia and hypertrophy of cortical parenchyma cells (Fig. 1.15c). This results in the formation of swelling, the root gall or knot at the site of juvenile penetration within 1–2 days of infection.

The galls formed by root-knot nematodes are often confused with rhizobium nodules in leguminous plants. Rhizobium nodules are side appendages, soft in texture, and can easily be separated from roots by slight disturbance. On the contrary, nematode galls are swellings of the root itself and cannot be detached from roots; these are harder in texture (Fig. 1.15d).

1.7.1 Assessment of Disease (Gall Index)

Greenhouse trials pertaining to a screening of crop germplasm to assess resistance against root-knot nematode or field trials on root-knot nematode management always require crucial observations on nematode disease suppression. Observations and categorization of infected roots based on root galling are routinely done. This is purely a visual observation that involves interperson errors. Nevertheless, with adequate experience, the parameter is extremely useful for judging the efficacy of treatments and the response of different germplasm lines while screening for resistance.

Various gall rating schemes are in vogue:

(i) Based on the number of galls on the root system; this is possible in greenhouse trials after the completion of one nematode generation.
(ii) giving a numerical score based on the per cent root area galled; this is usually called gall index (GI) or root-knot index (RKI) (Fig. 1.16).

RKI	Galling		Reaction
1.	No galls and egg masses	–	Highly resistant
2.	1–10 galls/egg masses	–	Resistant
3.	11–30 galls/egg masses	–	Moderately resistant
4.	31–100 galls/egg masses	–	Susceptible
5.	101 and above galls/egg masses	–	Highly susceptible

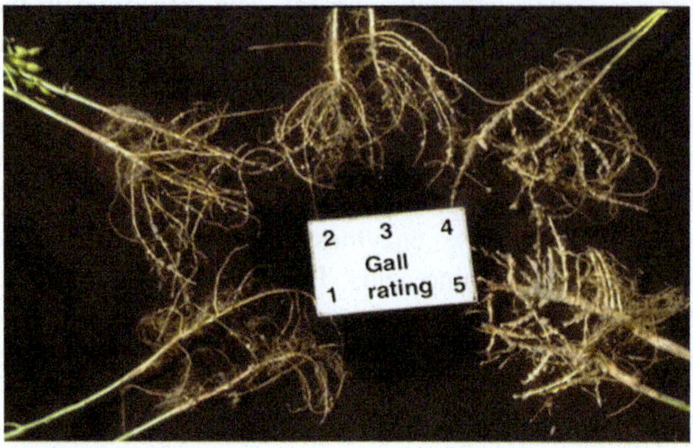

Fig. 1.16 Assessing infected roots for root-knot index (Source: MACTODE)

1.8 Symptoms of Damage in Different Crops (Figs. 1.17–1.21)

The above-ground symptoms due to root-knot nematodes, in general, are not diagnostic. The damage symptoms on shoots reflect root damage and appear as stunted plant growth, usually in patches, yellowing of foliage, less tillering, undersized fruits, etc. in annual crops. In perennials, usually, the dieback symptoms are common; bare twigs, poor seasonal flushes are indicative of nematode infection on roots. The severity of infection is more when nematode-infected planting material is used. Nutrient deficiency symptoms could be more appropriate to describe the above-ground expression of damage to roots by nematodes. Temporary wilting during hot days is common in broad leaves plants; such plants tend to recover by evening. However, gradually the wilting becomes permanent.

Below-ground symptoms are clearly discernible in the form of galls or knots. The initial infection leads to the formation of very small galls (primary galls). The subsequent infection leads to the fusion of primary galls; the bigger galls (secondary galls) are now visible more clearly. The pattern of galling, however, varies with crops.

- Vegetable crops like tomato, brinjal, and okra are highly susceptible and form heavy galling (Fig. 1.17a-d), but chilli has very small galls.
- Cucurbits usually have very big galls, so much so that the entire root may become swollen (Figs. 1.18–1.20). In many such crops, usually, egg masses are formed inside the galls.
- On tuberous crops like potatoes, besides roots, the infection may extend to tubers also. Infected tubers show pimple-like growth on the surface (Fig. 1.17g), significantly reducing their market value.

Fig. 1.17 Symptoms of root-knot nematode infection on (**a**) tomato, (**b**) brinjal, (**c**) okra field, (**d**) okra close-up, (**e**) beetroot, (**f**) carrot, (**g**) potato, (**h**) potato system infected by *Globodera* spp.

- Fleshy edible parts of the crops like beetroot, carrot, radish, and turnip bear small-sized galls on feeder roots, but tap roots frequently show forking as a result of nematode infection (Fig. 1.17f).

Fig. 1.18 Symptoms of root-knot nematode infection on (**a**) beetroot system, (**b**) cucumber, (**c**) bottle gourd, (**d**) poi (*Basella*), (**e**) cucumber, (**f**) cauliflower

- In leguminous plants, nematode galls are distinct from rhizobium nodules. While the bacterial nodules are side appendages, soft and can be detached easily, the nematode galls are axial swellings of the root itself, hard in consistency, and do not detach (Fig. 1.15d). But nematode infection hampers bacterial nitrogen fixation due to reduced root system, reduction in number and size of nodules, and infection of nodules themselves.

Fig. 1.19 Symptoms of root-knot nematode infection on (**a**) dolichos bean, (**b**) cowpea, (**c**) red amaranth, (**d**) Ash gourd

Fig. 1.20 Symptoms of root-knot nematode infection on (**a**) cucumber (infested polyhouse), (**b**) cucumber close-up, (**c**) capsicum, (**d**, **e**) pointed gourd (*parwal*)

Fig. 1.21 *Meloidogyne graminicola on onion* (**a**): Field symptoms, drying up of leaves from tip; (**b**) galls on roots in endemic areas: some parts of Karnataka, Haryana, West Bengal, Gujarat; *Meloidogyne* sp. on garlic (**c**): Field symptoms, drying up of leaves from tip; (**d**) galls on roots in endemic areas: some parts of Karnataka, Haryana, West Bengal, Gujarat; *Meloidogyne* sp. on potato in Gujarat (**e**); *Meloidogyne arenaria* on potato in Meghalaya (**f**)

1.9 Disease Complexes

The resident soil biota comprises fungi, bacteria, viruses, protozoans, etc., among microorganisms, and these numerically outnumber the nematodes. Nematodes, with their limited locomotion in thin water films surrounding the soil particles, come in contact with propagules of these microorganisms. For a long, it has been believed that due to repeated thrusting with stylet while feeding, nematodes create micro-punctures on the roots, thus paving the pave for other pathogens to invade the plant roots. Some of these microorganisms are vectored by nematodes externally (fungi and bacteria) or internally (viruses). But more important than this is physiological modifications induced by nematodes in the plant system. This has been conclusively proved using the split-root technique in some plant systems (Fig. 1.22). The qualitative changes in the nematode-infected plant root exudates have been reported to activate the dormant propagules of fungi/bacteria.

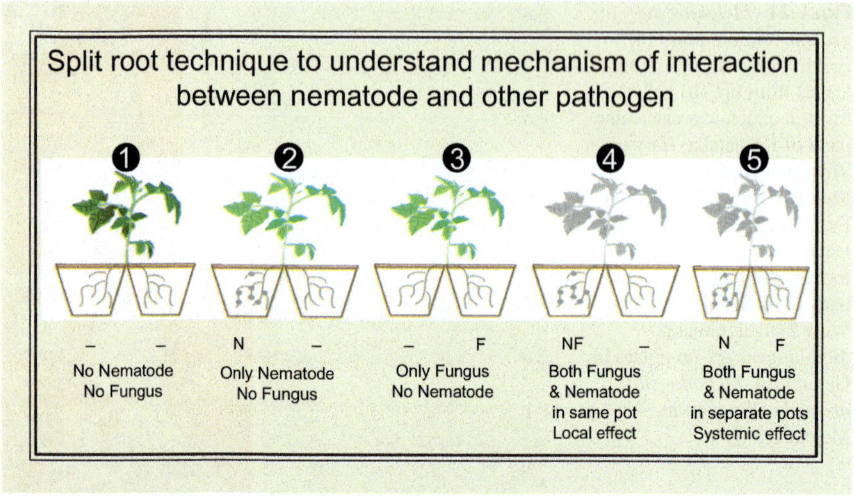

Fig. 1.22 Split root technique: A hypothetical model to explain the role of nematodes in disease complexes by rendering the plants more susceptible to other pathogens due to physiological modifications induced by nematodes by prior infection. ❶ No nematodes, no fungus, plants look healthy; ❷ Only nematode (N) inoculated in one pot, the systemic damage extends to the entire plant rendering it diseased; ❸ Only fungus (F) inoculated in one pot, the systemic damage extends to the entire plant rendering it diseased; ❹ Both N and F are inoculated in the same pot, N facilitating the F infection through stylet injuries; ❺ N inoculated in one pot brings about systemic physiological changes in the entire plant, and F inoculated in the other pot is able to cause damage. The combined damage is much more in intensity often killing the plant

Plant-parasitic nematodes, in general, and root-knot nematodes, in particular, have been the subject of investigations in such "disease complexes." Such associations are more common in field conditions than envisioned. They are known to predispose some plants to fungal pathogens (Atkinson 1892; Back et al. 2002). This type of interaction has been evaluated genetically in the *Meloidogyne incognita-Fusarium oxysporum* f. sp. *lycopersici* disease complex on tomato host. The interaction of root-knot nematodes with other organisms is known in many vegetables like tomato, potato, brinjal, cowpea, etc. However, the most common problem is the breakdown of disease resistance and wilting of healthy plants. Similarly, interaction with *Ralstonia* (=*Pseudomonas*) *solanacearum* is reported to cause "pseudomonas wilt" in tomato, brinjal, and potato.

In India, interactions of root-knot nematodes with other pathogens have been studied and in the majority of the cases, the association of nematodes is a predisposer of soil-borne pathogens. Root-knot nematodes interact synergistically with large numbers of root-infecting fungi (Khan 1993). Among the soil-borne fungal pathogens, *Fusarium, Rhizoctonia, Sclerotium, Pythium, Phytophthora, Macrophomina* etc., most frequently interact in the rhizosphere of different crops like vegetables, pulses, tobacco, potato, ginger, carnation, cardamom, betelvine, banana, jute, cotton, etc. Most of the interactions of fungi with *Meloidogyne* spp.

result in the root-rotting or wilt complex of crops. In some instances, root-knot nematode interferes physiological activity of *Rhizobium* nodule; they directly infect the nodule and reduce the number of nodules in leguminous crops (Upadhyay and Dwivedi 1987; Chahal and Chahal 1989; Sharma and Tiagi 1990; Siddiqui and Mahmood 1994; Jain and Trivedi 1995). More than one fungus, along with a root-knot nematode, is also known to be involved in the wilt-disease complex of different crops (Kavathiya and Pandey 2000; Parvatha Reddy 2008). The association of root-knot nematodes with wilt-causing bacteria (*Ralstonia solanacearum*) is more serious in solanaceous crops; in some instances, soil-borne fungi and bacteria jointly participate in the development of the disease complex. For example, in "Hooghly wilt" of jute in West Bengal, three pathogens, viz., *Meloidogyne incognita*, *Macrophomina phaseolina*, and *Ralstonia solanacearum* are involved in the development of the disease complex (Mishra et al. 1988; Mandal and Mishra 2001). In the presence of root-knot nematodes, plant viruses like tobacco ring spot, tobacco mosaic virus, tomato leaf curl virus, etc., produced early symptoms in cowpea and tomato (Swarup and Goswami 1969; Goswami et al. 1974; Goswami and Chenulu 1974; Mayee et al. 1974; Alam et al. 1990). Root-knot nematode (*M. javanica*) showed synergistic interactions with stem borer (*Stomopteryx nertaria*) in mungbean, and the nematode, together with the insect caused more damage (Prasad et al. 1971). Further, *M. incognita* also developed a disease complex in association with phytophagous mite (*Tyrophagus putrescentiae*) in tuberose (Ganguly et al. 1993).

1.10 Management

Historically speaking, the first attempts towards the management of nematodes in India were made by Ayyar and and his co-workers during 1926–33 on vegetable crops in Agriculture College, Coimbatore. He published the results of his experiments in the *Madras Agricultural Journal*. His experiments for controlling root-knot nematode on brinjal included: the burning of sorghum stalks and other materials; the use of knol-khol as a trap crop; the use of chemicals such as carbolic acid, kerosene, formalin, potassium cyanide, carbon disulfide, lime, sulphuric acid, and ranicide; and rotations with crops like *ragi*, maize, sorghum, or red gram. Serious efforts for developing nematode management technologies, however, were initiated during the late 1970s, when the economic importance of nematodes to agriculture was fully realized.

Root-knot nematodes usually attain damaging levels because the commonly occurring species have wide host ranges, are multivoltine with high fecundity and reproduction rates, and are vastly distributed. The predisposition of galled roots often leads to secondary infection, thus aggravating wilt and rot problems, ultimately leading to plant mortalities. Under these situations, a blend of effective control practices is employed to "manage" the nematode populations below damaging levels rather than depending on short-term strategies (Bernhard et al. 1985; Norris et al. 2003).

1.11 Effective Cultural/Agronomic Practices

1.11.1 Removal of Infected Materials

Destruction of nematode-infected roots after the crop is over can help tremendously remove significant nematode inoculums from the field for the next crop. In most instances, the shoot is cut, and roots are left in the field for decomposition till the field is prepared for the next crop. However, the practice can be particularly helpful in polyhouses where intensive cultivation is followed with little time between the crops. The roots of the previous crops can be pulled out along with the roots as much as possible. The galled roots containing millions and billions of nematode eggs can be piled in a corner outside, allowed to dry, and then destroyed.

1.11.2 Crop Rotation/Cropping Sequences

Crop rotation is a change of the main crop in one season only, whereas cropping sequence is a long-term (1–2 years) plan of the cropping sequence. The wide host ranges of common species of *Meloidogyne* spp. notwithstanding, most graminaceous crops do offer alternatives as non-host crops for main susceptible crops. However, the alternative graminaceous crop will find acceptability only if it (1) supports profitable production at par with the replaced crop; (2) is agronomically suitable to grow in that agro-climatic region; and (3) makes sure that the populations of other pests and diseases, including new nematode problems, do not develop/crop up. The practice will be redundant for established long-duration perennial crops, such as trees and vines. Use of non-hosts or poor hosts (such as graminaceous crops), or nematode antagonistic crops in rotation for 1–2 years effectively reduces the root-knot nematode populations (Sundresh and Setty 1977; Patel et al. 1979). Cropping sequences involving mustard, sesame, maize, wheat, etc. are also suppressive to root-knot nematodes (Alam et al. 1981; Haque and Gaur 1985; Siddiqui and Saxena 1987). The cropping sequence of tomato-onion-resistant tomato-okra was found best in managing the nematode population and giving the best economic returns out of the 15 cropping sequences studied by Kanwar (1990).

Four cropping sequences were tested at AAU Anand; the crop combinations included: susceptible (S) crops/cvs. of cowpea, chickpea, green gram; resistant (R) crops/cvs. of cowpea, onion, cowpea; non-host (NH) crop of cluster bean; and poor host (PH) crop of groundnut in different sequences, that is, Cowpea-Chickpea-Greengram (S-S-S) (check), Cowpea-Garlic-Cluster bean (S-R-NH), Cowpea-Onion-Cowpea (Veg.) (R-R-R), and Cowpea-Cabbage-Groundnut (S-R-PH). All three sequences with NH, R, PH crops/cvs. resulted in an approximately 50% reduction in gall index at the end of the third crop and yielded double the income compared to S-S-S check (Patel 2018).

1.11.3 Deep Summer Ploughing and Solarization

In North-West India, drastic reductions in root-knot nematode populations can be achieved by simple deep ploughing of infested fields during peak summer months (May-June) that leads to desiccating the soil populations of juveniles as well as eggs harboured in the leftover root tissues (Jain and Bhatti 1987). Polythene mulching can further enhance the efficiency of summer ploughing by trapping solar energy and retaining more heat (Gaur and Perry 1991). This practice also suppresses weeds, besides soil fungi and bacteria.

Solarization is widely practised in polyhouses during May/June in North Indian conditions. After harvesting the crop in April/May, the soil is levelled, given light irrigation, and covered with a 25 μm transparent polythene sheet in such a way that the edges are sealed. The shade nets are removed, and the polyhouse is virtually closed for about 3–4 weeks. The soil temperature in the top 15 cm layer reaches up to 62 °C. The practice is very effective for managing nematodes.

1.11.4 Planting Dates

Prevailing temperature plays a crucial role in the biology and pathogenicity of root-knot nematodes. In regions having wide fluctuations in the seasonal temperatures, such as North Indian plains (a low of 5 °C during winter to a high of 45 °C during summer), planting dates can be suitably changed to the disadvantage of nematode development, thus preventing crop losses. Overall, the populations of root-knot nematodes are generally high during *kharif* (summer) and low during *rabi* (winter). Therefore, delayed (mid-November instead of mid-October) sowing of chickpea and lentil prevented crop damage despite the high nematode population (Gaur et al. 1979; Mishra and Gaur 1979). Similarly, in southern California, *M. incognita* damage to carrots can be prevented by adjusting sowing dates (Roberts 1987).

1.11.5 Trap Crops, Antagonistic Crops

In certain crops like crotalaria, *Meloidogyne* J2 can infect the roots but is unable to develop further because of antagonistic response of the plant. Alternatively, a good host (e.g., a leguminous crop like *dhaincha,* and *Sesbania aculeata*) can be grown before the main crops. Even before the nematode completes one generation, the crop is ploughed back into the field for green manuring and allowed to decompose before the main crop is planted.

Antagonistic crops release certain root exudates in the rhizosphere that have nematoxic traits. Such crops, for example, African marigold, mustard, sesame, and asparagus (*Asparagus officinalis*) can be grown as intercrops with the main crop or in the basin areas of fruit crops to check the buildup of root-knot infection in the main crop (Gaur 1975; Haque and Gaur 1985).

The nematode antagonistic properties of marigold (*Tagetes* spp.) have long been documented (Gommers et al. 1980) because of the nematicidal action of α-terthienyl in their root exudates. Therefore, intercropping of marigold drastically reduced galling in the susceptible brinjal crop. Similarly, intercropping of onion and maize reduced galling due to *M. incognita* in potato. The cropping sequences of okra-cowpea-cabbage and okra-cucumber-mustard were most effective in suppressing the nematode population under West Bengal (India) conditions (Chandra and Khan 2011).

1.11.6 Use of Healthy Planting Materials

The problems are different for crops raised from true seed (e.g., white tip nematode of rice) and vegetatively propagated crops, and annual versus perennial crops. Root-knot nematode-infected seedlings of rice and vegetable crops pose a serious problem of disseminating nematode into new production areas. Rice nursery seedlings are raised in infested field sites, and there exist commercial rice nurseries facilitating the transport of *M. graminicola* (and rice root nematode also) infected seedlings to far-flung areas. Seedlings of vegetable crops such as tomato, brinjal, and chilli are still raised conventionally in most areas that aids in nematode dissemination, locally though. However, there is a significant shift towards raising vegetable seedlings in plug trays using sterilised medium (cocopeat, etc.) free from nematodes. Many other vegetatively propagated crops that facilitate the widespread spread of root-knot nematodes are nematode-infected tubers (potato), bulbs (onion, garlic, tuberose, gladiolus, etc.), rhizomes (ginger and turmeric), and corms (banana). However, the problem in such crops is restricted to more than one season. Using certified planting materials and hot-water treatment is a very effective means to clean the planting materials of nematode infection.

In banana, root-knot nematode-infected corms can be disinfected by peeling followed by hot-water treatment (53–55 °C for 20 min) (De Waele and Davide 1998). Hot treatment has been recommended for the ginger corms infected with *Meloidogyne* sp. in Madhya Pradesh.

Another serious aspect of this problem pertains to perennial crops, particularly fruit crops. The rooted cuttings are propagated by different methods, for example, grafting, cloning, air layering, ground layering, and tissue culture. However nematode free these materials may be, conventionally, the commercial nurserymen grow them further in nematode-infested soil (usually taken from the same orchard) for hardening. This is the hallmark of the problem that needs to be addressed because these nematode-infected saplings are transported all across the country without undergoing mandatory certifications.

1.11.7 Soil Amendments

Readily available organic materials, for example, animal manures, poultry litter, and crop residues, are typical examples of soil amendments used to manage root-knot nematodes. Farmyard manure (FYM) is often applied to fields since time immemorial to improve soil fertility. The increased microbial activity (including that of nematode antagonists) consequent to the incorporation of organic material in the soil leads to the suppression of plant-parasitic nematodes (Widmer et al. 2002), as well as due to the release of volatile fatty acids during this process. The improved plant health also imparts tolerance to plants to withstand nematode damage.

Deoiled cakes (e.g., neem cake) are concentrated organic materials that provide slow-releasing nitrogen to the plants. Neem seed kernel powder used as a seed treatment in pulse crops @ 5–10 g/kg seed combined with the application of bioagents is effective against root-knot nematode (Anonymous 2012).

Sundararaju et al. (2002) found that soil amendments with sewage sludge, spent compost, distillery sludge, and vermicompost, etc. are also effective for managing plant-parasitic nematodes. Some other organic products like composted horticultural waste or fresh poultry waste (manure and bedding material) also resulted in a reduction of *M. incognita* population in soil and an increase in vegetable yields in the USA (McSorley and Gallaher 1995; Riegel and Noe 2000).

Fortification of organic material, particularly FYM or vermicompost, with bioagents is being popularized nowadays to enhance the efficacy of bioagents. The bioagents, for example, *Pseudomonas fluorescens*, *Trichoderma harzianum/T. viride*, and *Purpureocillium lilacinum* are mixed in FYM @ 2 kg/2 l per ton of FYM separately on a cemented floor. The heaps are turned weekly, followed by light watering, and protected from direct sunlight and rain. The fortified organic material with bioagents is ready for application to the field or incorporation in beds after 3–4 weeks.

1.11.8 Phytotherapeutic Methods/Use of Botanicals

Several plant species are known to possess nematicidal properties naturally. The nematicidal attributes of such plants can be harnessed crudely by directly using their plant parts, products, or extracts (phytotherapeutic control); or using their synthesized/purified formulations (botanicals). Such materials are easily available in farm vicinity and are no/low cost, pollution free, biodegradable, non-toxic, and improve soil health. Leaf extracts of several plants, for example, *Allium sativum, Calotropis procera, Datura stramonium, Ricinus communis, Xanthium strumarium, Mentha viridis,* and *Cassia fistula*, proved nematicidal (Nandal and Bhatti 1983, 1986; Nath et al. 1982; Haseeb et al. 1982).

Neem products obtained from *Azadirachta indica* have been studied extensively (Akhtar 2000), and several synthetic products (nematicides, insecticides, fungicides, and miticides) are now available commercially. Nematicidal properties of neem are attributed to many chemical substances, for example, azadirachtin, kaempferol, nimbidin, nimbin, quercetin, salannin, and thionemone (Khan et al. 1974; Ferraz and de Freitas 2004). Incorporation of neem (*Azadirachta indica*) or *Subabool* (*Leucaena lucocephala*) left in tomato nursery beds @ 50 q/ha resulted in better seedling growth and reduced galling (Jain et al. 1988; Mojumder and Mishra 1993). Combined application of neem products and bioagents was more effective in nematode suppression in several crops (Rao 1997a, b, c; Reddy 1997).

Biofumigation with brassicaceous crops is successful in the reduction of *Meloidogyne* spp. (Monfort et al. 2007; Qing et al. 2007; Dutta et al. 2019). These crops possess glucosinolates (volatile sulphur-containing compounds) which, upon hydrolysis, transform into active fungicidal, bactericidal, and nematicidal isothiocyanates (Kirkegaard et al. 1996; Brown and Morra 1997).

Roots of *Tagetes* spp. contain polythienyls, and the formation of singlet oxygen by photoactivated α-terthienyl may be responsible for nematode mortality (Ferraz and de Freitas 2004). The application of undiluted extracts and chopped leaves of *Tagetes* spp. inhibited hatching of *M. incognita*, and reduced root galling, number of egg masses, and final nematode population (Walia 1997). Remnants of *T. patula* and *T. erecta* are frequently left in the field, and the integration of their plant biomass lowers the populations of *M. incognita* in the soil (Dutta et al. 2019).

Castor (*Ricinus communis*) contains ricin that is nematicidal (Ferraz and de Freitas 2004), and its incorporation in soil resulted in a significant suppression in *M. incognita* in davana (*Artemisia pallens*) (Pandey 1994); and in combination with karanj (*Pongamia pinnata*) and mahua (*Madhuca longifolia*) seed cake checked penetration of *M. incognita* J2 and gall formation on tomato (Poornima 1997). The cyanogenic glucoside linamarin present in cassava (*Manihot* spp.) roots is nematicidal in action (Sena and Ponte 1982; Ponte et al. 1996); and the application of cassava flour by-product known as manipueira or cassareep has been reported to provide some level of control of *Meloidogyne* spp. (Whitehead 1998). The introduction of nematode suppressive crops such as *Crotalaria* spp. (monocrotoline and pyrrolizidine), marigold (polythienyls and α-terthienyl), brassicas (isothiocyanates), sudan grass (cyanoglycoside dhurrin), rye (butyric acid and hydroxamic acid), velvet bean (1-tricontanol and triacontanyltetracosanate), and sesame (acetic acid) as cover crops in the crop sequence could be useful for root-knot nematode management.

Several products based on algae, fungi, and bacteria (Whitehead 1998; Chitwood 2002; Haydock et al. 2006) and crustacean chitin (Rodríguez-Kábana 1990; Ehteshamul-Haque 1997; Chitwood 2002; Ferraz and de Freitas 2004) are also antagonistic to root-knot nematodes. Furfural, a by-product of sugarcane, is currently registered for use against plant-parasitic nematodes (Haydock et al. 2006; Nel et al. 2007). Essential oils of several plant families and their components (citronellol, eugenol, geraniol, and linalool) were found to have nematcidal efficacy against root-knot nematodes (Oka et al. 2000; Oka 2001). Nematicidal activities of carvacrol (1,5-isopropyl-2-methylphenol) at doses of 250–1000 ppm showed strong effects on different life stages against *M. javanica* (Nasiou and Giannakou 2017). The orange and citronella oils were most effective for immobilization and killing of nematodes of *M. incognita* (Kundu et al. 2020).

1.12 Biological Control

1.12.1 Parasitic Bacteria

This gram-positive endospore-forming bacterium currently comprises six species; except *Pasteuria ramosa* which parasitizes water fleas (*Daphnia* spp.), all others are associated with plant-parasitic nematodes. *P. penetrans* is a well-known parasite of commonly occurring *Meloidogyne* species (Sayre and Starr 1985), although *P. hartismeri* is parasitic on *Meloidogyne ardenensis* (Bishop et al. 2007). The endospores of *P. penetrans* are resistant to desiccation and can tolerate high temperatures in field soils. The spores of *P. penetrans* attach to the cuticle of J2 of root-knot nematode only in soil. The spore-encumbered juveniles enter the roots and germination is triggered by the formation of giant cells by the juvenile. The nematode development continues unimpeded; the bacterium undergoes vegetative growth in the nematode pseudocoloemic fluid. As the nematode enters adulthood, the bacterium turns into the sporulation phase. The reproductive system of the females is atrophied; consequently, the nematode is not able to lay eggs or lays very limited eggs. The bacterium thrives on the nematode body fluid and produces spores that may be around two million per female.

A method for the mass production of endospores in vivo was first described by Stirling and Wachtel (1980). The initial greenhouse, microplot, and field experiments on *P. penetrans* against *Meloidogyne* spp. revealed that it is a highly promising biological control agent. However, being an obligate parasite, the lack of in vitro cultivation was a big impediment to its commercialization and field use. Spo0F, a key protein involved in the initiation of sporogenesis in *Bacillus subtilis,* also has a homologue in *P. penetrans* (Kojetin et al. 2005), and it has been suggested that cation concentrations may be prohibiting the vegetative forms of the bacterium from entering sporogenesis and forming endospores. Hewlett et al. (2004) successfully cultured it on synthetic media. This paved the way for commercial activity, and *Pasteuria* BioScience LLC, a US-based company, launched two products: ECONEM® for managing sting nematodes and CLARVIA® for soybean seed treatment against cyst nematodes. However, the strain parasitizing root-knot nematodes was not commercialized. Nevertheless, it is possible to mass multiply *P. penetrans* on in vivo production systems to raise small quantities of the product that can be applied to high-value horticultural crops under protective cultivation systems (Walia et al. 2011). A new strain of *Pasteuria* parasitizing rice root-knot nematode, *Meloidogyne graminicola* has been reported (Thakur et al. 2015), and a novel method to mass multiply this on a soil-less system on rice has been developed (Kumar et al. 2017).

1.12.2 Antagonistic Bacteria

Among the rhizospheric bacteria, *Bacillus subtilis, B. sphaericus,* and *P. fluorescens* have been reported to antagonize plant-parasitic nematodes (Sikora 1992; Tian et al.

2007). Direct antagonism through the production of toxins, enzymes, or other secondary metabolites, interference with plant-nematode recognition, competition for nutrients, plant growth promotion, and induced systemic resistance is recognized as the mechanism for nematode antagonism (reviewed in Tian et al. 2007). Deny®, a commercial nematicide based on *Burkholderia cepacia*; BioNemWP® and BioSafe®, two biological nematicides based on lyophilized *Bacillus firmus* are being marketed mainly for controlling *Meloidogyne* spp.

P. fluorescens (strain CHA0) produces several antibiotic compounds like phenazines, tropolone, pyrrolnitrin, pyocyanin, hydrogen cyanide, and 2-4-diacethylphlorooglucinol that play a role in nematode control. Extracellular protease aprA from *Pseudomonas fluorescens* strain CHA0 reduced egg hatching by 45%, reduced juvenile mobility, and enhanced juvenile mortality of *M. incognita* and *M. javanica* (Siddiqui et al. 2005). Endophytic bacteria, like endoparasitic nematodes, colonize the internal plant tissue, and their beneficial effects on plant-parasitic nematodes were demonstrated (Siddiqui and Mahmood 1999).

Transgenic plants expressing the Bt Cry6A protein have some potential for suppressing plant-parasitic nematodes. Commercial products containing Bt, such as Dipel and Turex, have been shown to reduce damage caused by root-knot nematodes (Radwan 2007). Li et al. (2007) expressed that *M. incognita* could ingest the 54-kDa Cry6A protein, and that Cry6A was toxic to the J2, as indicated by a decrease of up to fourfold in progeny production.

Another group of nematode antagonists is actinomycetes and exemplified by *Streptomyces avermitilis*. This species produces macrocyclic lactones (avermectins), which are highly nematicidal compounds. For example, the abamectin B1 is now commercialized under the name Avicta® as a seed treatment for cotton and vegetables against plant-parasitic nematodes.

1.12.3 Parasitic Fungi

Several fungi have been isolated from different nematode stages; these may be both obligate and facultative parasites. The facultative parasite *Purpureocillium lilacinum* (=*Paecilomyces lilacinus*) has been the most investigated. It also produces antibiotics such as leucinostatin and lilacin and enzymes such as protease and chitinase. Protease has nematicidal activity, causes degradation of the eggshell, and inhibits hatching. Originally isolated from eggs of *M. incognita* infecting potato in Peru by Jatala et al. (1979), the fungus has been demonstrated to parasite the eggs of major plant-parasitic nematodes, including those of root-knot nematodes. The major structural changes that occur in eggs treated with protease and chitinase from *P. lilacinum* strain 251 involve the loss of the lipid layer and disintegration of the vitelline layer, which contains proteins. *P. lilacinum* strain 251 is now available commercially under different trade names in several countries (Table 1.7) (EPA 2005; Kiewnick and Sikora 2006). *P. lilacinum* strain PL1 (T. Stanes & Co.) has been granted regular registration in India. Unlike many other *P. lilacinum* strains, the registered strains do not produce mycotoxins or paecilotoxins. *P. lilacinum* has been

Table 1.7 *Purpureocillium lilacinum* (= *Paecilomyces lilacinus*) products

P. lilacinum strain	Product name	Manufacturer	Region/country
P. lilacinum 251	BioAct WG	Bayer crop Science Prophyta	USA, Canada, Brazil, Colombia, Peru, Mexico, Argentina, Costa Rica, Bulgaria, New Caledonia, EU, UK, Spain, Greece, France, Switzerland, Italy, Bulgaria, Kenya, South Africa, Morocco, Turkey, China, the Philippines, Australia
P. lilacinum 251	Melcon WG	CertisInc	USA, Canada
P. lilacinum strain F18	MYTECH WP	Dudu Tech	USA, Africa, Kenya, Canada, Asia
P. lilacinum strain PL11 TGAI	BioStat WP	LAM international	USA, Canada
P. lilacinum strain BCP2	PL Gold	BASF	China, Asia, Africa, EU
P. lilacinum	Rem G[a]	Green Solutions	Italy
P. lilacinum strain	NemaxxionBiol[a]	Green Corp	Mexico
P. lilacinum strain BCC 19497 and 4119		TBRC	Thailand
P. lilacinum strain	Paecil	Australian Technological Innovation Corporation Pvt. Ltd.	Australia
P. lilacinum strain P56	–	–	Brazil
P. lilacinum Strain PL1	Bionematon	T. Stanes	India

[a] As Consortium with other fungal/bacterial spp. in formulation

tested extensively against root-knot nematodes in different crops in diverse agro-climatic conditions in India under AICRP (Nematodes). Many centers have included this in their Package of Practices.

Another facultative parasite, *Pochonia chlamydosporia* (=*Verticillium chlamydosporium*) is a promising biocontrol agent. It is also basically an egg parasite. *P. chalamydosporium* produces a branched mycelial network in close contact with the eggshell (Morgan-Jones et al. 1983; Lopez Llorca and Duncan 1988; Lopez Llorca and Claugher 1990). The penetration of the eggshell leads to the disintegration of the vitelline layer and the dissolution of the chitin and lipid layers (Segers et al. 1996; Morton et al. 2004). *P. chlamydosporia* also secretes aurovertin and phomalactone, which are toxic to both egg and juvenile stages of *M. incognita* (Khambay et al. 2000).

Obligate fungal parasites can infect nematodes through their spores, either by direct ingestion or adhering to the nematode cuticle. Parasitic fungi with adhesive spores belong to several classes, for example, biflagellate zoospores of *Catenaria anguillulae*, *Myzocytium lenticulare* (Oomycetes), spherical conidia of *Meristracum asterospermum* (Zygomycetes), club-like spores of *Meria coniospora* (Deuteromycetes), and adhesive spores of *Nematoctonus leiosporus* (Basidiomycetes). *Hirsutella rhossiliensis* (Hyphomycetes) has a density-dependent relationship with its host nematode (Jaffee et al. 1992), and, therefore, might be expected to be able to control plant-parasitic nematodes successfully.

1.12.4 Predacious Fungi

Predacious or trapping fungi were the earliest to be investigated for nematode control during 1950–70. These are common saprophytic fungi that can trap the larval or adult stages of the nematodes they feed on. They have a variety of fascinating nematode-trapping structures, for example, adhesive hyphae (*Stylopage* spp.), adhesive branches (*Monacrosporium cianopagum*), network traps (*Arthrobrotrys oligospora* and *A. superba*), constrictive rings (*A. anchonia*, *A. dactyloides*, *Dactylaria brochopaga*, etc.), or adhesive knobs (*Monacrosporium cianopagum* and *Dactylella lobata*). Some fungi, such as *Dactylaria candida* (Hyphomycetes), present two types of trapping mechanisms: (1) adhesive knobs and (2) constrictive but non-adhesive rings. Predacious fungi prey on and trap both plant-parasitic and free-living nematodes indiscriminately. Trap formation is induced with the production of peptides by extracellular proteases hydrolyzing the nematodes' cuticles (Huang et al. 2006). Interestingly, the earliest commercial bioagents of nematodes were based on this group of fungi (*Arthrobotrys irregularis*). Royal 300 and Royal 350 were launched in France to control mushroom and root-knot nematodes on tomatoes, respectively (Cayrol and Frankowski 1980).

1.12.5 Fungal Antagonists

Several species within the genus *Trichoderma*, such as *T. harzianum* and *T. viride* provide excellent control of root-knot nematodes (Sharon et al. 2001, 2007). *T. harzianum* is not able to grow on gelatinous matrices but colonizes isolated eggs and J2 of *M. javanica* (Sharon et al. 2001). The involvement of lytic enzymes such as chitinase, glucanases, and proteases in *Meloidogyne* parasitism was demonstrated in the case of *T. asperellum* carrying a fusion of the proteinase or chitinase promoters (Spiegel et al. 2005). Fungal metabolites (such as trichodermin, a nematicidal sesquiterpene) and induced resistance are other mechanisms involved in nematode control by *Trichoderma* spp. (Umamaheswari et al. 2004).

1.12.6 Endophytic Fungi

Sikora and Schonbeck (1975) first demonstrated the potential of arbuscular mycorrhizal (AM) fungi to reduce infestation by *Meloidogyne* spp. in vegetables. Further, Saleh and Sikora (1984) reported that 38% root colonization by *Glomus fasciculatum* was required to control *M. incognita* in cotton. AM fungi are beneficial in multiple ways – absorption and accumulation of nutrients like phosphorus leading to imparting tolerance in plants against nematode infection, competition with nematodes for food and space, and imparting resistance in mycorrhyzal feeder roots to nematode infection as well as other soil-borne pathogens (Diedhiou et al. 2003; Elsen et al. 2008).

1.13 Host Plant Resistance

Resistance offers the best option for nematode management because of its efficacy in nematode population reduction, cost-effectiveness, compatibility with other management tactics, and environmental safety. The first-ever nematode-resistant cowpea variety "Iron" was developed by Weber and Orton (1902), conferring resistance to root-knot nematode. Identifying several nematode-resistant genes has provided insights into the possible mechanism for achieving resistant phenotypes.

The tomato gene *Mi1* is the best-characterized nematode resistance gene that confers resistance against *M. incognita, M. javanica,* and *M. arenaria* (Williamson 1998). The *Mi1*-mediated resistance was initially discovered in *Lycopersicon peruvianum* (wild species) and introgressed to cultivated tomato (*L. esculentum*) by embryo rescue technique (Smith 1944). Genetically linked molecular markers, first the isozyme acid phosphatase and later PCR markers, were used as aids in introgression (Williamson 1998). However, the gene is ineffective at high soil temperatures (>28 °C). Additionally, some isolates of *M. incognita, M. javanica,* and *M. arenaria* virulent on *Mi1* tomato have been identified (Williamson and Kumar 2006; Bozbuga et al. 2020). J2 does not elicit an extensive HR while penetrating or migrating through the root tissue but while attempting to establish a feeding site (Paulson and Webster 1972; Ho et al. 1992). The resistant reaction is characterized by localized host cell death. *Mi1* was the first root-knot nematode resistance gene cloned (Milligan et al. 1998; Vos et al. 1998).

Breeding for resistance involves the same basic principles as are used in breeding for resistance to other pathogens: (1) identification of root-knot nematode species/race, (2) establishment of pure culture, (3) standardization of screening methods including marker-assisted selection, (4) sources of resistance and study on inheritance of resistance, (5) breeding commercially viable resistant lines through crossing/backcrossing, and (6) rigorous testing under field conditions.

Some of the crop varieties that proved to be resistant against major root-knot nematode species are given below (Table 1.8).

RNAi technology is a particularly promising tool for understanding virulence traits in the nematode and resistance pathways in the host (Williamson and Kumar

Table 1.8 Vegetable crop varieties identified/developed resistant to plant-parasitic nematodes in India

Crop	Root-knot nematodes	Resistant varieties
Tomato	*M. javanica*, *M. incognita*	PNR-7, NT-3, NT-12, Hisar Lalit
Chilli	*M. javanica*, *M. incognita*	NP-46A, Pusa Jwala, Mohini
Cowpea	*M. javanica*, *M. incognita*	GAU-1
Brinjal	*M. incognita*	Gachhabaigan, Azadkranti, Kantabaigen, Athagara Local, Kamaghara local, Utkal Madhuri (Nayak and Pandey 2015)

2006). Basic research in molecular plant nematology is expanding the knowledge that can be applied to provide crop resistance to parasitic nematodes in an economically and environmentally benign manner.

1.14 Chemical Control

It is difficult to cite and compile the voluminous work conducted on chemical control of root-knot nematodes in India in different crops. The experiments conducted under the aegis of AICRP (Nematodes) and other crop-based AICRPs have led to several recommendations on chemical control of root-knot nematodes in different crops. They are included in POPs of various universities.

The chronology of events on chemical control is similar to world history. During the 1960s, the first trials on chemical control of root-knot nematode included halogenated hydrocarbon fumigants like DD and EDB (Sen 1960; Nirula 1961). Methyl bromide was used for experimental purposes to control root-knot nematodes in tobacco to a limited extent (Hussaini 1985). Another class of soil fumigants relating to methyl isothiocyanates, that is, dazomet and metham sodium, were introduced in 1952 and 1956, respectively. All these fumigants were not used extensively in India because of their hazardous nature, high toxicity to non-target organisms, inherent difficulties involved in their applications, etc. DBCP was relatively easy to apply through irrigation and was used commercially in grapevine and tobacco but had to be withdrawn because of bromine residues.

The scenario shifted towards organophosphates and carbamates from the 1970s onwards. The granular formulations of these chemicals offered easier applications at much lower dosages (1–4 kg a.i./ha), besides being systemic in action. Bhatti and Jain (1984) reviewed the information on the chemical control of root-knot nematodes in India. Among the organophosphates, fensulfothion, phosphamidon, fenamiphos, and ethoprophos, and carbamates like aldicarb, carbofuran, oxamyl, etc. were tried and tested under AICRP (Nematodes). However, the cost factor was a major deterrent for overall field treatments. To save costs, the economic methods of their applications were standardized. These included nursery bed treatments and bare root dip treatment of seedlings for transplanted vegetables (Anonymous 1991; Haq et al. 1980; Jain and Bhatti 1978, 1983a, b; Tiyagi et al. 1986), and seed treatments for crops like cowpea, pea, French bean (Parvatha Reddy 1984). Seed coating with Posse (25ST) and UC54229 (100SP) at 3% and 6% (a.i. w/w) lowered root galling and promoted plant growth. Based on this work, mainly carbofuran emerged as the single most extensively used nematicide in vegetables and fruit crops till now. However, the imminent withdrawal of these toxic molecules is again leading to a big void, and replacements are urgently warranted.

Fortunately, some new developments have taken place recently that offer an exciting phase on the chemical control of nematodes in India. New chemical nematicidal molecules have been launched recently that offer better standards for nematode control with excellent safety profile and high efficacy with very low rates. Two products that have been granted registration by Central Insecticides Board & Registration Committee (Ministry of Agriculture, Government of India) recently are discussed below, along with their key features as per their manufacturers.

1.14.1 Fluopyram: Launched Under the Trade Name Velum® Prime by Bayer CropScience

- Chemical class: Pyridinyl ethyl benzamide.

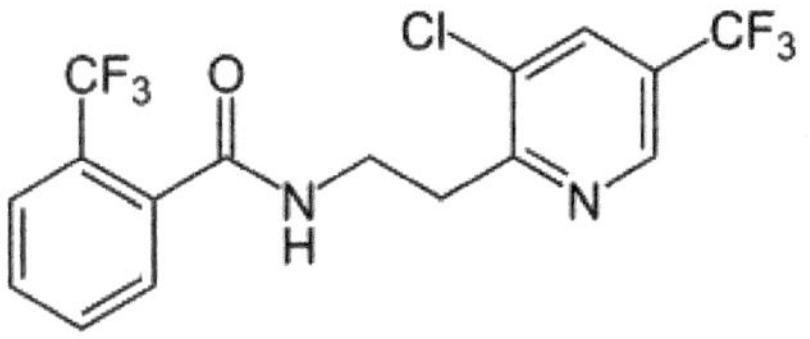

- Mode of action: Inhibition of the enzyme succinate dehydrogenase (SDH, complex II) within the nematode mitochondrial respiratory chain.
- Formulation: 400SC in India.
- Fast, effective, and long-term control of key nematodes in a wide range of crops.
- A revolutionary safety profile for operators and the environment.
- Efficacy at unprecedentedly low application rates, that is, 500 g a.i./ha (1250 ml/ha product) as a soil drench and drip for tomato. Since a sufficient level of soil moisture is required to activate Velum Prime, the beds must be applied with an adequate amount of water followed by light irrigation after 1–3 days of application.
- High application flexibility and convenient crop management.
- Wide profile of MRLs and import tolerance.
- Profitable and sustainable farm management.

1.14.2 Fluensulfone: Launched Under the Trade Name NIMITZ® by Adama

- Chemical name: 5-chloro-2- [(3,4,4trifluoro-3-buten-1-yl)sulfonyl]thiazole
- Chemical group: Heterocyclic fluoroalkenyl sulfones

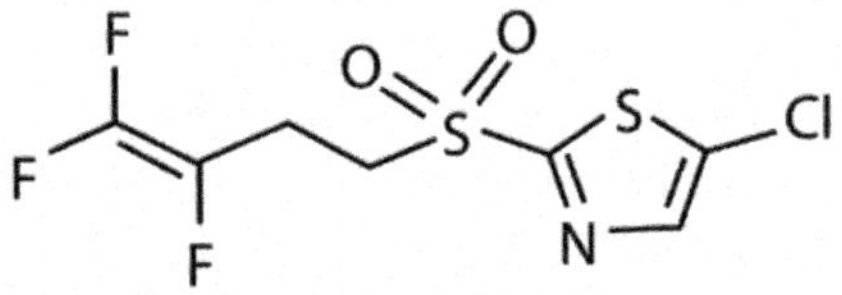

- Mode of action: Rapid irreversible paralysis
- Formulation: 2GR in India
- Innovative and safe chemistry
- Easy to handle granular formulation
- Highly effective against root-knot nematodes (cucumber, tomato, and okra)
  - Dose: 1.5 g per plant on tomato by ring method (two applications at 25 days interval)
- Non-fumigant, safe to applicator
- Very safe for humans and environment
- Long-duration control of nematodes

1.15 Integrated Management

Going by the definition, Integrated Nematode Management (INM) seeks to stabilize the pest population below damaging levels through the integration of various effective and unilateral practices leading to a long-term package programme based on ecologically sound, economically viable, and acceptable principles.

Unlike other pests and diseases, managing root-knot nematodes poses a serious challenge. The choice of unilateral management practices for integrating into INM based on the above-mentioned principles would vary according to crop, cropping system, and agro-climatic region. AICRP (Nematodes) embarked upon this programme quite early and has developed tangible INM programmes for various crops and regions. These INM technologies must be integrated with overall IPM modules for specific crops. In endemic areas (hot spots) where root-knot nematode is a major problem among other pests and diseases, nematode-centric IPM modules have been worked out and demonstrated in collaboration with ICAR-National Research Centre for Integrated Pest Management (NCIPM) in vegetables and rice successfully. The economic benefits to the farmers have been worked out.

A specific example of managing root-knot nematode problems in a protected cultivation system is cited as a case history.

1.15.1 Managing Root-knot Nematode Problem in Polyhouses

Respective governments are giving many subsidies for establishing protective cultivation systems. Consequently, polyhouses, in particular, have sprung up all across the country in a big way, and the area under protected cultivation is expanding. Polyhouses are subjected to intensive cultivation of high-value crops such as tomatoes, cucumber, and capsicum among vegetables, gerbera, carnations, lilies, and roses among ornamentals. All these crops are good to excellent hosts of root-knot nematodes. The microclimatic conditions inside the polyhouses provide continued favourable conditions for the multiplication of nematodes. The relatively higher temperature inside the polyhouse during winters, availability of optimum soil moisture due to drip irrigation systems, and growth of host crops continuously lead to nematode population explosion. This problem has been confronting the nematologists for the last few years, and in the absence of any sound management technology, many polyhouse growers were forced to abandon their facilities. Root-knot nematode problem came up in a big way in states like Punjab, Haryana, Rajasthan, Gujarat, Maharashtra, Karnataka, Andhra Pradesh, Telangana, Tamil Nadu, and Western Uttar Pradesh, besides Himachal Pradesh. Ironically, the polyhouse growers in the northern states particularly intend to grow cucumber after cucumber due to assured market price and short-duration crop.

It was soon realized that the basic problem of "constructing new polyhouse on nematode-infested land" leads to severe nematode infestation from the very first crop. It was, therefore, suggested to the state governments to enforce mandatory soil testing for nematode infestation in the proposed sites of new polyhouses and deny construction on root-knot nematode-infested lands. The Government of Haryana accepted the proposal. Prior soil testing has resulted in the containment of the nematode problem to a large extent, and the trend of closure of polyhouses has been reversed.

Many nematode management technologies relevant to vegetable crops in open field conditions become redundant under protected cultivation systems, for example, crop rotation with non-hosts. Polyhouse growers stick to cultivating cucumber, tomato, and capsicum only. No nematode-resistant cultivars are available that can be grown under polyhouse conditions. Much experimentation was done using fumigants like metham sodium, dazomet, silver nanoparticles, and formaldehyde. Ultimately, an INM protocol based on summer soil solarization, organic amendments, and bioagents was developed and tested at multilocations that have provided interim relief for the management of nematode problems in polyhouses. However, the recent developments with regard to the availability of newer chemical molecules (see Sects. 1.14.1 and 1.14.2) have solved the problem to a large extent.

1.15.2 Emerging Problems of Root-knot Nematodes (Source: Walia and Khan 2018)

Onion and garlic, hitherto considered antagonistic crops to root-knot nematodes, are being intercepted as susceptible to this nematode in several pockets of the country. In some cases, the species has been identified as *M. graminicola*. Similarly, some districts in north Gujarat exhibit large-scale root-knot nematode infestation on potato tubers, resulting in huge qualitative losses. Such infestations of potato have been recorded in northeast India.

1.16 Conclusions

Meloidogyne spp. constitutes the most formidable challenge for nematologists. The figures on recent estimations of crop losses presented in Table 1.2 reveal some interesting facts: (1) root-knot nematodes stand alone as the most damaging nematodes, causing 76% of the total losses inflicted by all plant-parasitic nematodes; (2) the monitory losses caused to field crops and horticultural crops are almost equal, although horticulture crops occupy very limited areas; (3) the mean per cent losses are higher in vegetable, fruits, and spices; and (4) the data is also a pointer to focus our management strategies on particular crops.

The agenda on basic aspects should focus on understanding the genetic diversity of rice root-knot nematode, *Meloidogyne graminicola*, rather "*M. graminicola* complex." The knowledge of the existence of races will never be complete, irrespective of species; however, *M. graminicola* deserves special attention because of its geographical distribution, economic importance, genetic diversity of its host crop (rice), and potential use of resistant planting material (*Oryza glaberrima*) for developing nematode-resistant rice varieties.

Prior soil testing for root-knot nematode at the proposed sites for constructing new polyhouses can greatly help. At least the polyhouse grower will not have to confront nematode problems for quite some time, and with careful planning long-term management of nematodes will be possible. The Government of Haryana has made it mandatory; others should follow suit.

We firmly believe that the problem of root-knot nematodes is bound to accentuate further due to the following main reasons:

1. *Unhindered and unchecked dissemination of nematodes through infected planting materials* from horticulture nurseries due to unawareness at all levels of stakeholders, that is, nurserymen (both private and public sector), farmers (orchardists), and government horticulture officers. Ironically, our legal framework is silent on the issues related to pests and diseases (including nematodes) applicable to horticultural nurseries and the interstate movement of planting materials. Periodical and random examination of nurseries for incidence of pests and diseases, especially nematodes, becomes redundant for want of trained human resources. Sensing the gravity of this situation, AICRP (Nematodes)

launched a series of initiatives to create awareness among all stakeholders by proposing the promulgation of nematode-centric nursery laws with the Government of India, a series of thematic workshops to sensitize government agriculture/horticulture officers as well as nurserymen and nursery managers, through print and electronic media, with tangible results. Such efforts need to be strengthened and carried forward more vigorously.

2. *Shift towards water-saving irrigation systems such as drip irrigation* and *sprinkler irrigation* that ensure continued availability of near-optimum soil moisture conditions for nematode infection. While crop husbandry and water scarcity necessitate such irrigation systems, we have to redesign our protocols to facilitate the application of chemical nematicides, bioagents, as well as other products for application through micro-irrigation systems. Already liquid formulations are being marketed by the corporate and public sectors.
3. *Intensive cultivation and shift towards horticultural cropping systems*. Growers are not inclined to go for lesser remunerative crops for nematode management. The problem is also beset by the non-availability of nematode-resistant cultivars of crops demanded by growers. Polyhouses pose a serious challenge of high magnitude in this category. Therefore, an intelligent blend of various methods must be developed to check the nematode population below ETL. The newer chemical molecules with better environmental safety properties are now available.

Acknowledgements Most of the photographs used in this chapter have been taken from resources and a rich collection of MACTODE and AICRP (Nematodes). The authors acknowledge the same.

References

Adam MAM, Phillips MS, Blok VC (2007) Molecular diagnostic key for identification of single juveniles of seven common and economically important species of root-knot nematode (*Meloidogyne* spp.). Plant Pathol 56:190–197

Akhtar M (2000) Nematicidal potential of the neem tree *Azadirachta indica* (A. Juss.). Integr Pest Manag Rev 5:57–66

Alam MM, Saxena SK, Khan AM (1981) Soil population of plant parasitic nematodes under various cropping sequences. Nematol Mediterr 9:49–56

Alam MM, Samad A, Anver S (1990) Interaction between tomato mosaic virus and *Meloidogyne incognita* in tomato. Nematol Mediterr 18:131–133

Anonymous (1991) Achievements (1977–87). Technical Bulletin No 1., All India Coordinated Research Project on Plant Parasitic Nematodes, ICAR, New Delhi, 27 pp

Anonymous (2012) Consolidated Annual Report, AICRP (Nematodes). In: Jain RK, Singh AU, Kumar V (eds) P.C. Cell, Division of Nematology. IARI, New Delhi, 160 pp

Anonymous (2014) AGMARKETNET Portal – Commodity-wise, month-wise arrival and price of horticulture produce. www.agmarketnet.nic.in/nhm/horticulture.aspx

Anonymous (2015) Agriculture statistics at a glance. Govt. of India, Ministry of Agriculture & Farmers' Welfare, Department of Agriculture & Cooperation, Directorate of Economics & Statistics, Krishi Bhawan, New Delhi, 507 pp

Anonymous (2016) Pocketbook of agricultural statistics. Govt. of India, Ministry of Agriculture & Farmers' Welfare, Department of Agriculture & Cooperation, Directorate of Economics & Statistics, Krishi Bhawan, New Delhi, 154 pp

Atkinson GF (1892) Some diseases of cotton. Alabama Polytech Instit Agric Exp Station Bull 41: 61–65

Ayyar PNK (1926) A preliminary note on the root gall nematode, *Heterodera radicicola* Miller and its economic importance in South India. Madras Agric J 14:113–118

Ayyar PNK (1933) Some experiments on the root gall nematode (*Heterodera radicicola* Greef) in South India. Madras Agric J 21:97–107

Ayyar PNK (1934) Further experiments on the root gall nematode, *Heterodera marioni* (Cornu) Goodey in South India. Indian J Agric Sci 3:1064–1071

Back MA, Haydock PPJ, Jenkinson P (2002) Disease complexes involving plant parasitic nematodes and soil–borne pathogens. Plant Pathol 51:683–697

Barber CA (1901) A tea eelworm disease in South India. Bull Depart Land Use Agric Madras 2: 227–234

Barker KR, Carter CC, Sasser JN (1985) An advanced Treatise on *Meloidogyne*. Vol. II. Methodology. North Carolina State University Graphics, Raleigh

Berge JB, Dalmasso A (1975) Biochemical characteristics of several populations *of Meloidogyne hapla* and *Meloidogyne* spp. *Cahiers O.R.S.T.O.M* Serie Biologie. Nematology 10:263–271

Berkeley MJ (1855) Vibrio forming cysts on the roots of cucumbers. Gardener's Chronicle Agric Gazette 14:220

Bernhard R, Bouquet A, Scotto La Massèse C (1985) Diversité des problèmesnématologiquesen vergers, solutions chimiques et génétiques. Création de variétésrésistantes aux nématodes des cultures: intérêt, possibilités et limites. C r séances Acad agric Fr 71:705–718

Bhatti DS, Jain RK (1984) Chemical control of root-knot nematode in India: a review. Haryana J Hort Sci 13:62–68

Bishop AH, Gowen SR, Pembroke B, Trotter JR (2007) Morphological and molecular characteristics of a new species of *Pasteuria* parasitic on *Meloidogyne ardenensis*. J Invertebr Pathol 96:28–33

Blok VC, Powers TO (2009) Biochemical and molecular identification. In: Perry RN, Moens M, Starr J (eds) Root-knot nematodes. CABI, Lincoln, pp 98–112

Blok VC, Wishart J, Fargette M, Berthier K, Phillips MS (2002) Mitochondrial differences distinguishing *Meloidogyne mayaguensis* from the major species of tropical root-knot nematodes. Nematology 4:773–781

Bozbuga R, Dasgan H, Akhoundnejad Y, İmren M, Cesur O, Toktay H (2020) Effect of Mi gene and nematode resistance on tomato genotypes using molecular and screening Assay. Cytol Genet 54:154–164

Brito J, Powers TO, Mullin PG, Inserra RN, Dickson DW (2004) Morphological and molecular characterisation of *Meloidogyne mayaguensis* isolates from Florida. J Nematol 36:232–240

Brown PD, Morra MJ (1997) Control of soil-borne plant pests using glucosinolate-containing plants. Adv Agron 61:167–231

Carneiro RMDG, Almeida MRA, Queneherve P (2000) Enzyme phenotypes of *Meloidogyne* spp. populations. Nematology 2:645–654

Castagnone-Sereno P, Esparrago G, Abad P, Leroy F, Bongiovanni M (1995) Satellite DNA as a target for PCR-specific detection of the plant parasitic nematode *Meloidogyne hapla*. Curr Gene 28:566–570

Cayrol JC, Frankowski JP (1980) New information on the nematophagus fungus *Arthrobotrys irregularis* (Royal 350). PHM revue horticole 203:33–38

Chahal PPK, Chahal VPS (1989) Effect of *Meloidogyne incognita* on leghaemoglobin, bacteroids and nitrogenase activity of nodules produced by *Rhizobium* spp., on mungbean (*Vigna radiata* L.). Pak J Nematol 7:21–25

Chandra B, Khan MR (2011) Dynamics of soil nematodes in vegetable based crop sequences in West Bengal, India. J Plant Prot Res 51:7–13

Chen YF, Wu JY, Hu XO Yu SF (1998) Using phast system for rapid identification of root-knot nematodes. Acta Phytopathol Sin 28:73–77

Chitwood BG (1949) Root-knot nematodes Part 1. A revision of the genus *Meloidogyne* Göldi, 1887. Proc Helminthol Soc Wash 16:90–104

Chitwood DJ (2002) Phytochemical based strategies for nematode control. Annu Rev Phytopathol 40:221–249

Cofcewicz ET, Carneiro RMDG, Castagnone-Sereno P, Queneherve P (2004) Enzyme phenotypes and genetic diversity of root-knot nematodes parasitizing *Musa* in Brazil. Nematology 6:85–95

Cornu M (1879) Etudes sur le *Phylloxera vastatrix*. Memoires Presentes Par Divers Savants A L'Academie Des Sciences de L'Institut de France 26:1–357

Curran J (1991) Application of DNA analysis to nematode taxonomy. In: Nickle WR (ed) Manual of agricultural nematology. Marcel Dekker, New York, pp 125–143

Curran J, Robinson MP (1993) Molecular aids to nematode diagnosis. In: Evans K, Trudgill DL, Webster JM (eds) Plant parasitic nematodes in temperate agriculture. CABI, Wallingford, pp 545–564

da Ponte JJ, Cavada BS, Silveira-Filho J (1996) Teste com lectina no controle de *Meloidogyne incognita* emtomateiro. Fitopatol Bras 21:489–491

Dalmasso A, Berge JB (1978) Molecular polymorphism and phylogenetic relationship in some *Meloidogyne* spp. application to the taxonomy of *Meloidogyne*. J Nematol 10:323–332

Dasgupta DR, Gaur HS (1986) The root-knot nematodes, *Meloidogyne* spp. in India. In: Swarup G, Dasgupta DR (eds) Plant parasitic nematodes of India – problems and progress. New Delhi, IARI, pp 139–171

De Giorgi C, Sialer MF, di Vito M, Lamberti F (1994) Identification of plant-parasitic nematodes by PCR amplification of DNA fragments. Bull OEPP 24:447–451

De Waele D, Davide RG (1998) The root–knot nematodes of banana. http://musalit.inibap.org/pdf/IN020258_en.pdf. Accessed 11 Feb 2008

Devappa V, Krishnappa K, Reddy BMR (1998) Estimation of avoidable losses due to root-knot nematode, *Meloidogyne incognita* in sunflower. Indian J Nematol 28:95

Devran Z, Sogut MA (2011) Characterizing races of *Meloidogyne incognita*, *M. javanica* and *M. arenaria* in the West Mediterranean region of Turkey. Crop Prot 30:451–455

Dickson DW, Sasser JN, Huisingh D (1970) Comparative disc-electrophoretic protein analyses of selected *Meloidogyne*, *Ditylenchus*, *Heterodera*, and *Aphelenchus* spp. J Nematol 2:286–293

Diedhiou PM, Hallmann J, Oerke EC, Dehne HW (2003) Effects of arbuscular mycorrhizal fungi and a non-pathogenic *Fusarium oxysporum* on *Meloidogyne incognita* infestation of tomato. Mycorrhiza 13:199–204

Dong K, Dean RA, Fortnum BA, Lewis SA (2001a) A species-specific DNA probe for the identification of *Meloidogyne hapla*. Nematropica 31:17–23

Dong KRA, Fortnum BA, Lewis SA (2001b) Development of PCR primers to identify species of root-knot nematodes: *Meloidogyne arenaria, M. hapla, M. incognita* and *M. javanica*. Nematropica 31:273–282

Dutta TK, Khan MR, Phani V (2019) Plant-parasitic nematode management via biofumigation using brassica and non-brassica plants: current status and future prospects. Curr Plant Biol 17: 17–32

Ehteshamul-Haque S (1997) Use of crustacean chitin and plant growth promoting bacteria for the control of *Meloidogyne javanica* root-knot nematode in chickpea. Pak J Nematol 15:89–93

Eisenback JD (1982) Morphological comparison of head shape and stylet morphology of second-stage juveniles of *Meloidogyne* species. J Nematol 14:339–343

Eisenback JD (1985) Diagnostic characters useful in the identification of the four most common species of root-knot nematodes (*Meloidogyne* spp.). In: Sasser JN, Carter CC (eds). An advanced treatise on *Meloidogyne*. Vol. 1. Biology and control. North Carolina State University Graphics, Raleigh, pp 95–112

Eisenback JD, Hirschmann H, Sasser JN Triantaphyllou AC (1981) A guide to the four most common species of root-knot nematodes, *Meloidogyne* species with pictorial keys. Department of Plant Pathology and Genetics, North Carolina State University and USAID, Raleigh, 47 pp

Elsen A, Gervacio D, Swennen R, de Waele D (2008) AMF-induced biocontrol against plant parasitic nematodes in *Musa* sp.: a systemic effect. Mycorrhiza 18:251–256

EPA US (Environmental Protection Agency) (2005) *Paecilomyces lilacinus* strain 251; exemption from the requirement of a tolerance. United States Federal Register 70:19278–19283

Esbenshade PR, Triantaphyllou AC (1985) Identification of major *Meloidogyne* species employing enzyme phenotypes as differentiating characters. In: Sasser JN, Carter CC (eds) An advanced treatise on *Meloidogyne*, Vol. I: biology and control. North Carolina State University Graphics, Raleigh, pp 135–140

Esbenshade PR, Triantaphyllou AC (1990) Isozyme phenotypes for the identification of *Meloidogyne* species. J Nematol 22:10–15

Fargette M (1987) Use of the esterase phenotype in the taxonomy of the genus *Meloidogyne*. 2. Esterase phenotypes observed in West African populations and their characterization. Rev Nematol 10:45–56

Ferraz S, de Freitas LG (2004) Use of antagonistic plants and natural products. In: Chen ZX, Chen SY, Dickson DW (eds) Nematology advances and perspectives, vol 2. CABI, Wallingford, pp 931–977

Ganguly S, Misra RL, Mishra SD (1993) New disease complex of tuberose (*Polianthes tuberosa*) involving root-knot nematode, *Meloidogyne incognita* and a mite species. Curr Nematol 4:113–114

Gaur HS (1975) Crop damage in relation to the density of nematode population and an integrated approach of nematode population management. Ph.D. Thesis, IARI, New Delhi, 184 pp

Gaur HS, Perry RN (1991) The biology and control of the plant parasitic nematode, *Rotylenchulus reniformis*. In: Evans K (ed) Agricultural zoological review, vol 4. Intercept, UK, pp 177–211

Gaur HS, Mishr SD, Sud UC (1979) Effect of date of sowing on the relation between the population density the root-knot nematode, *Meloidogyne incognita* and the growth of three varieties of chickpea, *Cicer arietinum*. Indian J Nematol 7:155–156

Gaur HS, Perry RN, von Mende N (1996) Differentiation of two groups of species of the genus *Meloidogyne* by polymerase chain reaction and restriction fragment length polymorphisms of ribosomal DNA. Afro-Asian J Nematol 6:50–54

Göldi EA (1892) Relatoriasôbre a molestia do cafeirona provincial da Rio de Janeiro. Arch Mus Nac (Rio de J) 8:1–112

Gommers FJ, Bakker J, Smith L (1980) Effect of singlet oxygen generated by the nematicidal compound ∞-terthienyl from *Tagetes* on the nematode *Aphelenchus avenae*. Nematologica 6: 369–375

Goodey T (1932) On the nomenclature of root-gall nematodes. J Helminthol 10:21–28

Goswami BK, Chenulu VV (1974) Interaction of root-knot nematodc, *Meloidogyne incognita* and tobacco mosaic virus in tomato. Indian J Nematol 4:69–80

Goswami BK, Singh S, Verma VS (1974) Interaction of a mosaic virus with root-knot nematode *Meloidogyne incognita* in *Vigna sinensis*. Nematologica 20:366–367

Gowda M, Sellaperumal C, Rai A, Singh B (2019) Root-knot nematodes menace in vegetable crops and their management in India: a review, ICAR-IIVR, Varanasi, Tech Bull No. 46:1–16

Greef R (1872) Uebernematoden in Wurtzelanschwellungen (Gallen) verschiedeneer Pflanzen. Sitzber Ges Naturw Marburg:169–174

Haq S, Saxena SK, Khan MW (1980) Inhibition of *Meloidogyne incognita* larvae penetration of tomato seedlings with different modes of application of Rogor. Acta Bot Indica 8:201–202

Haque MM, Gaur HS (1985) Effect of multiple cropping sequences on the dynamics of nematode population and crop performance. Indian J Nematol 15:262–263

Harris TS, Sandall LJ, Powers TO (1990) Identification of single *Meloidogyne* juveniles by polymerase chain reaction amplification of mitochondrial DNA. J Nematol 22:518–524

Hartman KM, Sasser JN (1985) Identification of *Meloidogyne* species on the basis of differential host test and perineal pattern morphology. In: Barker KR, Carter CC, Sasser JN (eds) An advanced treatise on *Meloidogyne*. Vol. II: Methodology. A Cooperative Publication of the Department of Plant Pathology & United States Agency for International Development, North Carolina State University Graphics, Raleigh, pp 69–77

Haseeb A, Khan AM, Saxena SK (1982) Toxicity of leaf extracts of plants to root-knot and reniform nematode. Indian J Parasitol 6:119–120

Haydock PPJ, Woods SR, Grove IG, Hare M (2006) Chemical control of nematodes. In: Perry RN, Moens M (eds) Plant nematology. CABI, Wallingford, pp 392–410

Hernandez A, Fargette M, Sarah JL (2004) Characterization of *Meloidogyne* spp. (Tylenchida: Meloidogynidae) from coffee plantations in Central America and Brazil. Nematology 6:193–204

Hewlett TE, Gerber JF, Smith KS (2004) In vitro culture of Pasteuria penetrans. In: Cook R, Hunt DJ (eds) Nematology monographs and perspectives, vol 2. Koninklijke Brill, Leiden, pp 175–185

Hinge VR, Patel BA, Vyas RV (2010) Differentiation among four *Meloidogyne* species from Gujarat by random amplified polymorphic DNA (RAPD–PCR). Indian J Nematol 40:167–170

Ho J-Y, Weide R, Ma HM, Wordragen MF, Lambert KN, Koornneef M, Zabel P, Williamson VM (1992) The root-knot nematode resistance gene (Mi) in tomato: construction of a molecular linkage map and identification of dominant cDNA markers in resistant genotypes. Plant J 2:971–982

Huang G, Allen R, Davis EL, Baum TJ, Hussey RS (2006) Engineering broad root-knot resistance in transgenic plants by RNAi silencing of a conserved and essential root-knot nematode parasitism gene. Proc Natl Acad Sci USA 103:14302–14306

Hunt DJ, Handoo ZA (2009) Taxonomy, identification and principal species. In: Perry RN, Moens M, Starr JL (eds) Root-knot nematodes. CABI, Wallington, pp 55–97

Hussaini SS (1985) Efficacy of durofume, rabbing and aldicarb against root-knot nematode and weeds in tobacco nurseries. Indian J Nematol 15:88–92

Hussey RS (1985) Host-parasite relationships and associated physiological changes. In: Sasser JN, Carter CC (eds) An advanced treatise on *Meloidogyne*, vol 1: Biology and control. North Carolina State University Graphics, Raleigh, pp 143–153

Hyman BC, Whipple LE (1996) Application of mitochondrial DNA polymorphism to *Meloidogyne* molecular population biology. J Nematol 28:268–276

Jaffee BA, Phillips R, Muldoon AE, Mangel M (1992) Density-dependent host-pathogen dynamics in soil microcosms. Ecology 73:495–506

Jain RK, Bhatti DS (1978) Effect of *Meloidogyne javanica* on the yield of brinjal and tomato. Nematol Mediterr 6:243–246

Jain RK, Bhatti DS (1983a) A note on seed treatment with systemic for the control of *Meloidogyne javanica* in okra. Haryana J Hort Sci 12:143–144

Jain RK, Bhatti DS (1983b) Chemical control of *Meloidogyne javanica* in tomato through nursery bed treatment. In: Proceedings of third nematology symposium, Himachal Pradesh Agricultural University, Solan, pp 37–38

Jain RK, Bhatti DS (1987) Population development of root-knot nematode (*Meloidogyne javanica*) and tomato yield as influenced by summer ploughings. Trop Pest Manag 33:122–125

Jain C, Trivedi PC (1995) Interaction between *Meloidogyne incognita* and *Rhizobium leguminosarum* on *Cicer arietinum*. Curr Nematol 6:155–159

Jain RK, Paruthi IJ, Bhatti DS (1988) Control of root-knot nematode (*Meloidogyne javanica*) in tomato through nursery bed treatment alone and in combination with application of carbofuran at transplanting. Indian J Nematol 18:340–341

Janati A, Berge JB, Triantaphyllou AC, Dalmasso A (1982) Nouvelles donnees sur l'utilisation des isoesterases pour l'identification des *Meloidogyne*. Revue Nematol 5:147–154

Jatala P, Kaltenbach R, Bocangel M (1979) Biological control of *Meloidogyne incognita acrita* and *Globodera pallida* on potatoes. J Nematol 11:303. (abstract)

Jepson SB (1987) Identification of root-knot nematode (*Meloidogyne* spp.). CABI, Wallingford, p 265

Jeyaprakash A, Tigano MS, Brito J, Carneiro RMDG, Dickson DW (2006) Differentiation of *Meloidogyne floridensis* from *M. arenaria* using high-fidelity PCR amplified mitochondrial AT–rich sequences. Nematropica 36:1–12

Jonathan EI, Balasubramanian TP, Wilfred MW (2001) Nematode menace in papaya. The Hindu. March, 2001. www.thehindu.com/2001/03/22/stories/0822042h.htm

Kanwar RS (1990) Studies on cropping systems for management of root-knot nematode, *Meloidogyne javanica* (Trueb, 1885) Chitwood, 1949, Ph.D. Thesis HAU Hisar India, 118 pp

Karssen G (2002) The plant-parasitic nematode genus *Meloidogyne* Göldi, 1892 (Tylenchida) in Europe. Brill Academic, Leiden

Karssen G, van Hoenselaar T, Verkerk-Bakker B, Janssen R (1995) Species identification of cyst and root-knot nematodes from potato by electrophoresis of individual females. Electrophoresis 16:105–109

Karssen G, Liao J, Kan Z, Van Heese EY, Den Nijs LGMF (2012) On the species status of the root-knot nematode *Meloidogyne mayaguensis* Rammah & Hirschmann, 1988. ZooKeys 181:67–77

Kavathiya YA, Pandey RN (2000) Interaction studies on *Meloidogyne javanica*, *Rhizobium* species and *Macrophomina phaseolina* in mungbean. J Mycol Plant Pathol 30:91–93

Khambay B, Bourne JM, Cameron S, Kerry BR, Zaki MJ (2000) A nematicidal metabolite from *Verticillium chlamydosporium*. Pest Manag Sci 56(12):1098–1099

Khan MW (1993) Mechanism of interactions between nematodes and other plant pathogens. In: Khan MW (ed) Nematode interactions. Chapman and Hall, London, pp 42–54

Khan MW, Haider SR (1991) Comparative damage potential and reproduction efficiency of *Meloidogyne javanica* and races of *Meloidogyne incognita* on tomato and eggplant. Nematologica 37:293–303

Khan MW, Alam MM, Khan AM, Saxena SK (1974) Effect of water soluble fractions of oilcakes and bitter principles of neem on some fungi and nematodes. Acta Bot Indica 2:120–128

Khan MR, Jain RK, Ghule TM, Pal S (2014) Root-knot nematodes in India – a comprehensive monograph. All India Coordinated Research Project on Plant Parasitic Nematodes with Integrated Approach for their Control. IARI, New Delhi, p 78

Khan MR, Pal S, Singh A, Patel AD, Patel BA, Ghule T, Phani V (2018) Further observations on *Meloidogyne indica* Whitehead 1968 from India. Pakistan J Zool 50(6):2009–2017

Khan MR, Poornima K, Somvanshi VS, Walia RK (2022) *Meloidogyne enterolobii*: a threat to guava (*Psidium guajava*) cultivation. Archv Phytopathol Pl Protec 55:1961–1997

Kiewnick S, Sikora R (2006) Evaluation of *Paecilomyces lilacinus* strain 251 for the biological control of the northern root-knot nematode *Meloidogyne hapla* Chitwood. Nematology 8:69–78

Kirkegaard JA, Wong PTW, Desmarchelier JM (1996) *In vitro* suppression of fungal root pathogens of cereals by brassica tissues. Plant Pathol 45:593–603

Kojetin DJ, Thompson RJ, Benson LM, Naylor S, Waterman J, Davies KG, Opperman CH, Stephenson K, Hoch JA, Cavanah J (2005) The structural analysis of divalent metals to the *Bacillus subtilis* response regulator Spo0F: the possibility for *in vitro* metalloregulation in the initiation of sporulation. Biometals 18:449–466

Kumar P, Walia RK, Kumar V, Mohan S (2017) A soil-less system for mass production of *Pasteuria penetrans* strain infecting rice root-knot nematode, *Meloidogyne graminicola*. In: Proceedings of national symposium on "Climate Smart Agriculture for Nematode Management", January 11–13, 2017, ICAR-CCARI, Goa, pp 130–131

Kumar V, Khan MR, Walia RK (2020) Crop loss estimations due to plant-parasitic nematodes in major crops in India. Natl Acad Sci Lett 43:409–412

Kundu A, Dutta A, Mandal A, Negi L, Malik M, Puramchatwad R, Antil J, Chikara A, Rao U, Saha S, Kumar R, Patanjali N, Manna S, Kumar GV, Anil DS, Singh PK (2020) A comprehensive *in vitro* and *in silico* analysis of nematicidal action of essential oils. Front Plant Sci 11. https://doi.org/10.3389/fpls.2020.614143

Lamberti F, Taylor CE (1979) Root-knot nematodes (*Meloidogyne* species) systematics, biology and control. Academic, New York, 489 pp

Li XQ, Wei JZ, Tan A, Arojan RV (2007) Resistance to root-knot nematode in tomato roots expressing a nematicidal *Bacillus thuringiensis* crystal protein. Plant Biotechnol J 5:455–464

Lopez Llorca LV, Claugher D (1990) Appressoria of the nematophagous fungus *Verticillium chlamydosporium*. Micron and Microscopica Acta 21:125–130

Lopez Llorca LV, Duncan GH (1988) A study of fungal endoparasitism of the cereal cyst nematode (*Heterodera avanae*) by scanning electron microscopy. Can J Microbiol 34:613–619

Mandal RK, Mishra CD (2001) Role of different organisms in inducing Hoogly wilt symptom in jute. Environ Ecol 19:969–972

Mayee CD, Mahajan R, Kanwar JS (1974) Relationship between leaf curl virus and root-knot nematode in some tomato hybrids. Nematol Mediterr 2:181–182

McSorley R, Gallaher RN (1995) Cultural practices improve crop tolerance to nematodes. Nematropica 25:53–60

Meher HC, Sharma SB, Singh G (2003) Genetic polymorphism in four geographically diverse *Meloidogyne incognita* populations in India. Ann Plant Prot Sci 11:96–100

Meng QP, Long H, Xu JH (2004) PCR assays for rapid and sensitive identification of three major root–knot nematodes, *Meloidogyne incognita, M. javanica* and *M. arenaria*. Acta Phytopathol Sin 34:204–210

Milligan S, Bodeau J, Yaghoobi J, Kaloshian I, Zabel P, Williamson VM (1998) The root-knot nematode resistance gene Mi from tomato is a member of the leucine zipper, nucleotide binding, leucine-rich repeat family of plant genes. Plant Cell 10:1307–1319

Mishra C, Som D, Singh B (1988) Interaction of *Meloidogyne incognita* and *Rhizoctonia bataticola* in white jute (*Corchorus capsularis*). Indian J Agric Sci 58:234–235

Mishra SD, Gaur HS (1979) Effect of planting schedule on population density of *Meloidogyne incognita* and growth of *Lens culinaris*. Bull Entomo 20:71–74

Mojumder V, Mishra SD (1993) Management of nematode pests. In: Parmar BS, Singh RP (eds) Neem in agriculture. Res Bull No. 4, IARI, New Delhi, pp 40–48

Molinari S (2001) Polymorphism of esterase isozyme zymograms of *Meloidogyne* populations detected by Phast system. Nematol Mediterr 29:63–66

Monfort WS, Csinos AS, Desaeger J, Seebold K, Webster TM, Diaz-Perez JC (2007) Evaluating *Brassica* species as an alternative control measure for root-knot nematode (*Meloidogyne incognita*) in Georgia vegetable plasticulture. Crop Prot 26:1359–1368

Morgan-Jones G, White JF, Rodríguez-Kábana R (1983) Phytonematode pathology: ultrastructural studies. I. Parasitism of *Meloidogyne arenaria* eggs by *Verticillium chlamydosporium* Nematropica 13:245–260

Morton CO, Hirsch PR, Kerry BR (2004) Infection of plant-parasitic nematodes by nematophagous fungi – a review of the application of molecular biology to understand infection processes and to improve biological control. Nematology 6:161–170

Nandal SN, Bhatti DS (1983) Preliminary screening of some weeds shrubs for their nematicidal activity against *Meloidogyne javanica*. Indian J Nematol 13:123–127

Nandal SN, Bhatti DS (1986) Influence of four plant extracts on the hatching of *Meloidogyne javanica* and invasion of host roots. Nematol Mediterr 14:291–294

Nasiou E, Giannakou IO (2017) The potential use of carvacrol for the control of *Meloidogyne javanica*. Eur J Plant Pathol 149. https://doi.org/10.1007/s10658-017-1191-z

Nath A, Sharma NK, Bhardwj S, Thapa CD (1982) Nematicidal properties of garlic. Nematologica 28:253–255

Nayak DK, Pandey R (2015) Screening and evaluation of brinjal varieties /cultivars against root-knot nematode, *Meloidogyne incognita*. Int J Adv Res 3:476–479

Nel A, Krause M, Khelawanlall N (2007) A guide for the control of plant pests, 39th edn. Department of Agriculture, Republic of South Africa. Government Printer, Pretoria

Nickle WR (1991) Manual of agricultural nematology. Marcell Dekker, New York, 1020 pp

Nirula KK (1961) Control of root-knot nematodes. Indian Potato J 3:72–75

Norris RF, Caswell-Chen EP, Kogan M (2003) Concepts in integrated pest management. Prentice Hall, Upper Saddle River

Oka Y (2001) Nematicidal activity of essential oil components against the root-knot nematode *Meloidogyne javanica*. Nematology 3:159–164. https://doi.org/10.1163/156854101750236286

Oka Y, Nacar S, Putievsky E, Ravid U, Yaniv Z, Spiegel (2000) Nematicidal activity of essential oils and their components against the root-knot nematode. Phytopathology 90:710–715

Pandey R (1994) Comparative performance of oil seed cakes and pesticides in the management of root-knot disease of davana. Nematol Mediterr 22:17–19

Parvatha Reddy P (1984) Efficacy of seed treatment with the nematicides for the control of *Meloidogyne incognita* infecting cowpea, frenchbean and pea. Indian J Nematol 14:39–40

Parvatha Reddy P (2008) Diseases of horticultural crops: nematode problems and their management. Scientific Publishers, Jodhpur, 379 pp

Patel AD (2018) Management of major nematode pests of pulse and fodder crops. In: Walia RK, Kranti KVVS, Kumar V (eds) Consolidated biennial report (2015–17). ICAR-AICRP (Nematodes), IARI, New Delhi, pp 83–92

Patel DJ, Patel BA, Patel SK, Patel RL, Patel RG (1999) Root-knot nematode, *Meloidogyne indica* on kagzi lime in north Gujarat. Indian J Nematol 29:197

Patel GJ, Patel DJ, Desai MV, Shah HM (1979) Effect of DD and DBCP on seed germination of some cereals and tobacco. Tobacco Res 5:143–146

Paulson RE, Webster JM (1972) Ultrastructure of the hypersensitive reaction in roots of tomato, *Lycopersicon esculentum* L., to infection by the root-knot nematode, *Meloidogyne incognita*. Physiol Plant Pathol 2:227–234

Petersen DJ, Zijlstra C, Wishart J, Blok V, Vrain TC (1997) Specific probes efficiently distinguish root-knot nematode species using signatures in the ribosomal intergeneric spacer. Fundam Appl Nematol 20:619–626

Peterson DJ, Vrain TC (1996) Rapid identification of *Meloidogyne chitwoodi, M. hapla* and *M. fallax* using PCR primers to amplify their ribosomal intergenic spacer. Fundam Appl Nematol 19:601–605

Phani V, Bishnoi S, Sharma A, Davies K Rao U (2018). Characterization of *Meloidogyne indica* (Nematoda: Meloidogynidae) parasitizing neem in India, with a molecular phylogeny of the species. J Nematol 50: 387-398.

Pokharel RR, Abawi GS, Zhang N, Duxbury JM, Smart C (2007) Characterization of isolates of *Meloidogyne* from rice-wheat production fields in Nepal. J Nematol 39:221–230

Pokharel RR, Abawi GS, Duxbury JM, Smat CD, Wang X, Brito JA (2010) Variability and the recognition of two races in *Meloidogyne graminicola*. Australas Plant Pathol 39:326–333

Poornima K (1997) Effect of some organic oils and their cakes on the infectivity of root-knot nematode, *Meloidogyne incognita* affecting tomato. Indian J Nematol 27:70–73

Poornima K, Suresh P, Kalairasan P, Subramanian S, Ramaraju K (2016) Root-knot nematode, *Meloidogyne enterolobii* in guava (*Psidium guajava* L), A new record from India. Madras Agric J 103(10/12):359–365

Poornima K, Walia RK (2017) Orchardists and nurserymen: beware of nematodes. Project Coordinating Cell, AICRP on Nematodes, 24 pp

Powers TO, Fleming CC (1998) Biochemical and molecular characterization. In: Perry RN, Wright DJ (eds) The physiology and biochemistry of free-living and plant parasitic nematodes. CABI, Wallingford, pp 335–379

Powers TO, Harris TS (1993) A polymerase chain reaction method for the identification of five major *Meloidogyne* species. J Nematol 25:1–6

Powers TO, Mullin PG, Harris TS, Sutton LA, Higgins RS (2005) Incorporating molecular identification of *Meloidogyne* spp. into large-scale regional nematode survey. J Nematol 37: 226–235

Prasad SK, Chawla ML, Kumar S, Saxena HP (1971) Root-knot nematode, *Meloidogyne javanica* (Treub, 1885) Chitwood, 1949 and stem borer, *Stomopteryx nertaria* Meyrick affecting green gram, *Phaseolus aureus*. Indian J Entomol 33:55–60

Qing Y, Tsao R, Chiba M, Potter J (2007) Elucidation of the nematicidal activity of bran and seed meal of oriental mustard (*Brassica juncea*) under controlled conditions. J Food Agric Environ 5: 374–379

Radwan MA (2007) Bioactivity of commercial products of *Bacillus thuringiensis* on *Meloidogyne incognita* infecting tomato. Indian J Nematol 37:30–33

Randig O, Bongiovanni M, Carneiro RMDG, Sarah JL, Castagnone-Sereno P (2002a) A species-specific satellite DNA family in the genome of the coffee root-knot nematode *Meloidogyne exigua* application to molecular diagnostics of the parasite. Mol Plant Pathol 3:431–437

Randig O, Bongiovanni M, Carneiro RMDG, Castagnone-Sereno P (2002b) Genetic diversity of root-knot nematodes from Brazil and development of SCAR markers specific for the coffee damaging species. Genome 45:862–870

Rao MS (1997a) Integration of *Paecilomyces lilacinus* with neem leaf suspension for the management of root-knot nematodes on egg plant. Nematol Mediterr 25:249–252

Rao MS (1997b) Management of root-knot nematode, *Meloidogyne incognita* on tomato by integration of *Trichoderma harzianum* with neem cake. Z Pflanzenkr Pflanzenschutz 104: 423–425

Rao MS (1997c) Management of root-knot nematodes, *Meloidogyne incognita* in tomato nursery by integration of endomycorrhiza, *Glomus fasciculatum* with castor cake. Pest Manage Hortic Ecsyst 3:31–35

Ray S, Mohanty S, Mahapatra SN, Patnaik PR, Ray P (1995) Yield loss in ginger and turmeric due to root-knot nematode, *Meloidogyne incognita*. J Spices Aromat Crops 4:67–69

Reddy PP (1997) Effect of integration of *Pasteuria penetrans*, *Paecilomyces lilacinus* and neem cake for the management of root-knot nematodes infecting tomato. Pest Manage Hortic Ecsyst 3:100–104

Riegel C, Noe JP (2000) Chicken litter soil amendment effects on soilborne microbes and *Meloidogyne incognita* on cotton. Plant Dis 84:1275–1281

Roberts PA (1987) The influence of planting date of carrot on *Meloidogyne incognita* reproduction and injury to roots. Nematologica 33:335–342

Robertson L, Diez-Rojo MA, Lopez-Perez JA, Piedra-Buena A, Escuer M, Lopez-Cepero J, Martinez C, Bello A (2009) New host races of *Meloidogyne arenaria, M. incognita,* and *M. javanica* from horticultural regions of Spain. Plant Dis 93:180–184

Rodríguez-Kábana R (1990) Chitinous materials from blue crab for control of root-knot nematode. II. Effect of soybean meal. Nematropica 20:153–168

Salalia R, Walia RK, Somavanshi VS, Kumar P, Kumar A (2017) Morphological, morphometric and molecular characterization of intraspecific variations within the Indian populations of rice root-knot nematode, *Meloidogyne graminicola* Golden and Birchfield, 1965. J Nematol 49:254–267

Saleh H, Sikora RA (1984) Relationship between *Glomus fasiculatum* root colonization and its effect on *Meloidogyne incognita*. Nematologica 30:230–237

Sasser JN (1952) Identification of root-knot nematodes (*Meloidogyne* spp.) by host reaction. Plant Dis Rep 36:84–86

Sasser JN, Carter CC (1985) An advanced treatise on *Meloidogyne*. Vol. I: Biology and control. A Cooperative Publication of the Department of Plant Pathology and the United States Agency for International Development, North Carolina State University Graphics, Raleigh, North Carolina, pp 79–93

Sasser JN, Kirby MF (1979) Crop cultivars resistant to root-knot nematodes, *Meloidogyne* species with information on seed sources. International *Meloidogyne* Project. A Cooperative Publication of the Department of Plant, Pathology, North Carolina State University and the United States agency for international development, Raleigh, p 24

Sayre RM, Starr MP (1985) *Pasteuria penetrans* (Ex Thorne, 1940) *nom-rev., comb-n., sp-n.*, a mycelial and endospore-forming bacterium parasitic in plant-parasitic nematodes. Proc Helminthol Soc Wash 52:149–165

Segers R, Butt TM, Kerry BR, Beckett A, Peberdy JF (1996) The role of the proteinase VCP1 produced by the nematophagous *Verticillium chlamydpsporium* in the infection process of nematode eggs. Mycol Res 100:421–428

Sen AK (1960) Preliminary studies on parasitic nematodes on vegetable crops in Bihar. Indian Agric 4:113–116

Sena ES, da Ponte JJ (1982) A manipueira no controle da meloidoginosa da cenoura. Rev Soc Brasil Nematol 6:95–98

Sharma RK, Tiagi B (1990) Effect of *Meloidogyne incognita* on nodulation and symbiotic nitrogen fixation in pea. Nematol Mediterr 18:15–17

Sharon E, Bar-Eyal M, Chet I, Herrera-Estrella A, Kleifeld O, Spiegel Y (2001) Biological control of the root-knot nematode *Meloidogyne javanica* by *Trichoderma harzianum*. Phytopathology 91:687–693

Sharon E, Chet I, Viterbo A, Bar-Eyal M, Nagan H, Samuels GJ, Spiegel Y (2007) Parasitism of *Trichoderma* on *Meloidogyne javanica* and role of the gelatinous matrix. Eur J Plant Pathol 118: 247–258

Siddiqi MR (2000) Tylenchida: parasites of plants and insects, 2nd edn. CABI, UK, p 848

Siddiqui ZA, Mahmood I (1994) Interactions of *Meloidogyne javanica, Rotylenchulus reniformis, Fusarium oxysporum* f. sp. *ciceri* and *Bradyrhizobium japonicum* on the wilt disease complex of chickpea. Nematol Mediterr 22:135–140

Siddiqui ZA, Mahmood I (1999) Role of bacteria in the management of plant parasitic nematodes: a review. Bioresour Technol 69:167–179

Siddiqui MA, Saxena SK (1987) Effect of interculture of margosa and Persian lilac with tomato and egg plant on root knot and reniform nematode. Int Nematol Netw Newsl 4:5–8

Siddiqui I, Haas D, Heeb S (2005) Extracellular protease of *Pseudomonas fluorescens* CHA0, a biocontrol factor with activity against the root-knot nematode *Meloidogyne incognita*. Appl Environ Microbiol 71:5646–5649. https://doi.org/10.1128/AEM.71.9.5646-5649.2005

Sikora RA (1992) Management of the antagonistic potential in agricultural ecosystems for the biological control of plant parasitic nematodes. Annu Rev Phytopathol 30:245–270

Sikora RA, Schonbeck (1975) Effect of vesicular-arbuscular mycorrhizae (*Endogone mosseas*) on population dynamics of root-knot neamtodes, *Meloidogyne incognita* and *Meloidogyne hapla*. In: Proceedings of 8th International Plant Protection Congress, Moscow, pp 158–166

Smith PG (1944) Embryo culture of a tomato species hybrid. Proc Am Soc Hortic Sci 44:413–416

Spiegel Y, Sharon E, Chet I (2005) Mechanisms and improved biocontrol of the root-knot nematodes by *Trichoderma* spp. Acta Hortic 698:225–228

Stirling GR, Wachtel MF (1980) Mass production of *Bacillus penetrans* for the biological control of root-knot nematodes. Nematologica 26:308–312

Subbotin SA, Burbridge J (2021) Sensitive, accurate and rapid detection of the northern root-knot nematode, *Meloidogyne hapla*, using recombinase polymerase amplification assays. Plants 10: 336. https://doi.org/10.3390/plants10020336

Sundararaju P, Cannayane I, Ramesh R (2002) Penetration and development of *Pratylenchus coffeae* and *Meloidogyne incognita* in susceptible cultivars of Banana. In: Abstracts of National Seminar of Nematological Research in India at C. S. Azad and University of Agric. and Tech. Kanpur, p 189

Sundresh HN, Setty KGH (1977) Crop rotation as an effective and practical means of controlling root-knot nematode *Meloidogyne incognita* Chitwood. Curr Res 6:157–158

Suresh P, Poornima K, Kalaiarasan P, Sivakumar M, Subramanian S (2017) Occurrence of barley root-knot nematode, *Meloidogyne naasi* in orange jasmine (*Cestrum aurantiacum* L.) in Nilgiris, Tamil Nadu, India: a new record. J Entomol Zool Stud 5(6):629–634

Swain SC, Ganguly AK, Sahoo P, Mohapatra T (1999) RAPD analysis distinguishes four races of root-knot nematode, *Meloidogyne incognita*. Indian J Nematol 29:1–7

Swarup G, Goswami BK (1969) Interrelationship of root-knot nematode and leaf curl virus in tomato. Indian J Exp Biol 7:64–65

Tanaka R, Kikuchi T, Aikawa T, Kanzaki N (2012) Simple and quick methods for nematode DNA preparation. Appl Entomol Zool 43:291–294

Taylor AL, Sasser JN (1978) Identification and control of root-knot nematodes (*Meloidogyne* spp.). Publication of Department of Plant Pathology, North Carolina State University and U.S. Agency International Development, 111 pp

Tesarova B, Zouhar M, Rysanek P (2003) Development of PCR for specific determination of root-knot nematode *Meloidogyne incognita*. Plant Prot Sci 39:23–28

Thakur S, Udaya Kumar M, Salalia R, Walia RK (2015) Studies on a new strain of *Pasteuria penetrans* infecting rice root-knot nematode, *Meloidogyne graminicola*. Indian J Nematol 45: 129–137

Tian B, Yang J, Zhang KQ (2007) Bacteria used in the biological control of plant-parasitic nematodes: populations, mechanisms of action, and future aspects. FEMS Microbiol Ecol 61: 197–213

Tiyagi SA, Siddiqui MA, Alam MM (1986) Control of *Meloidogyne incognita* on tomato by bare-root dip in dimethoate. Int Nematol Netw Newsl 3:4–5

Umamaheswari R, Sivakumar M, Subramanian S, Samiyappan R (2004) Induction of systemic resistance by *Trichoderma viride* treatment in greengram (*Vigna radiata*) against root-knot nematode *Meloidogyne incognita*. Curr Nematol 15:1–7

Umarao, Ahlawat S, Ganguly AK (2003) Molecular characterization of *Meloidogyne incognita* races based on DNA sequences of internal transcriber spacers of ribosomal DNA. Indian J Nematol 33:103–110

Upadhyay KD, Dwivedi BK (1987) Effect of interaction between *Meloidogyne javanica* and *Fusarium oxysporum* f. sp. *ciceri* on chickpea. Indian J Nematol 17:145–146

Uysal A, Yigit F, Yildirim AF (2017) Identification of *Meloidogyne* species and races infecting vegetables in Turkey. J Nematol 49(4):393–401

Vos P, Simons G, Jesse T, Wijbrandi J, Heinen L, Hogers R, Frijters A, Groenendijk J, Diergaarde P, Reijans M, Fierens-Onstenk J, de Both M, Peleman J, Liharska T, Hontelez J, Zabeau M (1998) The tomato Mi-1 gene confers resistance to both root-knot nematodes and potato aphids. Nat Biotechnol 16:1365–1369

Walia KK (1997) Management of root-knot nematode, *Meloidogyne javanica* on vegetable crops with *Tagetes* sp. Indian J Nematol 27:18–23

Walia RK, Bajaj HK (2014) Text book on introductory plant nematology, 2nd rev edn, ICAR, 240 pp

Walia RK, Khan MR (2018) A compendium of nematode diseases of crop plants. ICAR-All India Coordinated Research Project on Nematode in Cropping Systems, New Delhi, p 106

Walia RK, Kumar A, Mehta SK, Kapoor A (2011) An efficient *in vivo* system for mass production of *Pasteuria penetrans*. Int J Nematol 20(2):211–218

Walia RK, Kranti KVVS, Kumar V (eds) (2018a) Consolidated Biennial Report 2015-17. All India Coordinated Research Project on Nematodes in Cropping Systems, held at AAU Anand, February 15–17, 2018

Weber HJ, Orton WA (1902) A cowpea resistant to root-knot (*Heterodera radicicola*). USDA Bur Pl Industr Bull 17(2):23–38

Whitehead AG (1998) Plant nematode control. CABI, Wallingford

Widmer TL, Mitkowski NA, Abawi GS (2002) Soil organic matter and management of plant-parasitic nematodes. J Nematol 34:289–295

Williamson VM (1998) Root-knot nematode resistance genes in tomato and their potential for future use. Annu Rev Phytopathol 36:277–293

Williamson VM, Kumar A (2006) Nematode resistance in plants, the battle underground. Trends Genet 22:396–403

Williamson VM, Caswell Chen EP, Westerdahl BB, Wu FF, Caryl G (1997) A PCR assay to identify and distinguish single juveniles of *Meloidogyne hapla* and *M. chitwoodi*. J Nematol 29: 9–15

Yang BJ, Eisenback JD (1983) *Meloidogyne enterolobii* n. sp. (Meloidogynidae), a root-knot nematode parasitizing pacara earpod tree in China. J Nematol 15:381–391

Zijlstra C (1997) A reliable, precise method to differentiate species of root-knot nematodes in mixtures on the basis of ITS–RFLPs. Fundam Appl Nematol 20:59–63

Zijlstra C (2000) Identification of *Meloidogyne chitwoodi, M. fallax* and *M. hapla* based on SCAR-PCR a powerful way of enabling reliable identification of populations or individuals that share common traits. Eur J Plant Pathol 106:283–290

Zijlstra C, Lever AEM, Uenk BJ, Van-Silfhout CH (1995) Differences between ITS regions of isolates of root-knot nematodes *Meloidogyne hapla* and *M. chitwoodi*. Phytopathology 85: 1231–1237

Zijlstra C, Van-Hoof R, Donkers-Venne D (2004) A PCR test to detect the cereal root-knot nematode *Meloidogyne naasi*. Eur J Plant Pathol 110:855–860

Zubeyir D, Sogut MA (2011) Characterizing races of *Meloidogyne incognita, M. javanica* and *M. arenaria* in the West Mediterranean region of Turkey. J Crop Prot 30:451–455

Meloidogyne Species: Threat to Vegetable Produce

2

Amir Khan, Shahid Anwar Ansari, Mohammad Haris, Touseef Hussain, and Abrar Ahmad Khan

Abstract

Vegetables are the richest source of vitamins, essential elements, and minerals like calcium and iron. Most of the human population are vegetarians; they fulfil their daily nutrient requirements by consuming vegetables. However, the production of vegetables is seriously hampered by several biotic stresses, viz., bacteria, fungi, nematodes, and viruses, which pose a considerable challenge to meeting future demands for such a large population. Among several biotic stresses, root-knot nematodes (RKNs) (*Meloidogyne* spp.) are the major threat to vegetable production. RKNs are obligate and sedentary root endoparasites of almost all vegetable crops and are considered the most damaging pests in agriculture. Since RKNs target the root vascular system, they provoke host nutrient deprivation and defective food and water transport by forming galls in the roots. They also cause aboveground symptoms of growth stunting, wilting, chlorosis in patches, and reduced crop yields. Besides the direct damage, RKNs act as a predisposing agent to other soil-borne bacterial and fungal pathogens and aggravate the problem, further leading to development of disease complexes. Considering the difficulties,

A. Khan · M. Haris · A. A. Khan
Plant Pathology and Nematology Section, Department of Botany, Aligarh Muslim University, Aligarh, Uttar Pradesh, India

S. A. Ansari
Department of Plant Protection, Faculty of Agricultural Sciences, Aligarh Muslim University, Aligarh, Uttar Pradesh, India

T. Hussain (✉)
Plant Pathology and Nematology Section, Department of Botany, Aligarh Muslim University, Aligarh, Uttar Pradesh, India

Division of Plant Pathology, ICAR-Indian Agricultural Research Institute, New Delhi, India

F. Ahmad, G. Nombela (eds.), *Root-Galling Disease of Vegetable Plants*,
https://doi.org/10.1007/978-981-99-3892-6_2

researchers worldwide find eco-friendly approaches to protect vegetable production from such tiny and more damaging soil-borne pathogens.

Keywords

Disease development · Endoparasite · Economic damage · Production · Nematode · Vegetable

2.1 Introduction

Vegetables are the source of essential nutrients and are one of the paramount constituents of daily life by enabling us to energy-rich food. These are the components of our diet and the ample source of vitamins, minerals, and fibres for growth and development. Also possess phytochemicals having antioxidant, antifungal, antiviral, antibacterial, and anti-carcinogenic properties (Steinmetz and Potter 1996; Gruda 2005). In addition, short duration, economic viability, high yield, growing in every climate, and ability to create on-farm and off-farm employment have developed an advantage for the growers. In an agriculture-based country like India, large producers after China (leading producer), vegetables are grown in tropical, subtropical, and temperate regions. India's most commonly cultivated vegetables are potatoes, onions, cauliflowers, brinjal, cabbages, and other legumes. However, India occupies the top position in okra and ginger production (FAO 2020). In the global paradigm, India is the second largest producer of vegetables, producing 196.27 million tons of world vegetable production (Indian Horticulture Database 2020–21, Second Advanced Estimates). In India, West Bengal (30,330.77 million tons) is the leading vegetables producing state, followed by Uttar Pradesh (29,160.91 million tons) during the year 2020–21 (http://agricoop.nic.in/statistics/) (Table 2.1).

From the last decades, it has been observed that several soil-borne pathogens highly hamper vegetable production. This may be due to the climatic conditions that favour the development and reproduction of pathogens. Among the major obstacles, plant parasitic nematodes (PPNs) occupy the predominant position in restricting the productivity of vegetables (Sharma et al. 2006; Anwar and Mckenry 2007; Dhaliwal and Koul 2007). PPNs are ruinous and economically most important pests of many cultivated crops worldwide, damaging vegetables, particularly in tropical and subtropical countries (Trifonova et al. 2009; Sikora and Fernandez 2005). The importance of nematode as a constraint on vegetables production was realised long ago in our country. Since then, a nematode problem of national importance has appeared. In some production areas, the reduction in vegetable yield due to phytonematodes has reached as high as 30% (Anwar et al. 2009). At present, approximately 4100 PPNs species have been discovered. These species exhibit detrimental effects on the agricultural sector by degrading a variety of vegetable crops, such as cauliflower, cabbage, spinach, carrot, chilli, tomato, okra, eggplant, etc. (Chariou and Steinmetz 2017; Decraemer and Hunt 2013). Several nematologists came up with a list of the top 10 PPNs that affect the global economy. These are RKNs (*Meloidogyne* spp.),

Table 2.1 List of countries and their vegetables production based on FAO

Vegetables	Production	Countries	References
Okra (*Abelmoschus esculentus* L.)	6 million tons	India	FAO (2018)
Okra (*Abelmoschus esculentus* L.)	2 million tons	Nigeria	
Tomato (*Solanum lycopersicum* L.)	19,007,000 tons	India	FAO (2019)
Chilli (*Capsicum annuum* L.)	18,978,027	China, Mainland	
Tomato (*Solanum lycopersicum* L.)	62,869,502 tons	China	
Carrot (*Daucus carota* L.)	21,482,971 tons	China	
Carrot (*Daucus carota* L.)	2,259,000 tons	United states	
Tomato (*Solanum lycopersicum* L.)	12,841,990 tons	Turkey	
Carrot (*Daucus carota* L.)	2,769,613	Uzbekistan	
Sugar beet (*Beta vulgaris* L.)	51.366.830 tons	Russian Federation	
Eggplant (*Solanum melongena* L.)	12,777,000 tons	India	FAO (2020)
Eggplant (*Solanum melongena* L.)	36,557,611 tons	China	
Spinach (*Spinacia oleracea* L.)	28,507,829 tons	China	
Spinach (*Spinacia oleracea* L.)	367,433 tons	United states	
Cucumber (*Cucumus sativus* L.)	72,779,781 tons	China	

dagger nematode (*Xiphinema index*-the virus vector nematode), false root-knot nematode (RKN) (*Nacobbus aberrans*), rice white-tip nematode (*Aphelenchoides besseyi*), reniform nematode (*Rotylenchulus reniformis*), bulb and stem nematode (*Ditylenchus dipsaci*), pine wilt nematode (*Bursaphelenchus xylophilus*), burrowing nematode (*Radopholus similis*), cyst nematode (*Globodera* and *Heterodera*), and root lesion nematode (*Pratylenchus* spp.) (Jones et al. 2013). The RKNs (*Meloidogyne* spp.) are the most harmful and commercially significant among PPNs.

2.2 Root-Knot Nematodes as an Emerging Pest in Vegetable Crops

RKNs (*Meloidogyne* spp.) belong to the Phylum-nematoda and Order-tylenchida. They are considered highly adaptive, widespread and sedentary obligate endoparasites among all PPNs and dependent entirely on the host for their reproduction and survival (Khan 2008). Berkeley (1855) reported the presence of a nematode called "vibrios" in greenhouse-grown cucumbers in England. The name *Meloidogyne* comes from the Greek word that means "apple" or "pear" (Khan 2008). In the late twentieth century, this ubiquitous pest gained more attention as a result of its negative impact on global vegetable production. In India, Barber (1901) first reported RKNs from the tea plantation in Kerala. After that, Ayyar (1926) reported RKNs infesting a variety of vegetables (Reddy 2021).

Globally, >100 species of RKNs (*Meloidogyne* spp.) have been reported on more than 3000 host plants, including fruits and vegetables. (Khan et al. 2022). Four different species of *Meloidogyne*, such as *M. incognita*, *M. javanica*, *M. arenaria*, and *M. hapla*, are considered high pathogenic, which cause up to 90% of crop losses worldwide and make crops more susceptible to other soil-borne pathogens (Lunt et al. 2014; Hunt and Handoo 2009). In India, 14 species of RKNs have been discovered, according to Ghule et al. (2014). Among all species, *M. incognita* was the most prevalent in agricultural fields, causing significant damage and reducing the quality of crops like cabbage, tomato, eggplant, spinach, cauliflower, and okra (Khan and Khan 2021). However, the root-knot disease was frequently seen in vegetable fields across India, which detrimentally affects the quality and productivity of vegetables. Besides *M. javanica* and *M. arenaria*, *M. graminicola* is also the most common and pathogenic in rice crops (Ghule et al. 2014). But, *M. hapla* is commonly present in colder regions (Escobar et al. 2015).

2.3 Races of Root-Knot Nematodes

Using the North Carolina host differential test, researchers identified the physiological races of RKNs in the four most common species (Gorny et al. 2021). These include *M. incognita, M. javanica, M. arenaria*, and *M. hapla*. There are currently four confirmed *M. incognita* races (race 1, 2, 3, and 4), three confirmed *M. javanica* races (race 1, 2-pepper race and 3-groundnut race), and one confirmed *M. arenaria* (race 2) (Qiu et al. 2021). Before breeding resistant varieties of any root-knot nematode species, it is crucial to determine which races are present in the target population (Table 2.2).

Table 2.2 Root-knot nematode races and their distribution status in India (Source: Khan et al. 2014)

Meloidogyne species	Race	Distribution
M. incognita	1	India
	2	Uttar Pradesh, Himachal Pradesh, Punjab, Bihar, Assam, Andra Pradesh, Karnataka, Maharashtra, Kerala, Mizoram, West Bengal, Odisha
	3	Uttar Pradesh, Haryana, West Bengal, Assam, Gujarat, Karnataka, Maharashtra, Tamil Nadu
	4	Uttar Pradesh, Haryana, Himachal Pradesh
	5	Maharashtra, Tripura, Haryana, Tamil Nadu
	6	Manipur
M. arenaria	2	Uttar Pradesh, Haryana
M. javanica	1	Haryana
	2	Uttar Pradesh, West Bengal
	3	Andhra Pradesh
	4	Gujarat, Andaman and Nicobar Islands
	5	Uttar Pradesh, West Bengal
	6	Haryana

2.4 Biological Cycle of Root-Knot Nematodes on Vegetable Crops

RKNs life cycle, from egg to adult female, generally takes 25–28 days. However, the life cycle duration depends (Fig. 2.1) on the host, temperature, and soil properties, including moisture (Khan et al. 2022). The life cycle starts from the second-stage juveniles (J2s) after hatching from the egg; it is known as an infectious stage because it starts the infection process in the host roots (Rawal 2020). On the other hand, CO_2 is a crucial diffusate that plays a vital role in attracting J2s of *M. incognita* and other PPNs (Robinson 2002). After penetration inside the roots of the host, J2s release some proteins and enzymes such as pectate lyases, polygalacturonases, cellulases, and endo-xylanases from the sub-ventral glands inside the roots (Davis et al. 2011; Wieczorek et al. 2014; Perry and Moens 2011). These released enzymes alter the host plant's cell wall constituents or interrupt the host plant's cell cycle. Additionally, they accelerate the degradation of proteins in the host cell while simultaneously slowing down defence mechanisms and transcriptional regulation (Eves-van den Akker and Birch 2016). Few infectious enzymes break down the host plant's cellular components, establish a nematode-feeding site inside the roots of the host, and assist

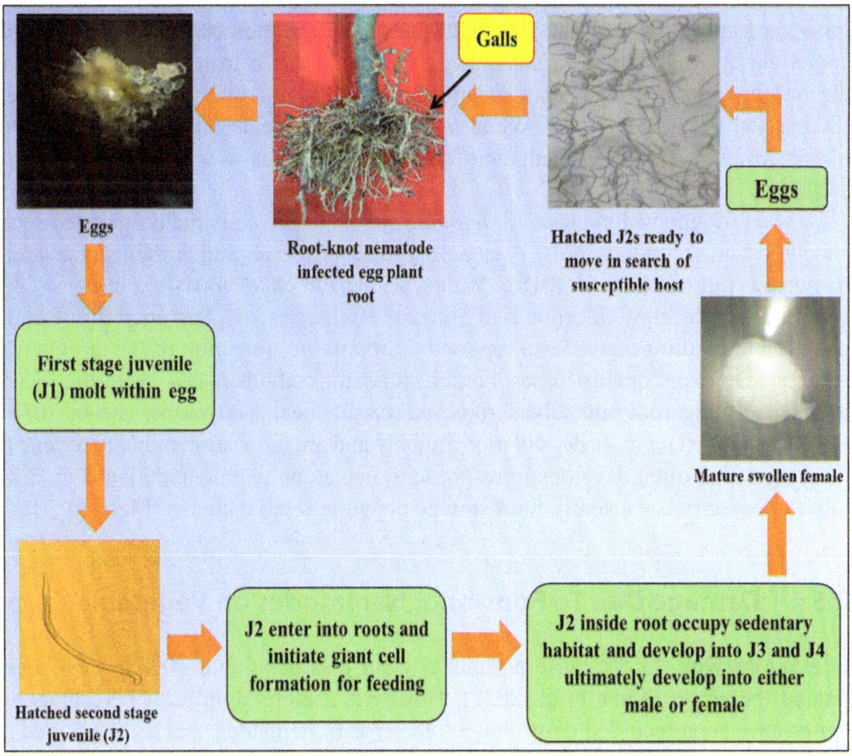

Fig. 2.1 Life cycle of root-knot nematode on vegetable crops

in forming giant cells (GCs) (Shakeel et al. 2020). By hypertrophy and hyperplasia, GCs turn into galls or knots and provide food for nematodes till reproduction. When galls form on the plant's roots, they damage the vascular system and make it difficult to get water and mineral nutrients from the soil, which causes the plant to wilt and turn yellow (Jamal et al. 2017; Lee and Kim 2016; Gao et al. 2016). These plant infections make the plants more susceptible to other pathogens in the soil and cause disease complexes with other harmful soil microorganisms (Zhou et al. 2016).

2.5 Expression of Both Above- and Below Ground Symptoms on Vegetable Crops

Aboveground symptoms due to *Meloidogyne* infestation are not distinct from other factors that cause root malfunction but may often occur clustered (Ahmad et al. 2022a, b). *Meloidogyne* infection is characterised by galling of the root system. Root galling can be mild, moderate, or severe depending on the nematode species, population density, and plant cultivar. The species, number of nematodes in the tissue, host, and age of the plant all play a role in the gall's size and appearance. Whereas galls on most other vegetables tend to be small to medium in size, cucurbit galls can be enormous in both dimensions (Sikora et al. 2018). In the tropics, galls grow more prominent in the lowlands due to warmer temperatures than higher, cooler elevations. Typically, isolated egg masses protrude from the root surface as the only symptom of root-knot on monocotyledonous crops like onion and leek (Sikora and Fernandez 2005). While most nematode species cause galling, others, like *M. artiellia*, do not. Instead, their females and egg masses are visible on the root surface.

A severely infected plant with *Meloidogyne* will have a few heavily galled roots and almost no secondary roots. Their ability to absorb water and nutrients is severely hampered (Sikora et al. 2018). Plants wither quickly when exposed to dry conditions, becoming chlorotic and stunted. High rates of plant mortality due to infection in seedlings have been reported. Furthermore, infection of the taproot can result in significant quality losses. Forked and deformed tubers can cause significant losses in selling root and tuber crops like carrot, beet, and radish (Khan 2015). Because infected taproots dry out more quickly and are more susceptible to rot due to the fungus that often develops in response to nematode gall degradation (Fig. 2.2), tuber infections also severely limit storage potential (Hallmann and Meressa 2018).

2.6 Damage Due To Root-Knot Nematodes on Vegetable Crops

Infestations with RKNs cause an annual yield loss of around 10% in vegetables around the world (Khan et al. 2022). However, a more significant percentage of losses have been recorded depending on the type of nematode species, area, type of crop, and soil nematode population. Losses in India due to RKNs in major vegetable crops were estimated at 5131.80 million rupees per year. In addition to causing

Fig. 2.2 Root samples of various vegetable crops infected with root-knot nematodes, (**a**) okra root, (**b**) beet root, (**c**) eggplant root, (**d**) cucurbit root, and (**e**) chilli plant root

damage directly, RKNs also serve as a "predisposing agent" for the entry of soil-borne pathogens such as bacterial and fungal (Gowda et al. 2019), exacerbating the problem and leading to the development of disease complexes and severe yield losses of 40–70% in vegetable crops grown in different regions of the country (Gowda et al. 2019). Root-knot nematode (*M. incognita* and *M. javanica*) and reniform nematode (*Rotylenchulus reniformis*) cause enormous crop loss (up to 80%) in selected crops and constitute a significant problem even in greenhouses (Table 2.3) where tomatoes, chillies, watermelon, muskmelon, okra, gherkins, and flower crops like carnations, gerbera, and roses are grown (Phani et al. 2021).

2.7 Strategies to Manage Root-Knot Nematodes and Sustain Vegetable Production

High reproductive potential, polyphagous nature, and unique survival mechanism made managing RKNs more difficult under intensive vegetable cultivation. Once the RKNs are located in the field, it is challenging to eradicate them from the soil. However, eliminating nematodes is neither economically nor ecologically sound unless there is a regulatory requirement for total control of nematodes. Hence the

Table 2.3 Root-knot nematodes and vegetables loss across the world

Root-knot nematodes	Loss	Countries	Host vegetables	References
Meloidogyne incognita	35.09%	India	Ivy gourd (*Coccinia indica* L.)	Basumatary et al. (2018)
Meloidogyne spp.	24–38%	Pakistan	Tomato (*Solanum lycopersicum* L.)	Mukhtar (2018)
Melodogyne spp.	8–20%	India	Radish (*Raphanus sativus* L.)	Gowda et al. (2017)
Melodogyne spp.	8–23%	India	Chilli (*Capsicum annuum* L.)	Gowda et al. (2017)
Meloidogyne spp.	10%	Egypt	Potato (*Solanum tuberosum* L.)	Shaltoot (2001)
Meloidogyne incognita	44%	India	Pointed gourd (*Trichosanthes dioica* Roxb.)	Verma and Anwar (1996)
Meloidogyne spp.	45%	United States	Carrot (*Daucus carota* L.)	Widmer et al. (1999)
Meloidogyne spp.	80%	Turkey	Tomato (*Solanum lycopersicum* L.)	Kaskavalci (2007)
Meloidogyne spp.	480.00 million/ annually	India	Okra (*Abelmoschus esculentus* L.)	Jain et al. (2007)
Meloidogyne incognita	66.84% in poly house	India	Cucumber (*Cucumis sativus* L.)	Bhati and Baheti (2021)
Meloidogyne incognita	2.3%	Ethiopia	Tomato (*Solanum lycopersicum* L.)	Sikora and Fernandez (2005), Wesemael et al. (2011)
Meloidogyne spp.	Rs. 547.5 billion annually	India	Sponge gourd (*Luffa aegyptiaca* Mill.)	Chandra et al. (2010), Jain et al. (2007)

concept of “living with nematodes” (Tyler 1933) has been strengthened, and nematologists have made sustained efforts to develop ideal approaches against RKNs. Several eco-friendly approaches, cultural, physical, biological, and genetics-based methods, are applied to manage RKNs.

2.7.1 Cultural Method

In the soil environment, biotic and abiotic factors highly govern the nematode behaviour (Khan et al. 2023). Soil moisture and temperature, chemical and physical composition of soil, and antagonistic flora and fauna influence the nematode behaviour and their effects on hosts. Therefore, for effective management, it is imperative to disturb the harmonious relationship between the host plant and nematode by altering the soil ecosystem with strategic approaches (Gaur 2006). Major cultural practices such as sanitation, crop rotation, deep summer ploughing, utilisation of trap crops, cover crops and antagonistic crops, organic amendments, weed plants, and crop residues play a major role in managing the nematode population, including RKNs.

2.7.1.1 Sanitation

Sanitation is the solitary principle to prevent new-area infestations and to avoid the secondary spreading of RKNs in the vegetable field. Generally, RKNs are easily spread through vegetative propagules (Collange et al. 2011; Rao et al. 2015). In the vegetable cropping system, several weeds like *Tithonia rotundifolia, Solanum nigrum, Chenopodium album,* and other unknown weeds are known to act as excellent alternate hosts for maintaining the RKNs (Khan et al. 2014). In addition, they can also survive in crop residues. Crop residues increase the rate of nematode survival by reducing the rate of decomposition and providing mechanical protection under extreme conditions. Thus, timely removal and destruction of weeds and crop residues generally help to minimise the inoculum level of RKNs under field conditions.

2.7.1.2 Summer Ploughing

Two or three deep summers ploughing in the hot summer months expose the nematodes and infected tissue to solar heat and dehydrate them. This practice has effectively managed RKNs (Jain and Bhatti 1987). Fallowing and summer ploughing during hot summer months in eggplant fields significantly reduce *M. incognita* (Singh 2013). Soil solarisation can increase the effectiveness of summer ploughing by capturing and holding more heat underneath polyethylene mulching rather than direct exposure (Gaur and Perry 1991).

2.7.1.3 Crop Rotation and Cropping Sequence

Rotation of crops is a widely used effective cultural practice to reduce the RKNs population in the soil. One- to two-year crop rotation with graminaceous poor hosts and specific antagonistic crops efficiently reduce RKN (Patel et al. 1979; Sundresh and Setty 1977). RKNs levels can be reduced by rotating non-host crops such as onions, garlic, mustard, and cereals for at least 2 to 3 years in an appropriate cropping strategy (Khan et al. 2010). The cropping sequence in vegetable-based cropping systems plays a key role in nematode management. However, sometimes vegetable-based cropping sequence predominantly increases nematode damage potential in vegetable crops. Chandra and Khan (2011) found that the sequence of okra-brinjal-okra stimulated the RKN population under field conditions. In contrast, cropping sequences such as okra-cucumber-mustard and okra-cowpea-cabbage also effectively suppressed the population of *M. incognita* in field conditions.

2.7.1.4 Trap, Antagonistic, and Cover Crops

Trap crops, cover crops, and antagonistic crops are typically termed nematode suppressive crops. These crops inhibit or reduce the nematode population by their planting or incorporation. *Crotalaria spectabilis* is most commonly used as a trap crop against RKNs. The crops with major nematode antagonistic properties from their root exudates can be used as rotation, cover, or intercrops to retard the nematode attack on host crops. Crops like marigolds, mustard, sesame, and asparagus have nematode-suppressive activity by releasing nematotoxic compounds through root exudates (Gaur 1975; Haque and Gaur 1985). Of these, marigold is the most studied crop, which can suppress nematode activity by releasing polythienyls toxic compounds (Umashankar et al. 2005). Cover crops are generally exploited to manage the nematodes because nematodes move slowly and cover a very short distance, and cannot migrate to the neighbouring field. If a cover crop is not a host of nematodes, some populations may starve, which helps to reduce the initial population density to the next crop.

2.7.1.5 Organic Amendments

Organic amendments are often used in farming for a long time in Indian agriculture to enhance soil fertility, soil physical condition, recycling of nutrients, and soil biological activity. However, several studies evident that organic amendments also utilised for the management of PPNs, including RKNs (Ahmad et al. 2021a). Generally, organic amendments include organic manures (animal and poultry), plant parts and their extracts, plant products, industrial wastes, green manures from cover crops, vermicompost, etc. (Collange et al. 2011; Hussain et al. 2019). Three major biological processes in organic amendments which helps in nematode suppression are (a) enhance soil microbial activities against nematode that feed and kill nematodes in the soil during decomposition; (b) decomposition led to secretions of antinemic compounds; and (c) increase soil capacity to hold elements, which enhance plant tolerance to nematodes (Akhtar and Mahmood 1996; Bridge 1996; Oka 2010). This endeavour examines relevant studies on different forms of organic amendments for managing RKNs in vegetable crops.

2.7.1.6 Green Manure

Incorporating chopped leaves of plants, including *Brassica* species (*Brassica juncea*, *B. napus*, *B. rapa*), into the soil also limits the reproduction of nematodes. They produce volatile compounds, isothiocyanates acting as an antagonist to PPNs, including RKNs. Ahmad et al. (2010) demonstrated that leaf extracts of *Lantana camara* were highly nematostatic, where juveniles were paralysed entirely after 12 h of exposure, and 96% of juvenile's mortality was observed at 48 h after exposure.

2.7.1.7 Leaf Extract

Rather et al. (2008) found that pot application of 100 g chopped leaves of neem, *Persian lilac*, and marigold considerably reduced the incidence of RKNs with enhancing plant growth in tomato. Similarly, the application of madar (*Calotropis procera*) and neem (*Azadirachta indica*) chopped leaves considerably reduced the population of nematode in soil with lesser root-knot index under pot experiment (Singh and Patel 2015).

2.7.1.8 Oil Cake

Oil cakes have been widely used and recommended for the suppression of the nematode population in the soil. Goswami and Meshram (1991) found that applying mustard and karanja cake as soil amendments reduced the RKNs juvenile's penetration on tomato roots. Similar investigations showed that neem cake (*A. indica*) successfully lowers the incidence of RKN in vegetable crops, even at small dosages (1–2 t/ha) (Devi and Das 2016).

2.7.2 Physical Method

Physical methods rely on heat, as nematodes have different maximum and minimum temperature thresholds for their survival, activity, infection, and growth. Hence, the key abiotic factor, the temperature, can be exploited for the management of nematodes. Among important physical methods, soil solarisation effectively manages RKNs infesting vegetables (Gowda et al. 2019).

Soil solarisation is a method of heating moist soil by covering it with transparent plastic sheets to trap solar radiation during the summer or hottest period of the year. The solarisation period between 2 and 9 weeks has been reported to be effective for nematode suppression (Gowda et al. 2019). Gaur and Dhingra (1991) revealed that 4 to 6 weeks of solarisation in the mid-summer period had been effective under tropical and subtropical conditions in reducing nematode incidence. Soil solarisation through 100 gauzes (25 μm) and linear low-density polyethylene (LLDPE) transparent plastic film for 15 days during May month reduced the root-knot disease and weeds by 66% and 93%, respectively (Walia et al. 2016).

2.7.3 Biological Method

Among all control measures, bio-control is considered as a most effective and eco-friendly method to manage nematodes. Application of biotic agents such as bacteria, fungi, or other microbes is considered (Ahmad et al. 2021b). Bio-control agents (BCAs) consist of substances or factors that enhance plant growth and induce resistance against PPNs, including RKNs. It is another best tool to manage nematodes, replace the chemicals in agriculture, and sustain vegetable production (Forghani and Hajihassani 2020). Many researchers have suggested that synthetic chemicals and pesticides cause environmental problems. In the soil, large amounts of chemicals lower its fertility, increase the rate of soil disintegration, and have a detrimental effect on the health of humans. Bio-agents cause the plants to reduce or control the damaging effects of soil-borne pathogens such as root-knot nematode (*M. incognita*), which is accomplished by interacting with the roots (Forghani and Hajihassani 2020). Several others activities, such as assisting the resources accumulation, plant hormones (cytokinin and gibberellins) production, antibiotics, and lytic enzymes, are also done by beneficial microorganisms in the soil (Glick 2012). A few mechanisms contributed to the endogenous defence at the gene levels through activating pathogenesis-related genes (PR-genes), PR-1, PR-1b, PR-3, 5, and salicylic acid (SA)-dependent genes related to the pathogenesis of the systemic acquired resistance (SAR) and other genes responsible for killing the *M. incognita*. Some enzymatic activities occur in pre-treated infected plant's roots, such as glucanases and endochitinase. These activities help to retard nematode activities (Molinari and Leonetti 2019).

2.7.3.1 Arbuscular Mycorrhizal Fungi (AMF) as Bio-Control Agents Against Root-Knot Nematodes

AM fungi are found in more than 80% of almost all soil plant species as obligate root symbionts. They promote plant development, reduce plant stress, and increase the intake of mineral elements in their host plant in exchange for carbon (abiotic and biotic stress) (Vos et al. 2012; Smith et al. 2010). The protective effects of AM fungus against RKNs in various plants were discovered by several in vitro and in vivo studies, such as in coffee (against *M. exigua* and *M. cofeeicola*), tomato (against *M. incognita*), and banana (against *X. index*) (Koffi et al. 2013; Vos et al. 2012). Additionally, AM fungi helped the plants create specific compounds harmful to PPNs and interfered with the development of root diffusates which attract PPNs (Teillet et al. 2013).

2.7.3.2 Nematophagous Fungi as Bio-control Agents Against Root-Knot Nematodes

Different fungi that feed and grow on nematodes are called nematophagous fungi. Most are facultative nematode saprophytes, and other fungi are obligatory nematode parasites (Lopez-Llorca et al. 2008). Nematophagous fungi can be divided into four main categories based on their modes of action against nematodes: (1) nematode-trapping fungi (predatory fungi), (2) egg-parasitic fungi, (3) endoparasitic fungi, and (4) toxin-producing fungi.

2.7.3.2.1 Nematode Trapping Fungi

Fungi, such as *Arthrobotrys* spp. and *Monacrosporium* spp., are soil-borne pathogens which capture moving stages of *M. incognita* by using various trapping systems of different shapes and sizes (Khan et al. 2022). It was discovered that some fungi were responsible for capturing the J2s of *M. incognita* and are responsible for the release of certain compounds such as nematicidal and antimicrobial properties, pleurotin (*Nematoctonus concurrens* and *N. robustus*), or linoleic acid (*Arthrobotrys conoides* and *A. oligospora*), viz., *A. superba also* relies a compound for trapping J2s of *M. incognita* (Hallmann et al. 2009). Some other fungi, like *A. dactyloides,* efficiently trapped pathogenic juveniles of *M. graminicola* compared to *Monacrosporium eudermatum* and *Dactylella brochopaga* (Hallmann et al. 2009).

2.7.3.2.2 Egg-Parasitic Fungi

Fungi that attack eggs, females, and different stages of PPNs got more attention due to their potent approach to controlling economically important nematodes such as RKNs and cyst nematodes (*Heterodera* spp.). These fungi can infect the nematodes by specialised structures called zoospores, appresoria, penetration peg, and lateral mycelial branches (Lopez-Llorca et al. 2008). *Lecanicillium psalliotae*, *Pochonia chlamydosporia*, and *Paecilomyces lilacinus* are the most important bio-control among all egg-parasitic fungi that manage *M. incognita* (Li et al. 2015).

2.7.3.2.3 Endoparasitic Fungi

Drechmeria coniospora is an endoparasitic obligate parasite that parasitises the nematodes by their conidia and exists as a conidial form in the environment. These conidia adhere to the nematode's cuticle by using hyphae and kill them. Different nematode species, such as *Pratylenchus penetrans*, *Ditylenchus* spp., *H. schachtii*, and *Cephalenchus* sp. parasitises by the conidial attachment of *D. coniospora* (Lebrigand et al. 2016; Zhang et al. 2020).

2.7.3.2.4 Toxin-Producing Fungi

The fungi that make toxins are the most common and dangerous to RKNs. These fungi produce toxins that have nematicidal characteristics and paralyse the nematode juveniles before the penetration of fungal hyphae through nematode cuticle (Lopez-Llorca et al. 2008). Trans-2-decenedioic acid, a potent toxin with nematicidal characteristics both in vivo and in vitro, was produced by the fungus *Pleurotus ostreatus* and rapidly paralysed the RKNs (Luo et al. 2004). *Trichoderma*, which produces a range of enzymes, is an effective bio-control agent against *M. incognita* in addition to the four different types of nematophagous fungi and also produces compounds which help in plant growth (Agrawal and Kotasthane 2012; Haris et al. 2021). According to Ahmad et al. (2022a, b), combined applications of *T. harzianum* with fly ash manage *M. incognita* and improve the chilli plant's growth, yield, and biochemical substances. *Trichoderma* spp. produce extracellular hydrolytic enzymes such as serine protease (SprT), trypsin like chitinolytic (chi18–5 and chi18–12), and proteas (PRA1) which can parasitise the J2s and the eggs of RKNs (Szabo et al. 2012).

2.7.3.3 Bacterial Bio-control Agents Against Root-Knot Nematodes

Bacteria are single-celled, microscopic organisms that live in different environmental conditions, ranging from water, soil, acidic conditions, and radioactive waste (Fredrickson et al. 2004). They also exist as symbionts and parasites of numerous plants and animals. Several products obtained from bacteria are also considered BCAs to inhibit the PPNs (Hallmann et al. 2009). *Bacillus subtilis* is an important rhizobacterium that gained worldwide attention as a bio-pesticide against phytonematodes (Yu et al. 2015; Prakob et al. 2009; Hussain et al. 2020). According to Li et al. (2015) few rhizospheric bacteria like *Bacillus*, *Pasteuria*, and *Pseudomonas* are potent BCAs against PPNs, including RKNs. They are considered nematophagous soil-borne bacteria. They reduce the population of nematodes, including RKNs, through competition for space and nutrients, direct parasitism, and antibiosis (Lee and Kim 2016). Among all mechanisms against PPNs, antibiosis is considered the most important and widely used tool because of the production of volatile organics (VOCs), antibiotics, and toxins (Saraf et al. 2014). According to

Bharali et al. (2019), bacterial bioagents have more potential to control RKNs than fungal bioagents on black gram under in vivo conditions through seed treatments (Hussain and Khan 2020). From the last two decades of bacterial genera, *Serratia*, *Bacillus*, and *Pseudomonas* showed the maximum efficiency as BCAs against PPNs and plant growth enhancers (Radhakrishnan et al. 2017; Raymaekers et al. 2020).

2.7.3.3.1 Spore-Forming Bacteria Against Root-Knot Nematodes

Pasteuria penetrans is a gram-positive obligates parasitic and endospore-forming bacteria widely distributed in agricultural soils worldwide. Many studies proved their potential against RKNs infecting different vegetable crops (Swaranakumari and Sivakumar 2012; Swarnakumari 2017).

2.7.3.3.2 Cry Protein-Forming Bacteria Against Root-Knot Nematodes

Crystal protein or cry protein is an important bacterial compound that is secreted by ubiquitous and spore-producing bacterium *B. thuringiensis* (Bt), in response to RKNs. According to Prasad et al. (1972), the populations of *M. incognita* were significantly decreased by the application of *B. thuringiensis* var. *thuringiensis*. Currently, three families of cry protein known as Cry55, Cry6, and Cry5 cause inhibition in the growth and killing of nematodes larvae (Luo et al. 2013) (Table 2.4).

2.7.4 Genetics-Based Methods

Genetics-based methods of nematode control include the use of resistant/tolerant varieties developed by classical plant breeding and genetic engineering. Identifying the source of resistance and its utilisation is the best option for nematode management because of resistant varieties compatible with other management methods. In the recent decades, genetic engineering, a new approach, began in the field of nematology, which provides a strategy to design effective, durable, and resistant crops against economically important PPNs. Many studies were started in the country on the application of biotechnological approaches such as RNA interference and proteinase inhibitors to combat the RKNs.

Breeding for nematode-resistant cultivars is essential as an effective and environmentally safe alternative to chemical nematicides. Roberts (1995) reviewed that several wild plant species have a natural source of resistance against RKNs, *Meloidogyne* spp. In this context, studies were initiated to identify sources of resistance; for example, Williamson (1998) and Milligan et al. (1998) found resistance Mi gene from the wild tomato species, such as *Solanum peruvianum,* which conferred resistance to three economic species of RKNs, *M. incognita, M. javanica, and M. arenaria*. Similarly in pepper Me3 gene (Djian-Caporalino et al. 2001) and peanut Mae and Mag genes (Garcia et al. 1996). Globally, the Mi gene has been commercially utilised to develop root-knot-resistant in tomato cultivars. In India, Reddy et al. (2018) reported H-88-78-1, an advanced tomato breeding line is

Table 2.4 Application of bio-control agents against root-knot nematodes infesting vegetables

Bio-control agents	Vegetables	Root-knot nematodes	References
Lecanicillium muscarium	Tomato (*Solanum lycopersicum* L.)	*Meloidogyne incognita*	Hussain et al. (2018)
Bacillus cereus	Tomato (*Solanum lycopersicum* L.)	*M. incognita*	Li et al. (2019)
Purpureocillium lilacinum	Tomato (*Solanum lycopersicum* L.)	*M. enterolobii*	Silva et al. (2017)
Xylaria grammica	Tomato (*Solanum lycopersicum* L.)	*M. incognita*	Kim et al. (2018)
Trichoderma longibrachiatum	Cucumber (*Cucumis sativus* L.)	*M. incognita*	Zhang et al. (2015)
B. firmus	Tomato (*Solanum lycopersicum* L.)	*M. incognita*	d'Errico et al. (2019)
Paecilomyces lilacinus	Eggplant (*Solanum melongena* L.)	*M. incognita*	Nisha and Sheela (2016)
Trichoderma spp.	Pepper (*Capsicum annuum* L.)	*M. incognita*	Herrera-Parra et al. (2017)
Pochonia chlamydosporia	Carrot (*Daucus carota* L.)	*M. incognita*	Bontempo et al. (2017)
Pseudomonas fluorescens and *B. subtilis*	Cowpea (*Vigna unguiculata* L.)	*M. incognita*	El-Nagdi et al. (2019)
T. harzianum with Fly ash	Chilli (*Capsicum annuum* L.)	*M. incognita*	Ahmad et al. (2022a, b)
B. licheniformis	Tomato (*Solanum lycopersicum* L.)	*M. incognita*	Colagiero et al. (2018)
Pasteuria fluorescens	Sugar beet (*Beta vulgaris* L.)	*M. incognita*	Kavitha et al. (2007)
T. asperellum	Pineapple (*Ananas comosus* L.)	*M. javanica*	Kiriga et al. (2018)
Penicillium chrysogenum	Cucumber (*Cucumis sativus* L.)	*M. incognita*	Sikandar et al. (2019)
P. lilacinum	Tomato (*Solanum lycopersicum* L.)	*M. incognita*	Hore et al. (2018)

resistant to *M. incognita.* Further, molecular screening with Mi gene-linked markers Pmi and Mi2.3 indicated the presence of the Mi gene in H-88-78-1 (Table 2.5).

2.8 Conclusions and Future Perspectives

RKNs that have become a growing concern for vegetable producers due to their negative impact on growth and yield. The phase-out of fumigant nematicides due to human health and environmental pollution, the problem of RKNs still further intensified and become a major stumbling block for the successful cultivation of vegetables in fields as well as protected cultivation. Country-wise distribution of

Table 2.5 List of vegetables and their resistance varieties against root-knot nematodes

Vegetables	Resistant varieties	Root-knot nematodes	References
Tomato (*Solanum lycopersicum* L.)	Hisar Lalit, LA 3471, LA 2823, and H-88-78-1	*Meloidogyne incognita*	Reddy et al. (2018)
Brinjal (*Solanum melongena* L.)	Azadkranti, Gachhabaigan, Athagara Local, Kantabaigen, PBR 129-5, BB1-3, Utkal madhuri, LB-44, LB-5, ARU-1	*M. incognita*	Nayak and Pandey (2015)
Chilli (*Capsicum annuum* L.)	Surajmukhi, Pant Chilli-4, Brahmpur, Roshni, CA-960, ZCH-3025	*M. incognita* race 1	Ravishankar (2007)
Tomato (*Solanum lycopersicum* L.)	Hisar Lal, PAU Acc.-1, EC531804, EC631955, EC119197, IC117012, EC520075	*M. incognita*	Kaur et al. (2014)
Carrot (*Daucus carota* L.)	Golden Rosy	*M. incognita*	Khan et al. (2018)
Cabbage (*Brassica oleracea* var. *capitata*)	Parvati super cross	*M. incognita*	Khan and Khan (2021)
Okra (*Abelmoschus esculentus* L. Moench)	Arka Anamika, Sanam, Ikra-2, Ikra-1, and Dikshah	*M. incognita*	Mukhtar et al. (2014)
Cucumber (*Cucumis sativus* L.)	Marketmore, Dynasty, Long Green, and Pioneer-II	*M. incognita*	Mukhtar and Kayani (2019)

root-knot nematodes in the vegetable ecosystem cause substantial economic losses in vegetable production. Thus, the presence of RKNs becomes more challenging in intensive vegetable cultivation due to their polyphagous nature, high reproductive potential, and unique survival mechanism. However, an approach is warranted to provide basic information and awareness to stakeholders through academia, scientific publications, and extension activities related to management practices for keeping population of RKNs below damage threshold in the vegetable ecosystem. Each management approach has its limitations; therefore, integrated nematode management (INM) approach involving the combination of two or more suitable methods by holistically exploiting locally available resources is necessary to combat the menace of RKNs in the vegetable ecosystem. In addition, use of novel biotechnology tools is required to develop cost-effective and green reliable nematode management approaches. Whatever strategies/methods are invented in the future that should focus on the essential aspects like sustainable approaches which manage the RKNs and enhance vegetable production without affecting soil, human health, and the environment.

Acknowledgements All the authors are very thankful to the Chairman, Department of Botany, AMU, Aligarh, India, for moral encouragement, upliftment, and support throughout this chapter.

References

Agrawal T, Kotasthane AS (2012) Chitinolytic assay of indigenous *Trichoderma* isolates collected from different geographical locations of Chhattisgarh in Central India. Springer Plus, vol 1, pp 1–10. https://doi.org/10.1186/2193-1801-1-73

Ahmad F, Rather MA, Siddiqui MA (2010) Nematicidal activity of leaf extracts from *Lantana camara* L. against *Meloidogyne incognita* (Kofoid and White) chitwood and its use to manage roots infection of *Solanum melongena* L. Braz Arch Biotechnol 53:543–548

Ahmad G, Nishat Y, Ansari M, Khan A, Haris M, Khan AA (2021a) Eco-friendly approaches for the alleviation of root-knot nematodes. In: Plant growth-promoting microbes for sustainable biotic and abiotic stress management. Springer, Cham, pp 557–575

Ahmad G, Khan A, Khan AA, Ali A, Mohhamad HI (2021b) Biological control: a novel strategy for the control of the plant parasitic nematodes. A V Leeu 114:885–912

Ahmad G, Khan A, Ansari S, Khan AA, Elhakem A, Sami R, Mohamed HI (2022a) Management of root-knot nematode infection by using fly ash and *Trichoderma harzianum* in *Capsicum annum* plants by modulating growth, yield, photosynthetic pigments, biochemical substances, and secondary metabolite profiles. Notulae Botanicae Horti Agrobotanici Cluj-Napoca 50:12591–12591

Ahmad S, Rehman F, Adnan M, Ahmad I, Ahmad S, Iqbal Z, Ashraf E, Kalsoom M, Ehetisham ul Haq M (2022b) Rice nematodes and their integrated management. In: Modern techniques of rice crop production. Springer, Singapore, pp 517–543

Akhtar M, Mahmood I (1996) Control of plant-parasitic nematodes with organic and inorganic amendments in agricultural soil. Appl Soil Eco 4:243–247

Anwar SA, McKenry MV (2007) Variability in reproduction of four populations of *Meloidogyne incognita* on six cultivars of cotton. J Nematol 39:105

Anwar SA, Zia A, Javed N, Shakeel Q (2009) Weeds as reservoir of nematodes. Pak J Nematol 27: 145–153

Ayyar PNK (1926) On root knot nematodes infesting vegetables and other crops in south India. Madras Agric J 49:113–118

Barber CA (1901) Dept land records and agriculture. Madras agricultural branch 2. Bull No 45: 237–234

Basumatary B, Mahanta B, Borah A, Dutta P (2018) Assessment of yield losses due to *Meloidogyne incognita* on ivy gourd (*Coccinea indica* L.). Ind J Nematol 48:119–121

Berkeley MJ (1855) Vibrio forming excrescences on the roots of cucumber plants. Gard Chron 14: 220

Bharali A, Bhagawati B, Uday K (2019) Bio-efficacy of native bio-agents and bio-fertilisers for the management of root-knot nematode *Meloidogyne incognita* infecting black gram *Vigna mungo*. Int J Curr Microbio Appl Sci 8:1484–1501

Bhati SSB, Baheti BL (2021) Estimation of avoidable losses caused by *Meloidogyne incognita* infecting cucumber in poly-house. J Agric Appl Biol 2:35–40

Bontempo AF, Lopes EA, Fernandes RH, Freitas LGD, Dallemole-Giaretta R (2017) Dose-response effect of *Pochonia chlamydosporia* against *Meloidogyne incognita* on carrot under field conditions. Revista Caatinga 30:258–262

Bridge J (1996) Nematode management in sustainable and subsistence agriculture. Annu Rev Phytopathol 34:201–225

Chandra B, Khan MR (2011) Dynamics of soil nematodes in vegetable-based crop sequences in West Bengal. In J Pl Protec Res 51:7–13

Chandra P, Sao R, Gautam SK, Poddar AN (2010) Initial population density and its effect on the pathogenic potential and population growth of the root-knot nematode *Meloidogyne incognita* in four species of cucurbits. Asian J Plant Patho 4:1–15

Chariou PL, Steinmetz NF (2017) Delivery of pesticides to plant parasitic nematodes using tobacco mild green mosaic virus as a Nano carrier. ACS Nano 11:4719–4730

Colagiero M, Rosso LC, Ciancio A (2018) Diversity and bio-control potential of bacterial consortia associated to root-knot nematodes. Biol Control 120:11–16

Collange B, Navarrete M, Peyre G, Mateille T, Tchamitchian M (2011) Root-knot nematode (*Meloidogyne*) management in vegetable crop production: the challenge of an agronomic system analysis. Crop Prot 30:1251–1262

d'Errico G, Marra R, Crescenzi A, Davino SW, Fanigliulo A, Woo SL, Lorito M (2019) Integrated management strategies of *Meloidogyne incognita* and *Pseudopyrenochaeta lycopersici* on tomato using a *Bacillus firmus*-based product and two synthetic nematicides in two consecutive crop cycles in greenhouse. Crop Prot 122:159–164

Davis EL, Haegeman A, Kikuchi T (2011) Degradation of the plant cell wall by nematodes. In: Genomics and molecular genetics of plant-nematode interactions. Springer, Dordrecht, pp 255–272

Decraemer W, Hunt DJ (2013) Structure and classification. In: Perry RN, Moens M (eds) Plant nematology, 2nd edn. CABI, Wallingford, pp 3–39

Devi S, Das D (2016) Effect of organic amendments on root-knot nematode (*Meloidogyne incognita*) in cucumber. Pest Manag Hort Ecosyst 22:176–181

Dhaliwal GS, Koul O (2007) Biopesticides and pest management: conventional and biotechnological approaches. Kalyani Publishers, New Delhi

Djian-Caporalino C, Pijarowski L, Fazari A, Samson M, Gaveau L, O'Byrne C, Lefebvre V, Caranta C, Palloix A, Abad P (2001) High-resolution genetic mapping of the pepper (*Capsicum annuum* L.) resistance loci Me3 and Me4 conferring heat-stable resistance to root-knot nematodes (*Meloidogyne* spp.). Theor Appl Genet 103:592–600

El-Nagdi WMA, Youssefi MMA, Abd-El-Khair H, Abd Elgawad MMM, Dawood MG (2019) Effectiveness of *Bacillus subtilis*, *B. pumilus*, *Pseudomonas fluorescens* on *Meloidogyne incognita* infecting cowpea. Pak J Nematol 37:35–43

Escobar C, Barcala M, Cabrera J, Fenoll C (2015) Overview of root-knot nematodes and giant cells. Adv Bot Res 73:1–32

Eves-van den Akker S, Birch PR (2016) Opening the effector protein toolbox for plant–parasitic cyst nematode interactions. Mol Pl 9:1451–1453

FAO (2018) Food and Agricultural Organization of the United Nations, Statistics Division (FAOSTAT, 2018)

FAO (2019) Food and Agricultural Organization of the United Nations, Statistics Division (FAOSTAT, 2019)

FAO (2020) Food and Agricultural Organization of the United Nations, Statistics Division (FAOSTAT, 2020)

Forghani F, Hajihassani A (2020) Recent advances in the development of environmentally benign treatments to control root-knot nematodes. Front Pl Sci 11:1125

Fredrickson JK, ZacharaJM KDW, Kukkadapu RK, McKinley JP, Heald SM, Plymale AE (2004) Reduction of TcO4-by sediment-associated biogenic Fe (II). Geochim Cosmochim Acta 68: 3171–3187

Gao H, Qi G, Yin R, Zhang H, Li C, Zhao X (2016) *Bacillus cereus* strain S2 shows high nematicidal activity against *Meloidogyne incognita* by producing sphingosine. Sci Rep 6:1–11

Garcia GM, Stalker HT, Shroeder E, Kochert G (1996) Identification of RAPD, SCAR and RFLP markers tightly linked to nematode resistance genes introgressed from *Arachis cardenasii* into *Arachis hypogaea*. Genome 39:836–845

Gaur HS (1975) Crop damage in relation to the density of nematode population and an integrated approach of nematode population management. Ph.D. Thesis, IARI, New Delhi, pp 184

Gaur HS (2006) Approaches for alternation of soil environment for management of plant parasitic nematodes. In: Gaur HS, Jain RK, Ganguly S, Sirohi A, Kamra A (eds) Nematode biodiversity, identification and role of agriculturally useful nematodes in soil health, dynamics, of nematode community structure in major cropping systems and relevance in integrated nematode management. Division of Nematology, Indian Agricultural Research Institute, New Delhi pp 147–159

Gaur HS, Dhingra A (1991) Management of *Meloidogyne incognita* and *Rotylenchulus reniformis* in nursery beds by soil solarisation and organic amendment. Revue de nematologie 14:190–197

Gaur HS, Perry RN (1991) The biology and control of the plant parasitic nematode, *Rotylenchulus reniformis*. In: Evans K (ed) Agricultural zoological review. Intercept Ltd., London, vol 4, pp 177–211

Ghule TM, Singh A, Khan MR (2014) Root knot nematodes: threat to Indian agriculture. Pop Kheti 2:126–130

Glick BR (2012) Plant growth-promoting bacteria: mechanisms and applications. Scientifica 2012: 1

Gorny AM, Ye W, Cude S, Thiessen L (2021) Soybean root-knot nematode: a diagnostic guide. Pl H Prog 22:164–175

Goswami BK, Meshram N (1991) Studies on comparative efficacy of mustard and karanj oil seed cakes with a nematicide, carbofuran against root-knot nematode, *Meloidogyne incognita* in tomato. In J Nematol 21:66–70

Gowda TM, Rai AB, Singh B (2017) Technical bulletin No. 76. IIVR, Varanasi, p 32

Gowda MT, Rai AB, Singh B (2019) Root knot nematodes menace in vegetable crops and their management in India: a review. Veg Sci 46:1–16

Gruda N (2005) Impact of environmental factors on product quality of greenhouse vegetables for fresh consumption. Crit Rev Pl Sci 24:227–247

Hallmann J, Meressa BH (2018) 10 Nematode parasites of vegetables. Plant parasitic nematodes in subtropical and tropical agriculture, 346

Hallmann J, Davies KG, Sikora R (2009) Biological control using microbial pathogens, endophytes and antagonist. In: Root-knot nematodes, p 380

Haque MM, Gaur HS (1985) Effect of multiple cropping sequences on the dynamics of nematode population and crop performance. In J Nematol 15:262–263

Haris M, Shakeel A, Ansari, MS, Hussain T, Khan AA, Dhankar R (2021) Sustainable crop production and improvement through bio-prospecting of fungi. In: Fungi bio-prospects in sustainable agriculture, environment and nano-technology. Academic Press, pp 407–428

Herrera-Parra E, Cristóbal-Alejo J, Ramos-Zapata JA (2017) *Trichoderma* strains as growth promoters in *Capsicum annuum* and as bio-control agents in *Meloidogyne incognita*. Chil J Ag Res 77:318–324

Hore J, Roy K, Maiti AK (2018) Evaluation of bio-nematon (*Purpureocillium lilacinum* 1.15% WP) against root-knot nematode (*Meloidogyne incognita*) in tomato. J Entomo Zoo Stud 6: 1700–1704

Hunt DJ, Handoo ZA (2009) Taxonomy, identification and principal species. In: Perry RN, Moens M, Starr JL (eds) Root-knot nematodes. Cap. 3. CAB International, Wallingford, pp 55–88

Hussain M, Zouhar M, Rysanek P (2018) Suppression of *Meloidogyne incognita* by the entomopathogenic fungus *Lecanicillium muscarium*. Pl Dis 102:977–982

Hussain T, Haris M, Shakeel A, Ahmad G, Khan AA, Khan MA (2020) Bio-nematicidal activities by culture filtrate of Bacillus subtilis Hussain T-AMU: new promising biosurfactant bioagent for the management of root galling caused by Meloidogyne incognita. Vegetos 33:229–238

Hussain T, Khan AA (2020) Bacillus firmus HussainT:Lab.66, a new biosurfactant bioagent having potential bio-nematicidal activity against root knot nematode Meloidogyne incognita. National Conference on Recent Advances in Biological Science (NCRABS 2020) held on 5th March, 2020, organized by Dept. of Biosciences, Faculty of Natural Sciences, Jamia Millia Islamia University, New Delhi, OP 25, p 42

Hussain T, Shakeel A, Haris M, Ahmad G, Khan AA (2019) First report on induction of resistance to root-knot nematode (Meloidogyne incognita) by DL-beta aminobutyric acid under abiotic stress (Fly Ash). Agricultura 109(1–2):121–125

Jain RK, Bhatti DS (1987) Population development of root-knot nematode (*Meloidogyne javanica*) and tomato yield as influenced by summer ploughings. Trop Pest Manage 33:122–125

Jain RK, Mathur KN, Singh RV (2007) Estimation of losses due to plant parasitic nematodes on different crops in India. In J Nematol 37:219–221

Jamal Q, Cho JY, Moon JH, Munir S, Anees M, Kim KY (2017) Identification for the first time of cyclo (d-Pro-l-Leu) produced by *Bacillus amyloliquefaciens* Y1 as a nematocide for control of *Meloidogyne incognita*. Molecules 22:1839

Jones JT, Haegeman A, Danchin EG, Gaur HS, Helder J, Jones MG, Perry RN (2013) Top 10 plant-parasitic nematodes in molecular plant pathology. Mol Pl Pathol 14:946–961

Kaskavalci G (2007) Effects of soil solarisation and organic amendment treatments for controlling *Meloidogyne incognita* in tomato cultivars in Western Anatolia. Turk J Agric For 31:159–167

Kaur SG, Dhaliwal MS, Cheema DS, Jindal SK, Gaikwad AK (2014) Screening of tomato (*Solanum lycopersicum* L.) germplasm for root-knot nematode resistance using conventional and molecular marker techniques. In J Nematol 44:56–61

Kavitha J, Jonathan EI, Umamaheswari R (2007) Field application of *Pseudomonas fluorescens*, *Bacillus subtilis* and *Trichoderma viride* for the control of *Meloidogyne incognita* (Kofoid and White) Chitwood on sugar beet. J Bio Cont 21:211–215

Khan MR (2008) Current options for managing nematodes pest of crops in India. Department of Agricultural Entomology, West Bengal

Khan MR (2015) Nematode diseases of crops in India. In: Recent advances in the diagnosis and management of plant diseases. Springer, New Delhi, pp 183–224

Khan A, Khan AA (2021) Evaluation of cabbage cultivars through screening against root-knot Nematode, *Meloidogyne incognita*. Res J Ag Sci 12:1767–1772

Khan MR, Jain RK, Singh RV, Pramanik A (2010) Economically important plant parasitic nematodes ATLAS. Directorate of Information and Publications of Agriculture Krishi Anusandhan Bhavan 1, Pusa New Delhi, pp 1–145

Khan MR, Jain RK, Ghule TM, Pal S (2014) Root knot nematodes in India-a comprehensive monograph. All India Coordinated Research Project on Plant Parasitic nematodes with Integrated approach for their Control, Indian Agricultural Research Institute, New Delhi, pp 1–78

Khan F, Asif M, Khan A, Tariq M, Siddiqui MA (2018) Screening of carrot cultivars against root-knot nematode *Meloidogyne incognita*. Phytopathol 71:415–421

Khan A, Ahmad G, Haris M, Khan AA (2022) Bio-organics management: novel strategies to manage root-knot nematode, *Meloidogyne incognita* pest of vegetable crops. Gesunde Pflanzen 75:193–209. https://doi.org/10.1007/s10343-022-00679-2

Khan A, Khan A, Ali A, Fatima S, Siddiqui MA (2023) Root-knot nematodes (Meloidogyne spp.): biology, plant-nematode interactions and their environmentally benign management strategies. Gesunde Pflanzen:1–19. https://doi.org/10.1007/s10343-023-00886-5

Kim TY, Jang JY, Yu NH, Chi WJ, Bae CH, Yeo JH, Kim JC (2018) Nematicidal activity of grammicin produced by *Xylaria grammica* KCTC 13121BP against *Meloidogyne incognita*. P Manag Sci 74:384–391

Kiriga AW, Haukeland S, Kariuki GM, Coyne DL, Beek NV (2018) Effect of *Trichoderma* spp. and *Purpureocillium lilacinum* on *Meloidogyne javanica* in commercial pineapple production in Kenya. Biol Control 119:27–32

Koffi MC, Vos C, Draye X, Declerck S (2013) Effects of *Rhizophagus irregularis* MUCL 41833 on the reproduction of *Radopholus similis* in banana plantlets grown under in vitro culture conditions. Mycorrhiza 23:279–288

Lebrigand K, He LD, Thakur N, Arguel MJ, Polanowska J, Henrissat B, Ewbank JJ (2016) Comparative genomic analysis of *Drechmeria coniospora* reveals core and specific genetic requirements for fungal endoparasitism of nematodes. PLoS Genet 12:e1006017

Lee YS, Kim KY (2016) Antagonistic potential of *Bacillus pumilus* L1 against root-knot nematode, *Meloidogyne arenaria*. J Phytopathol 164:29–39

Li J, Zou C, Xu J, Ji X, Niu X, Yang J, Huang X, Zhang KQ (2015) Molecular mechanisms of nematode-nematophagous microbe interactions: basis for biological control of plant-parasitic nematodes. Annu Rev Phytopathol 53:67–95

Li X, Hu HJ, Li JY, Wang C, Chen SL, Yan SZ (2019) Effects of the endophytic bacteria *Bacillus cereus* BCM2 on tomato root exudates and *Meloidogyne incognita* infection. Pl Dis 103:1551–1558

Lopez-Llorca LV, Macia-Vicente JG, Jansson HB (2008) Mode of action and interactions of nematophagous fungi. In: Integrated management and bio-control of vegetable and grain crops nematodes. Springer, Dordrecht, pp 51–76

Lunt DH, Kumar S, Koutsovoulos G, Blaxter ML (2014) The complex hybrid origins of the root not nematodes revealed through comparative genomics. PeerJ 2:e356

Luo H, Mo M, Huang X, Li X, Zhang K (2004) Coprinus comatus: a basidiomycete fungus forms novel spiny structures and infects nematode. Mycologia 96:1218–1224

Luo H, Xiong J, Zhou Q, Xia L, Yu Z (2013) The effects of *Bacillus thuringiensis* Cry6A on the survival, growth, reproduction, locomotion, and behavioural response of *Caenorhabditis elegans*. App Microbio Biot 97:10135–10142

Milligan BS, Bodeau J, Yaghoobi J, Kaloshian I, Zabel WVM (1998) The root-knot nematode resistance gene *Mi* from tomato is a member of the Leucine Zipper, nucleotide binding, leucine-rich repeat family of plant genes. Plant Cell 10:1307–1319

Molinari S, Leonetti P (2019) Bio-control agents activate plant immune response and prime susceptible tomato against root-knot nematodes. PLoS One 14:213230

Mukhtar T (2018) Management of root-knot nematode, *Meloidogyne incognita*, in tomato with two *Trichoderma* species. Pak J Zoo 50:1589–1592

Mukhtar T, Kayani MZ (2019) Growth and yield responses of fifteen cucumber cultivars to root-knot nematode (*Meloidogyne incognita*). Acta Sci Pol Hortorum Cultus 18:45–52

Mukhtar T, Hussain MA, Kayani MZ, Aslam MN (2014) Evaluation of resistance to root-knot nematode (*Meloidogyne incognita*) in okra cultivars. Crop Prot 56:25–30

Nayak DK, Pandey R (2015) Screening and evaluation of brinjal varieties /cultivars against root-knot nematode, *Meloidogyne incognita*. Int J Adv Res 3:476–479

Nisha MS, Sheela MS (2016) Effect of fungal egg parasite, *Paecilomyces lilacinus* (Thom.) samson on *Meloidogyne incognita* in brinjal. In J Nematol 46:157–159

Oka Y (2010) Mechanisms of nematode suppression by organic soil amendments a review. Appl Soil Ecol 44:101–115

Patel GJ, Shah HM, Patel DJ (1979) Reaction of some tomato cultivars to root-knot nematode. In J Nematol 9:60–61

Perry RN, Moens M (2011) Introduction to plant-parasitic nematodes; modes of parasitism. In: Genomics and molecular genetics of plant-nematode interactions. Springer, Dordrecht, pp 3–20

Phani V, Khan MR, Dutta TK (2021) Plant-parasitic nematodes as a potential threat to protected agriculture: current status and management options. Crop Prot 144:105573

Prakob W, Nguen-Hom J, Jaimasit P, Silapapongpri S, Thanunchai J, Chaisuk P (2009) Biological control of lettuce root knot disease by use of *Pseudomonas aeruginosa*, *Bacillus subtilis* and *Paecilomyces lilacinus*. J Agric Technol 5:179–191

Prasad SSV, Tilak KVBR, Gollakota RG (1972) Role of *Bacillus thuringiensis* var. *thuringiensis* on the larval survivability and egg hatching of *Meloidogyne* spp. the causative agent of root-knot disease. J Inver Pathol 20:377–378

Qiu S, Maquilan MAD, Chaparro JX, Brito JA, Beckman TG, Dickson DW (2021) Susceptibility of Flordaguard peach rootstock to a resistant-breaking population of *Meloidogyne floridensis* and two populations of *Meloidogyne arenaria*. J Nematol 53:1

Radhakrishnan R, Hashem A, AbdAllah EF (2017) *Bacillus*: a biological tool for crop improvement through bio-molecular changes in adverse environments. Front Physiol 8:667

Rao MS, Umamaheswari R, Priti K, Rajinikanth R, Vidyashree P, Prabhu KM (2015) Nematode management in vegetable crops. IIHR Technical Bulletin No.: 47. Published by Director ICAR-IIHR, Bengaluru

Rather MA, Ahmad F, Siddiqui MA (2008) Nematicidal effect of chopped leaves of some selected plants against root-knot nematode, *Meloidogyne incognita* on tomato. Int J Pl Sci 3:339–341

Ravishankar M (2007) Nematode development and biochemical changes in genotypes of chilli (*Capsicum* spp.) infected with root knot nematode (*Meloidogyne incognita* race 1). Ph.D. Thesis, IARI New Delhi, pp 60

Rawal S (2020) A review on root-knot nematode infestation and its management practices through different approaches in tomato. Trop Agroecosyst 1:92–96

Raymaekers K, Ponet L, Holtappels D, Berckmans B, Cammue BP (2020) Screening for novel bio-control agents applicable in plant disease management–a review. Biol Control 144:104240

Reddy PP (2021) Nematode diseases of crops and their management. Springer Singapore

Reddy YS, Sellaperumal C, Prasanna HC, Yadav A, Kashyap SP, Singh S, Rai N, Singh M Singh B (2018) Screening of tomato genotypes against root-knot nematode and validation of Mi 1 gene linked markers. Proc Natl Acad Sci India, Sect B Boil Sci 88: 65–72

Roberts PA (1995) Conceptual and practical aspects of variability in root-knot nematodes related host-plant resistance. Annu Rev Phytopathol 33:199–221

Robinson AF (2002) Soil and plant interactions' impact on plant-parasitic nematode host finding and recognition. The behavioural ecology of parasites. CAB International, Wallingford, pp 89–110

Saraf M, Pandya U, Thakkar A (2014) Role of allelochemicals in plant growth promoting 8 rhizobacteria for biocontrol of phytopathogens. Microbiol Res 169:18–29

Shakeel A, Khan AA, Haris M (2020) Multifaceted strategies used by root-knot nematodes to parasitise plants-a review. Phyton 89:205

Shaltoot A (2001) Economic loss resulting from nematode-infected vegetables and fruits. In: Disease problems caused by nematodes in horticultural farms in Egypt Proceedings of a workshop (In Arabic), Menoufia Univ, pp 43–50

Sharma A, Haseeb A, Abuzar S (2006) Screening of field pea (*Pisum sativum*) selections for their reactions to root-knot nematode (*Meloidogyne incognita*). J Zhej Uni Sci B 7:209–214

Sikandar A, Zhang MY, Zhu XF, Wang YY, Ahmed M, Iqbal MF, Javeed A, Xuan YH, Fan HY, Liu XY, Chen LJ, Duan YX (2019) Efficacy of *Penicillium chrysogenum* strain SNEF1216 against root-knot nematodes (*Meloidogyne incognita*) in cucumber (*Cucumis sativus* L.) under greenhouse conditions. Appl Eco Envi Res 17(5):2451–12464

Sikora RA, Fernández E (2005) Nematodes parasites of vegetables. In: Liuc M, Sikora RA, Bridge J (eds) Plant parasitic nematodes in subtropical and tropical agriculture. CAB International, Wallingford, pp 319–392

Sikora RA, Coyne D, Hallmann J, Timper P (eds) (2018) Plant parasitic nematodes in subtropical and tropical agriculture. CABI

Silva SD, Carneiro RM, Faria M, Souza DA, Monnerat RG, Lopes RB (2017) Evaluation of *Pochonia chlamydosporia* and *Purpureocillium lilacinum* for suppression of *Meloidogyne enterolobii* on tomato and banana. J Nematol 49:77–85

Singh S (2013) Integrated approach for the management of the root-knot nematode, *Meloidogyne incognita*, on eggplant under field conditions. Nematol 15:747–757

Singh T, Patel BA (2015) Management of root-knot nematode (*Meloidogyne incognita*) in bottle gourd using different botanicals in pots. J Parasit Dis 39:441–445

Smith SE, Facelli E, Pope S, Smith FA (2010) Plant performance in stressful environments: interpreting new and established knowledge of the roles of arbuscular mycorrhizas. Plant Soil 326:3–20

Steinmetz KA, Potter JD (1996) Vegetables, fruit, and cancer prevention: a review. J Amer Diet Assoc 96:1027–1039

Sundresh HN, Setty KGH (1977) Crop rotation as an effective and practical means of controlling root-knot nematode *Meloidogyne incognita* Chitwood. Curr Res 6:157–158

Swaranakumari N, Sivakumar CV (2012) Bio-efficacy of obligate bacterial parasite, *pasteuria penetrans* against root-knot nematode, *Meloidogyne incognita* infestation in chilli. In J Nematol 42:42–45

Swarnakumari N (2017) Role of bacterial bioagent, *Pasteuria penetrans* in the management of root knot nematode, *Meloidogyne incognita* by altering the lifecycle. Pest Manag Hort Ecosyst 23: 76–80

Szabo M, Csepregi K, Galber M, Viranyi F, Fekete C (2012) Control plant-parasitic nematodes with *Trichoderma* species and nematode- trapping fungi: The role of chi18-5 and chi18-12 genes in nematode egg-parasitism. Biol Control 63:121–128

Teillet A, Dybal K, Kerry BR, Miller AJ, Curtis RH, Hedden P (2013) Transcriptional changes of the root-knot nematode *Meloidogyne incognita* in response to *Arabidopsis thaliana* root signals. PLoS One 8:61259

Trifonova Z, Karadjova J, Georgieva T (2009) Fungal parasites of the root-knot nematodes *Meloidogyne* spp. in southern Bulgaria. Est J Eco 58

Tyler J (1933) Development of the root-knot nematode as affected by temperature. Hilgardia 7:389–415

Umashankar KN, Krishnappa K, Reddy BMR, Ravichandra NG, Karuna K (2005) Intercropping for the management of root-knot nematode, *Meloidogyne incognita* in vegetable-based cropping systems. In J Nematol 35:46–49

Verma AC, Anwar A (1996) Assessment of yield loss due to *Meloidogyne incognita* in pointed gourd (*Trichosanthes dioica* Roxb.). Af Asian J Nematol 6:92–93

Vos CM, Tesfahun AN, Panis B, DeWaele D, Elsen A (2012) Arbuscular mycorrhizal fungi induce systemic resistance in tomato against the sedentary nematode *Meloidogyne incognita* and the migratory nematode *Pratylenchus penetrans*. Appl Soil Ecol 61:1–6

Walia RK, Kumar V, Kumar P (2016) Major nematode problems and technologies generated for their management. Project coordinating cell, ICAR-IARI, New Delhi 110012:1–34

Wesemael W, Viaene N, Moens M (2011) Root-knot nematodes (*Meloidogyne* spp.) in Europe. Nematology 13:3–16

Widmer TL, Ludwig JW, Abawi GS (1999) The northern root-knot nematode on carrot, lettuce, and onion in New York. New York State Agricultural Experiment Station

Wieczorek K, Elashry A, Quentin M, Grundler FMW, Favery B, Seifert GJ, Bohlmann H (2014) A distinct role of pectate lyases in the formation of feeding structures induced by cyst and root-knot nematodes. Mol Pl Microbe Int 27:901–912

Williamson VM (1998) Root knot resistance genes in tomato and their potential for future use. Annu Rev Phytopathol 36:277–293

Yu Z, Xiong J, Zhou Q, Luo H, Hu S, Xia L, Yu Z (2015) The diverse nematicidal properties and bio-control efficacy of *Bacillus thuringiensis* Cry6A against the root-knot nematode *Meloidogyne hapla*. J Inve Pathol 125:73–80

Zhang S, Gan Y, Xu B (2015) Bio-control potential of a native species of *Trichoderma longibrachiatum* against *Meloidogyne incognita*. App Soil Eco 94:21–29

Zhang Y, Li S, Li H, Wang R, Zhang KQ, Xu J (2020) Fungi–nematode interactions: diversity, ecology, and bio-control prospects in agriculture. J Fungi 6:206

Zhou L, Yuen G, Wang Y, Wei L, Ji G (2016) Evaluation of bacterial biological control agents for control of root-knot nematode disease on tomato. Crop Prot 84:8–13

Chemotaxis in Root-Knot Nematodes

3

William César Terra, Letícia Lopes de Paula, Daniele de Brum, Vicente Paulo Campos, Denilson Ferreira de Oliveira, and Jorge Teodoro De Souza

Abstract

For a long time, chemotaxis in root-knot nematodes has received scant attention. In recent years, however, this topic has captured the attention of several researchers worldwide. Chemotaxis refers to the movement of living organisms towards or away from a chemical gradient. Second-stage juveniles (J2s) hatching from eggs are the only infective stage of *Meloidogyne* spp., and they locate their host through chemotaxis by sensing host-secreted chemoattractants. Despite its importance in the host location process, the structures and properties of compounds that are attractive to *Meloidogyne* spp. J2s are not well understood. This chapter will present a compilation of information on the attractiveness of volatile and non-volatile compounds identified in emissions from plant roots and microorganisms. The obstacles in chemotaxis studies, which include the characterization of compounds that attract or repel, the limitations of in vitro methodologies, such as Petri dishes filled with agar and the challenges of studies using soil, will be presented. On the other hand, the advances achieved in the recent years and how chemotaxis can be manipulated to manage these important soil-borne pathogens will also be discussed.

Keywords

Attraction · Chemical gradients · *Meloidogyne* spp. · Repellence

W. C. Terra · L. Lopes de Paula · D. de Brum · V. P. Campos · J. T. De Souza (✉)
Department of Phytopathology, Federal University of Lavras, University Campus, Lavras, Minas Gerais, Brazil
e-mail: jorge.souza@ufla.br

D. F. de Oliveira
Department of Chemistry, Federal University of Lavras, University Campus, Lavras, Minas Gerais, Brazil

F. Ahmad, G. Nombela (eds.), *Root-Galling Disease of Vegetable Plants*,
https://doi.org/10.1007/978-981-99-3892-6_3

3.1 Introduction

Root-knot nematodes (*Meloidogyne* spp.) are the most widespread and damaging among the plant-parasitic nematodes. These pathogens cause losses in agriculture that are estimated to be around US$157 billion per year (Coyne et al. 2018). These nematodes are obligate biotrophic parasites that penetrate into the roots of host plants to obtain food. They molt once inside the eggs, and the second-stage juveniles (J2s) hatch and move through the soil to find suitable hosts before their energy reserves are depleted. They enter just behind the root tips and establish the feeding site at the vascular tissue, known as giant cells and the external symptom as gall or root-knot. The nematode feeds and molts three more times before it reaches maturity when females lay eggs in a gelatinous matrix. The eggs hatch, and J2s, the only infective stage, will spread in the soil again, searching for new penetration sites in the same host or new hosts.

Molecules produced by one organism with the property of influencing the behaviour of other organisms are called semiochemicals or signaling molecules (Robinson and Perry 2006). When these interactions involve members of different species, they are named allelochemicals (Perry 1996). Semiochemicals influence all relationships among living organisms in nature. The process by which *Meloidogyne* spp. J2s follow chemical gradients to find a suitable host plant is known as chemotaxis. Nematodes use chemotaxis to locate food, for mating, to avoid predators and many other behavioural responses (Zuckerman and Jansson 1984). The most important semiochemicals that attract or repel *Meloidogyne* spp. are the ones produced by plants (Kihika et al. 2017; Murungi et al. 2018; Sikder and Vestergård 2020). Factors such as the presence of microorganisms, root zone and age, soil composition and texture heavily influence the attractiveness to *Meloidogyne* spp. J2s (Perry 1996; Rocha et al. 2016). Water-soluble compounds are used for short distance, whereas volatile organic compounds (VOCs) are used in long range chemotaxis (Čepulytė et al. 2018; Wang et al. 2019; Sikder and Vestergård 2020). Chemotaxis in *Meloidogyne* spp. has been extensively studied since its first demonstration (Lindford 1939), but only recently, due to the use of modern techniques, the compounds that exert chemotaxis are being revealed (Van Dam and Bouwmeester 2016).

Our objective in this chapter is to review the information on chemotaxis in *Meloidogyne* spp. J2s towards or away from the emitting source, with emphasis on chemicals produced by plants and microorganisms. The possible applications of chemotaxis in managing these pathogens are also discussed.

3.2 Perception of Environmental Stimuli by *Meloidogyne* spp.

In order to find suitable hosts, nematodes need to assimilate information from their external environment via sensing organs or sensilla (Perry 1996), most of which are located in the anterior end of the nematode body. Of all the nematode sensilla, the amphids are considered to be the primary chemosensilla. These organs are situated on either side of the nematode mouth, open to the exterior via a prominent pore (Bargmann 2006). Each amphid contains sensory cilia, dendrites of chemosensory neurons, that are exposed to the environment via a pore in the cuticle (Siddique et al. 2022). Axonal processes from these neurons project into the circumpharyngeal nerve ring, the main mass of the nematode central nervous system, where much of the sensory integration takes place. Sensory organs in the tail region are known as phasmids, and they are similar in general structure to the amphids, each consisting of an external pore. Anatomy and chemosensation in functional studies implicate amphid and phasmid neurons in chemosensation (Robinson and Perry 2006).

Migration of the nematode is enabled by separate innervation of dorsal and ventral muscle trunks by their respective nerve chords along most of the body length. Innervation is achieved via somatic muscle arms that extend to and synapse only with their respective dorsal or ventral nerve chords (Robinson and Perry 2006).

3.3 Rhizosphere Gradients

Meloidogyne species chemotaxis can be defined as the migration oriented with respect to a chemical stimulus gradient. The soil volume affected by roots—the rhizosphere—establishes several chemical gradients that affect the *Meloidogyne* spp. J2 movement (Fig. 3.1). It is certain that some of these gradients constitute cues that allow the migration of nematodes towards the root region.

Several authors have shown that most gradients in the rhizosphere extent for 0.5–4 mm, but gases may exceed this limit (Kuzyakov and Razavi 2019). The following are some of the gradients formed in the rhizosphere that are thought to help J2s find roots and establish a feeding site before their energy reserves are completely depleted (Rocha et al. 2010).

3.3.1 Carbon dioxide (CO_2)

The most frequently suggested attractant for plant-parasitic nematodes has been CO_2 (Klingler 1965; Pline and Dusenbery 1987). Carbon dioxide was long regarded as the most common and potent nematode attractant in nature (Robinson and Perry 2006).

By using planar optodes, a non-destructive visualization technique, gradients of CO_2 were clearly visible around root tips but less pronounced around mature root parts, probably due to high root respiration and microbial activity around the tips (Holz et al. 2020). The mean CO_2 concentration at the root center of young roots was 0.26 μmol L^{-1}, which was higher than in bulk soil. This CO_2-sensitive sensor revealed a CO_2 rhizosphere range of 1.5–3 mm (Holz et al. 2020). This seems to be a relatively short distance considering the gaseous nature of carbon dioxide. It is important to note that *Meloidogyne* spp. J2s only penetrate at a region just after the root tip.

Fig. 3.1 Gradients in the rhizosphere that affect the chemotaxis of second-stage juvenile (J2s) of *Meloidogyne* spp. towards the root system. These gradients include root exudates, volatile organic compounds (VOCs), organic compounds, CO_2 and pH, all of them under the influence of the microbes inhabiting the rhizosphere

3.3.2 pH

The release of H^+ by roots into slightly acidic, neutral and alkaline soils (without N fertilization) is one of the dominant mechanisms of plants to mobilize nutrients and maintain the electrochemical potential on the root surface (Kuzyakov and Razavi 2019). The common distance of root-induced pH changes is about 2–3 mm (Blossfeld et al. 2010).

Meloidogyne hapla was shown to be attracted to pH gradients between 4.5 and 5.4 formed by acetic acid and several other Brønsted acids (Wang et al. 2009). This observation is consistent with the idea that low pH is an attractant for nematodes. As mentioned above, root-knot nematodes have been reported to be attracted to CO_2; however, the study suggested that this attraction may be due to CO_2-acidified solutions rather than to CO_2 itself.

3.3.3 Organic Compounds

The organic compounds released by living roots into the soil are collectively referred to as rhizodeposits. It is estimated that approximately 3% of the assimilated C is released by plants as rhizodeposits, including the continuously and passively released exudates and the dynamically and actively released mucilage, secretions and enzymes from various root zones (Pausch and Kuzyakov 2018). Most root exudation takes place at the root tips, and two main mechanisms decrease the concentration of organic compounds in soil solution: (1) microbial uptake and utilization/modification and (2) sorption on surfaces of minerals or organic matter (Kuzyakov and Razavi 2019). The rhizosphere extent measured by ^{14}C imaging of exudates is usually only 2–3 mm (Holz et al. 2018).

In recent years, a variety of volatile and non-volatile organic compounds released by roots of host plants have been identified as attractants or repellents to *Meloidogyne* spp. J2s (Kirwa et al. 2018; Tsai et al. 2021). Oota et al. (2019), using cryo time-of-flight secondary ion mass spectrometry/scanning electron microscopy (cryo-TOF-SIMS/SEM) analyzes, techniques used to visualize the distribution of water-soluble compounds in freeze-fixed samples at microscopic resolution level, demonstrated that propane-1,3-diamine, putrescine and especially cadaverine (Fig. 3.2), are potent attractants to J2s of *M. incognita*. These compounds are produced and released by soybean root tips and form a gradient up to 250 μm from the root surface.

The evaluation of rhizosphere extent and shape are more complicated for signalling compounds like secondary metabolites and other chemoattractants because most of them are volatile and are not strongly absorbed by soil minerals. Consequently,

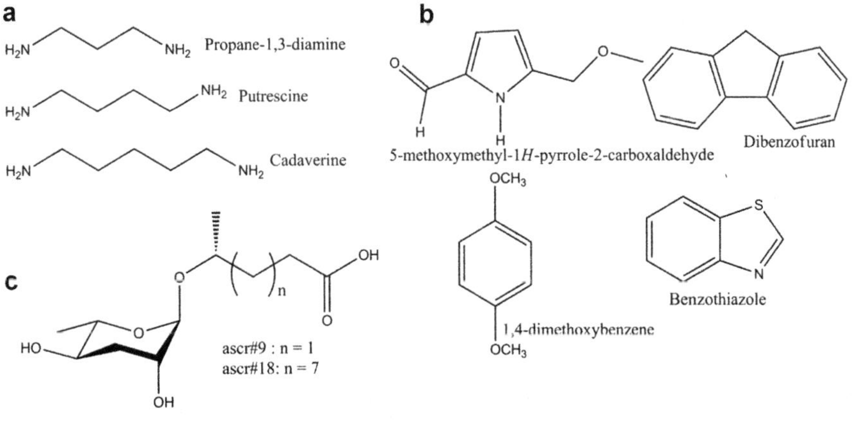

Fig. 3.2 Chemical structures of semiochemicals shown to influence *Meloidogyne* spp. chemotaxis. (**a**) Diamines produced by soybean roots that attract J2s of *M. incognita*. (**b**) Chemical structures of heterocyclic organic compounds produced by microorganisms. (**c**) Ascarosides produced by *Meloidogyne* spp. affect chemotaxis towards plant roots and nematode-trapping fungi

the travel distances and concentration gradients of some signalling compounds are very dynamic and dependent on soil properties (Kuzyakov and Razavi 2019).

3.4 Distances Root-Knot Nematodes Move

After hatching from the egg, *Meloidogyne* spp. J2s have to find a suitable host plant root to penetrate, otherwise, they will starve to death in approximately 7 days (Rocha et al. 2010; Campos et al. 2011). After the perception of chemical signals through the sensory organs, J2s start moving towards attractive gradients or in the opposite direction of repellent gradients. An issue still not well understood is the distance that J2s can migrate before losing their infective capacity.

Studies on the distances *Meloidogyne* spp. J2s move have generated a wide range of results. While some studies indicated that *Meloidogyne* spp. J2s were able to migrate more than 50 cm and infect the host plant; other studies showed a drastic reduction in migration and infectivity when J2s were placed 5 cm away from the host (Prot 1976; Rocha et al. 2016).

Nematode migration depends on the relation between pore size and J2 body diameter and the thickness of water films adhered to soil particles (Wallace 1968), among many other factors. Soil moisture has been kept close to ideal for the nematode movement in migration studies. On the other hand, soil texture and the three-dimensional environment in which J2s are inserted have varied. Vertical and horizontal migration of *Meloidogyne* spp. J2s have been studied mainly in three-dimensional systems using columns filled with sand (Prot 1976; Prot and van Gundy 1981; Pinkerton et al. 1987; Oliveira et al. 2020; Leitão et al. 2021a, b). In these apparatuses, the test nematode is placed at one end of the column and a bait plant at the opposite end, where J2s can migrate over different distances and periods of time (Leitão et al. 2021a).

Using columns with a diameter of 1.2 cm, Prot (1976) observed that J2s of *M. javanica* placed 75 cm vertically and 50 cm horizontally from tomato plants were capable of penetrating the roots in large numbers. Using the same apparatus, Prot and Van Gundy (1981) reported that up to 34% of *M. incognita* J2s were able to penetrate tomato roots after migrating 20 cm from the infestation point. Probably the small diameter used in these studies restricted nematode horizontal dispersal and imposed a vertical migration. In vertical columns with 4 cm of diameter assembled with metal or PVC rings, approximately 40% of *M. enterolobii* (Oliveira et al. 2020), 5% of the *M. floridensis* (Leitão et al. 2021b) and 1.6% of *M. incognita* (Leitão et al. 2021a) J2s were able to migrate 13 cm upwards after 9 days of infestation. By using a similar apparatus, Eo et al. (2007) reported that less than 10% of the *M. incognita* J2s migrated more than 7.5 cm 10 days after soil infestation. On the other hand, Pinkerton et al. (1987), using columns with a larger diameter (8.25 cm), filled with soil containing 16% clay plus silt, observed that less than 0.1% of the J2s of *M. chitwoodii* were able to migrate 45 cm and penetrate tomato roots.

After reaching the roots, only a small percentage will effectively penetrate and this percentage is highly dependent on the energy reserves. For example, when J2s of

M. javanica were placed 7.5 cm away from soybean roots in plastic pots, only 0.2% of them were able to penetrate the roots in a period of 5 days (L. Andrade-Souza, unpublished data).

These studies were performed with different set-ups, nematode species and soil characteristics and therefore are difficult to compare. Species such as *M. marylandi* and *M. javanica* are more motile than *M. incognita* (Oka 2020; Leitão et al. 2021b) and are expected to move longer distances. Nematodes appear to move longer distances in clayey than in sandy soils (Rocha et al. 2016). In addition to the *Meloidogyne* species and soil textures, migration distances are also influenced by the presence of bait plants, soil humidity, nutrients and salts, microorganisms and the amount of lipid reserves in the J2 body (Rocha et al. 2010, 2016). Probably, although there is no information on this topic, the amounts of reserves influence the capacity of these J2s to perceive and respond to chemical cues.

3.5 Compounds that Influence *Meloidogyne* Chemotaxis

The search for attractants and repellents to phytonematodes has been an ongoing endeavour. The chemical composition and identity of the plant-derived compounds that elicit nematode responses are mostly unknown. However, the precise and high-throughput detection and identification of semiochemicals from soils and rhizospheres have improved in recent times due to the development and higher sensitivity of scientific instrumentation (Torto et al. 2018). Interest in such molecules has increased with the need for new technologies to control nematodes (Oka 2021).

3.5.1 Plant Exudates

The main source of chemoattractants are exudates released by plants and metabolites secreted by microorganisms. Exudates are composed of high-molecular-weight polysaccharides and lower-molecular-weight organic compounds such as sugars, amino acids, flavonoids, tannins and other phenolic compounds, enzymes, fatty acids, growth regulators, nucleotides, carbohydrates, steroids, terpenes, alkaloids, polyacetylenes and vitamins (Bertin et al. 2003). They are released as a product of the interaction of the plant or microorganism with the environment that surrounds them (Kihika et al. 2017; Oota et al. 2019). The molecules perceived by nematodes include carbohydrates, amino acids, flavonoids, thiazoles, benzoxazinoids, terpenoids, alkaloids and many others (Sikder and Vestergård 2020; Sikder et al. 2021; Tsai et al. 2021).

Studies on the attractiveness and repellence of chemical compounds require specific tools. In vitro studies are carried out in Petri plates (Fig. 3.3), using water agar, agarose or pluronic F-127 gel (Williamson et al. 2009; Shivakumara et al. 2018; Liu et al. 2019; Oota et al. 2019; Oka 2020) or in adapted olfactometers filled with sand (Reynolds et al. 2011; Kihika et al. 2017; Murungi et al. 2018; Kirwa et al.

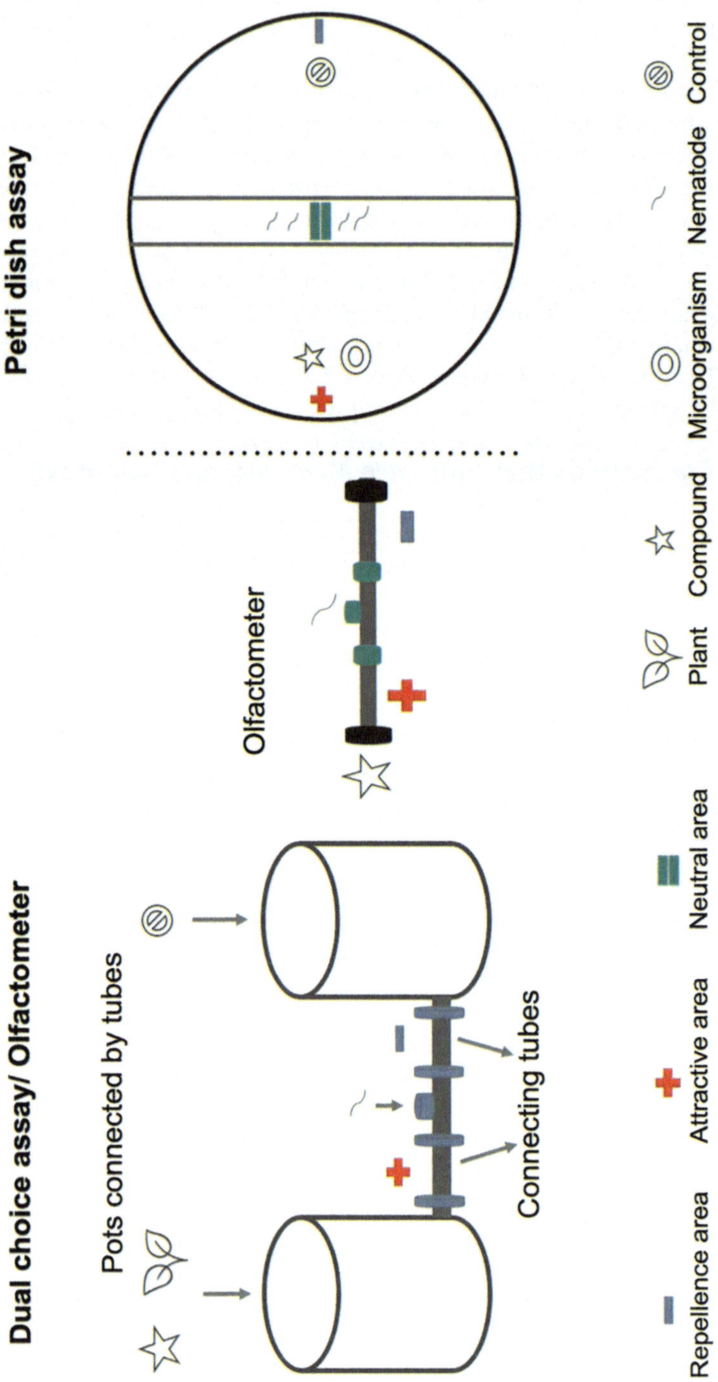

Fig. 3.3 Main techniques used to study the attraction and repellency of phytonematodes to chemical compounds or microorganisms

2018; Torto et al. 2018). Evaluations include most commonly counting the number of J2s that migrate to determined zones in the plates or olfactometers (Pacheco et al. 2021), number of stylet thrusts in selected specimens (Dutta et al. 2012; Kirwa et al. 2018) and time-lapse photographic evaluations of nematode tracks (Wuyts et al. 2006). Experiments with plants are generally carried out using pots connected by tubes (Kihika et al. 2017; Wang et al. 2019; Pacheco et al. 2021; Fig. 3.3), where recovery of nematodes from soil or sand may be challenging due to the low efficacy of the extraction methods.

VOCs are among the metabolites that compose exudates and are currently one of the most explored. They have up to 20 carbon atoms in their chemical structures and tend to present high vapour pressure, being easily released and dispersed in the environment (Dudareva et al. 2006). Several VOCs from different chemical groups had their toxicity to nematodes verified, and recently their attractiveness and repellence potential have been studied (Murungi et al. 2018; Oka 2021; Pacheco et al. 2021).

Nematode responses to plants are complex, and to illustrate this point, Wang et al. (2018a) measured the attractiveness of root tips, root exudates and extracts of marigold, a known trap plant and of soybean and pepper. They found that the root tips of all three species attracted *M. incognita* J2s, but only soybean root tips attracted *Heterodera glycines*. On the other hand, these three species' root exudates and root extracts attracted *H. glycines*, but repelled *M. incognita*. Although the chemoattractants were fractionated and found to be polar in their chemical nature, they were not identified. Similar species-dependent responses were also found for root border cells of different plant species to *M. incognita* (Zhao et al. 2000).

Susceptible and resistant cultivars of *Capsicum annum* and tomato showed that root exudates and VOCs emitted by susceptible plants are more attractive to *M. incognita* J2s than those emitted by resistant cultivars (Yang et al. 2016; Kihika et al. 2017). In addition to VOCs, some carbohydrates and proteins were related to the attractiveness of root-knot nematodes. *Arabidopsis* seeds attract *M. incognita* J2s, but it was dependent on the composition and presence of the seed-coat mucilage. Mutants that did not produce mucilage did not attract. Mucilage itself was not able to attract J2s, other components, such as carbohydrates and proteins, were determinant (Tsai et al. 2019).

3.5.2 Pure Chemical Compounds

Root-knot nematode species are among the most used in chemotaxis studies, especially *M. incognita*. A common approach adopted by many authors is the detection and identification of plant-derived chemicals by different techniques, such as gas chromatography (GC) coupled with mass spectrometry (MS) for volatiles and high-performance liquid chromatography (HPLC) coupled with MS for non-volatiles, followed by testing the pure chemicals in chemotaxis bioassays. Many compounds derived from plants were tested in their pure form, and their effects on chemotaxis have been confirmed (Table 3.1). These studies are difficult to compare because they

Table 3.1 Pure synthetic compounds or chemicals derived from plants and their activity on the chemotaxis of root-knot nematodes

Compound	Source	*Meloidogyne* species	Chemotaxis Evaluation[a]	Assays[b]	Attractant +[c]	Repellent −[c]	References
p-Coumaric acid	Pure compounds from the Phenylpropanoid pathway	*M. incognita*	Chemotaxis factor (Cf) done in plates with water agar and time-lapse photographs of nematode tracks			Repellent	Wuyts et al. (2006)
Caffeic acid						Repellent	
Ferulic acid						Repellent	
Kaempferol						Repellent	
Quercetin						Repellent	
Myricetin						Repellent	
Salicylic acid					Attractant		
Lauric acid – 0.5–2 mM	Crown daisy exudates	*M. incognita*	Chemotaxis index (CI) in agar plates	3	0.17–0.22		Dong et al. (2014)
Lauric acid – 4 mM				1		0.08	
Dibutyl phthalate	Tomato root exudates	*M. incognita*	CI in Petri plates with agar water	3		0.3–0.49	Yang et al. (2016)
Methyl salicylate	*Capsicum annum* root volatiles	*M. incognita*	CI determined in dual choice olfactometer filled with sand	3	0.4–0.62		Kihika et al. (2017)
α-Pinene				3	0.1–0.24		
(+)-Limonene				3	0.22–0.28		
Tridecane				3	0.12–0.2		
2-Methoxy-3-(1-methylpropyl)-pyrazine				2	0.05–0.1		
Thymol				3		0.2–0.64	
Ethephon[d]	Synthetic plant phytochemicals	*M. incognita*	CI in agar plates		0.6		Fleming et al. (2017)
Salicylic acid					0.42		
Mannitol					0.31		
Indole-3-acetic acid					0.26		
Gibberellic acid					0.47		

6-Dimethylallylamino purine					0.46		
Vanillic acid					0.45		
Coumaric acid						0.63	
trans-Cinnamic acid						0.31	
Palmitic acid	Castor bean exudates	*M. incognita*	CI in water agar			0.08–0.18	Dong et al. (2018)
Linoleic acid						0.02–0.12	
Methyl salicylate	Tomato root exudates	*M. incognita*	CI in sand	3	0.16–0.52		Kirwa et al. (2018)
Zeatin				5	0.14–0.44		
Luteolin				4		0.1–0.36	
Quercetin – low conc.[e]				3	0.08–0.38		
Quercetin – high conc.[e]				2		0.04–0.24	
Solasodine				4		0.06–0.16	
Tomatidine				3		0.12–0.2	
3-Methylbutan-1-ol	Pure compounds	*M. incognita*	CI in pluronic gel	6	0.4–0.9		Shivakumara et al. (2018)
Butan-1-ol				6	0.23–0.83		
Benzaldehyde – high conc.				6		0.44 (1)[f]	
Benzaldehyde – low conc.				6	0.1–0.64 (5)[f]		
Butan-2-one				6	0.25–0.8		
Octan-1-ol				6		0.1–0.42	
Methyl salicylate	Exudates from tomato and spinach	*M. incognita*	CI in olfactometer with sand		0.2–0.48		Murungi et al. (2018)
Tridecane					0.02–0.28		
Sabinene					0.04–0.2		
2-Isopropyl-3-methoxypyrazine					0.04–0.2		
Cadaverine		*M. incognita*	CI with pluronic gel		0.81		

(continued)

Table 3.1 (continued)

Compound	Source	*Meloidogyne* species	Chemotaxis: Evaluation[a]	Assays[b]	Attractant +[c]	Repellent −[c]	References
Putrescine	Soybean and tomato root exudates				0.71		Oota et al. (2020)
Propane-1,3-diamine					0.58		
Ethylenediamine					0.22		
Propylamine					0.07		
Spermidine					0.3		
Spermine					0.27		
Octane-1,8-diamine					0.11		
Heptane-1,7-diamine					0.07		
Hexane-1,6-diamine					0.06		
Nonane-1,9-diamine						0.02	
trans-Cinnamic acid	Pure compounds	*M. javanica/M. marylandi/M. hapla*	Relative density (RD) of J2 in pre-defined zones in agar plates. RD > 2.0 was defined as attractant		8.3/7.0/ 3.7[f]		Oka (2020)
Salicylic acid					6.0/13.4/ 2.3		
4′-Hydroxy-3′-methoxyacetophenone					3.0/4.1/ 5.2		
O-vanillin					5.1/11.6/ 5.0		
Carvacrol					9.6/11.3/ 4.2		
2-Methoxybenzaldehyde					6.1/8.3/ 6.6		
3-Methoxybenzoic acid					7.4/10.3/ 4.9		

3-Methoxybenzyl alcohol					3.0/6.8/ 5.3		
2-Methoxycinnamaldehyde					10.6/8.8/ 5.2		
trans-p-methoxycinnamaldehyde					2.8/10.9/ 2.5		
4-Methoxy-3-methylbenzaldehyde					11.0/7.4/ 5.6		
2-Methoxy-4-propenylphenol					7.3/5.0/ 4.9		
Thymol					3.2/10.0/ 2.5		
Salicylic acid	Pure compounds	*M. javanica, M. Marylandi, M. hapla, M. incognita*	CI calculated with the number of J2s trapped in tubes placed in sterile sand dune		0.96/0.95/ 0.083/ 0.09[g]		Oka (2021)
Carvacrol					0.31/0.88/ 0.78/0.74		
O-vanillin					–/0.90/ 0.64/–		
trans-Cinnamic acid					0.66/ /–/–/0.91		
2-Methoxycinnamaldehyde					0.85/0.95/ 0.65/0.65		
3-Methoxybenzoic acid					0.89/0.73/ 0.81/0.76		
4-Methoxybenzoic acid					0.96/0.85/ 0.63/0.83		
2-Methoxybenzaldehyde					0.90/0.94/ 0.85/0.80		

(continued)

Table 3.1 (continued)

Compound	Source	*Meloidogyne* species	Chemotaxis				References
			Evaluation[a]	Assays[b]	Attractant +[c]	Repellent −[c]	
Tridecane	Root exudates of tomato, radish, cucumber, alfalfa, lettuce and pepper	*M. incognita*	CI in pluronic gel	9	0.03–0.22 (5)	0.02–0.25 (4)	Wang et al. (2021)
Octadec-1-ene				9	0.04–0.22 (9)		
2-Hexyldecan-1-ol				9	0.04–0.26 (7)*	0.03–0.20 (2)	
Docos-1-ene				9	0.23 (1)	0.04–0.24 (8)*	
Malic acid				5	0.01–0.23 (4)*	0.03 (1)	
Tartaric acid				5	0.05–0.15 (4)*	0.07 (1)	
Maleic acid				5	0.07–0.10 (4)*	0.05) (1)	
Oxalic acid				5	0.23 (1)	0.07–0.43 (4)*	
Lactic acid				5	0.04–0.20 (4)*	0.02 (1)	
Glycolic acid				5	0.06–0.16 (4)*	0.05 (1)	
4-aminobenzoic acid				5	0.03–0.22 (4)*	0.16 (1)	
Ferulic acid				5	0.02–0.08 (4)*	0.03 (1)	
Amygdalic acid				5	0.10 (1)	0.01–0.19 (4)*	

Rhamnogalacturonan-I (RG-I)	From flaxseed mucilage	*M. incognita*	CI in Petri plates with pluronic gel		0.47		Tsai et al. (2021)
L-Gal(α1–3)-L-Rha(1) – RG-I side chain					0.35		

[a]The chemotaxis index (CI) is calculated with the formula: (Number of J2s in the test area – number of J2s in the control area)/number of J2s in both test and control areas. When the CI was not given by the authors, the numbers in the figures were extracted with the program WebPlotDigitizer (https://automeris.io/WebPlotDigitizer) and the CI index calculated to facilitate comparisons

[b]Number of assays with different concentrations done in the study

[c]Classification of the compounds as attractant when the CI is shown in this column or repellent when the CI is presented in the column. A dash between two numbers indicates a range of values

[d]Ethephon is a synthetic product used as growth regulator, therefore cannot be considered a plant-derived hormone

[e]High and low concentrations of the compound

[f]The numbers between parenthesis indicate the number of assays in which the response is either attractant or repellent to J2s

[g]The numbers separated by slashes refer to the respective *Meloidogyne* species. A dash just before a slash indicates that the species was not tested for this compound

*Asterisks indicate the predominant activity of the compound on the basis of the number of assays with different concentrations performed in the study

were done with different methods, nematode species and populations and chemicals, without any standardized controls across studies. In this chapter, we made an effort to compile the studies with purified chemicals tested in chemotaxis of *Meloidogyne* species in a quantitative way, whenever possible (Table 3.1). The determination of a chemotaxis index (CI) is the most common way of presenting the data. This is a convenient way to make comparisons, especially when the methods are the same, but one should always keep the differences in mind. For example, salicylic acid was used in four different studies and in only one of them, it did not attract *M. incognita*, although it did not repel (Table 3.1). In these four studies, the chemotaxis index varied from 0.09 to 0.42 and four different methods were used to determine CI (Table 3.1), illustrating the difficulties of comparing these data. Nevertheless, when the methods are the same, there is value in comparing the CIs obtained in different studies. As an example, the CI of methyl salicylate (MeSA) in sand varied from 0.16 to 0.52 in one study and from 0.2 to 0.48 in another, both in the same range (Table 3.1).

Although studies on chemotaxis are done with pure compounds, semiochemicals are not expected to exert their activities isolated, but in complex mixtures. In some studies, this aspect was taken into consideration. For example, MeSA was detected in tomato roots and shown to contribute to the attractiveness of tomato to *M. incognita*, whereas 2-isopropyl-3-methoxypyrazine and tridecane contributed to the attractiveness of spinach. MeSA exerted a stronger attraction even when mixed with other compounds and was responsible for the preference of tomato over spinach by *M. incognita* (Murungi et al. 2018). The blend composed of α-pinene + limonene +2-methoxy-3-(1-methylpropyl)-pyrazine + tridecane + MeSA was highly attractive to J2s of *M. incognita*. However, when MeSA was removed from the blend, the attractiveness was drastically reduced. Similarly, thymol induced negative chemotaxis (repellence) when it was added in any blend (Kihika et al. 2017).

There is an effect of the concentration for many of these chemical compounds, where lower concentrations attract nematodes and higher concentrations repel them and vice versa (Li et al. 2019; Tables 3.1 and 3.2). This is another factor that makes comparisons across studies difficult because there is no standardization among studies. Additionally, some compounds detected in root exudates might be contaminants from soil, microorganisms or the extraction process. One possible example is dibutyl phthalate, a common plasticizing agent, that was detected in tomato root exudates (Yang et al. 2016). Although its origin is unknown, it has been reported to be produced by filamentous fungi in nature (Tian et al. 2016).

It appears that there is no universal chemical that will function in the same way for all *Meloidogyne* spp. However, some chemical characteristics gave some hints in determined systems. For example, Oota et al. (2019) found that only diamines with a backbone containing three to five carbons, including cadaverine, putrescine and propane-1,3-diamine attracted J2s of *M. incognita* among the 376 compounds tested. Cadaverine was the most attractive compound to J2s of *M. incognita*, but it had no effect on *M. arenaria* and *M. enterolobii*, showing that this specificity may determine the host range of different *Meloidogyne* spp. (Oota et al. 2019). According to the authors, cadaverine is released by stressed plants, leading nematodes to potential

Table 3.2 Pure compounds from microorganisms with activity on *Meloidogyne* chemotaxis

				Chemotaxis			
Compound	Source	*Meloidogyne* species	Evaluation[a]	assays[b]	Attractant +[c]	Repellent -[c]	References
Acetone	VOCs from *Paenibacillus polymixa* KM2501–1	*M. incognita*	CI in agar plates	5	0.10–0.27		Cheng et al. (2017)
Decan-2-ol				5	0.06–0.17		
Furfural acetone				5	0.29–0.47		
Undecan-2-one				5		0.09–0.44	
4-Acetylbenzoic acid				5	0.09–0.24 (2)[d]	0.06–0.23 (3)[d]	
(*Z*)-Hexen-1-ol acetate	VOCs from *Pseudomonas putida* 1A00316	*M. incognita*	CI in 2% agar plates	5		0.42–0.58	Zhai et al. (2018)
Octan-2-one				5		0.41–056	
Undec-1-ene				5		0.33–0.56	
1-(Ethenyloxy)octadecane				5		0.23–0.49	
Dimethyl-disulfide				5		0.45–054	
Undecan-2-one				5		0.49–0.73	
Nonan-2-one				5		0.42–0.48	
3,3-Dimethyloctane	Tomato exudates with *Bacillus cereus* BCM2	*M. incognita*	CI in agar plates	3	0.46 (1)	0.18 (1)	Li et al. (2019)
Tridecane				3	0.08 (1)	0.06–0.08 (2)	
2,4-Di-tert-butylphenol				3	0.2–0.26		
Benzothiazole	VOCs from *Streptomyces plicatus* G	*M. incognita*	CI in plates with 1% agarose	3		0.04–0.4	Wang et al. (2019)
Dibenzofuran				3	0.22–0.46		
Benzothiazole + dibenzofuran				3	0.06–0.38		
1,4-Dimethoxybenzene	VOC from *Purpureocillium chlamydosporia* Pc-10	*M. incognita*	CI in agar plates	5	0.29–0.8		Pacheco et al. (2021)
5-Methoxymethyl-1*H*-pyrrole-2-carboxaldehyde—high conc.[e]	From *Purpureocillium lavendulum* YMF1.00683	*M. incognita*	CI in plates with agarose	3		0.07–0.38	Bao et al. (2022)

(continued)

Table 3.2 (continued)

				Chemotaxis			
Compound	Source	*Meloidogyne* species	Evaluation[a]	assays[b]	Attractant +[c]	Repellent -[c]	References
5-Methoxymethyl-1*H*-pyrrole-2-carboxaldehyde—low conc.[e]				3	0.02–0.13		

[a]The chemotaxis index (CI) is calculated with the formula: (Number of J2s in the test area – number of J2s in the control area)/number of J2s in both test and control areas. When the CI was not given by the authors, the numbers in the figures were extracted with the program WebPlotDigitizer (https://automeris.io/WebPlotDigitizer) and the CI index calculated to facilitate comparisons

[b]Number of assays with different concentrations done in the study

[c]Classification of the compounds as attractant when the CI appears in this column or repellent when the CI is presented in the column

[d]The numbers between parenthesis indicate the number of assays in which the response is either attractant or repellent to J2s

[e]High and low concentrations of the compound

hosts with a compromised immunity. In another study, Oka (2020) found that the most attractive chemicals to three different *Meloidogyne* spp. in a screening of 60 pure compounds contained a methoxy group (OCH_3) and postulated that its presence may play a role in attraction. Although the methoxy group was present in the attractants reported by Oka (2020), it is absent from widely known list of semiochemicals such as salicylic acid and carvacrol (attractants) and thymol and *trans*-cinnamic acid (repellents).

Non-volatile compounds from tomato root exudates were fractionated and the phytohormone zeatin (cytokinin) was shown to be attractive to the *M. incognita* J2s, whereas the flavonoid quercetin elicited concentration-dependent responses, being attractive at low concentrations and repellent at high concentrations (Kirwa et al. 2018). These results indicate that the concentration of certain chemicals and the ratio among compounds in mixtures determine the complex responses of *Meloidogyne* spp. (Kirwa et al. 2018). Furthermore, zeatin was shown to be secreted by *M. incognita* and is probably used in the manipulation of plant hormone balance in the initial stages of invasion for the establishment of feeding sites (Dowd et al. 2017; Kirwa et al. 2018). It appears that most phytohormones are somehow involved in the attractiveness of *Meloidogyne* to plants, including indolacetic acid (IAA), salicylic acid, jasmonic acid and ethylene (Wuyts et al. 2006; Bhattarai et al. 2008; Curtis 2008; Fudali et al. 2013; Fleming et al. 2017; Zinovieva et al. 2021). Salicylic acid was shown to be an attractant of *M. incognita* J2s, but it also inhibited egg hatching and had nematicidal effects (Wuyts et al. 2006). Foliar or drench applications of salicylic acid suppressed *M. incognita* (Maheshwari and Anwar 1990; Nandi et al. 2003), probably by increasing the level of plant resistance. However, exogenous application of IAA decreased the resistance of plants to *M. incognita* (Curtis 2008). Mutants deficient in the accumulation of salicylic acid and ethylene attracted more J2s than the wild type (Fudali et al. 2013; Čepulytė et al. 2018), whereas the role of jasmonic acid in chemotaxis is less understood (Bhattarai et al. 2008). In addition to VOCs and phytohormones, *Meloidogyne* spp. also responds to fatty acids, such as lauric acid that was found in exudates of crown daisy (Dong et al. 2014) and palmitic and linoleic acid from roots of castor bean (Dong et al. 2018).

In a relatively large-scale screening, Oka (2020) tested 60 pure aromatic compounds against *M. incognita*, *M. javanica*, *M. marylandi* and *M. hapla* and found that none of the compounds was repellent, even the ones with nematicidal activity, such as carvacrol. *Meloidogyne incognita* did not respond to any of the compounds and 35 of them attracted at least one of the three other species, and 13 were considered highly attractive (Table 3.1). Although *M. javanica* and *M. hapla* are considered species with a broad host range, the specialist *M. marylandi* was attracted to more chemicals. In this study, thymol and salicylic acid, previously found to be repellent and attractant, respectively, by other authors (Fleming et al. 2017; Kihika et al. 2017; Wuyts et al. 2006), did not elicit any response from *M. incognita*. These results raise awareness to the fact that either the methodology used by Oka (2020) needs to be further evaluated or populations of *M. incognita* are responding differently to the same chemicals as implied by Wang et al. (2009). In a follow-up study, Oka (2021) used a bioassay with trap tubes filled with sand. In contrast with the other study (Oka 2020), the author was able to show attraction of *M. incognita* J2s to salicylic acid and less attractiveness of all species of *Meloidogyne* to carvacrol (Table 3.1). Differential responses are known to occur among *Meloidogyne* species and their nature is still unknown. More investigations in this area will uncover if there is any link between chemotaxis and host range. Additionally, the concentrations used in laboratory assays are not always realistic in the field.

3.5.3 Nematode-Derived Compounds

The semiochemical compounds described up to now are produced either by plants or by microorganisms in soil or in the rhizosphere. However, there is a large class of glycosidic hormones called ascarosides, universally conserved among nematodes that function in mate location, aggregation and regulation of development (Choe et al. 2012; Schroeder 2015). Ascarosides seem to be devoid of antimicrobial activity and sometimes may act against parasitic nematodes as they are also perceived by other microorganisms such as nematophagous fungi, that are induced to produce trapping structures to capture nematodes moving in soil (Hsueh et al. 2013). These molecules are also perceived by plant roots at pico to nano molar concentrations and elicit systemic resistance to nematodes and other pathogens, in plants as diverse as tomato, *Arabidopsis* and barley (Manosalva et al. 2015).

Ascaroside ascr#18 (Fig. 3.2), the most common in *Meloidogyne* spp. and other nematodes, is a weak attractant to nemadodes (Hamada et al. 2020). This compound was shown to be metabolized by plants and transformed into ascr#9 (Fig. 3.2), which in mixtures with asc#18 repelled J2s of *M. incognita* (Manohar et al. 2020). It has also been shown that repellence, rather than systemic resistance, was mainly responsible for the reduced infection by *M. incognita* (Manohar et al. 2020). Therefore, these mixtures of ascarosides seem to interfere with the plant-nematode interaction by reducing the level of infection.

3.5.4 Inorganic Compounds

Inorganic salts and ions were investigated for their effect on the chemotaxis of *M. incognita* J2s and most of them were found to be repellent. No salt was found to be a consistent attractant to the J2s of this species. In some cases, higher concentrations resulted in stronger repellence (Qi et al. 2015). Salts of nitrate (NO_3^-), ammonium ($\left(NH_4^+\right)$), thiocyanate (SCN^-), cesium (Cs^+), potassium (K^+) and sodium (Na^+) were among the most repellent (Castro et al. 1990; Le Saux and Quénehervé 2002; Qi et al. 2015). Salts of chloride (Cl^-), sulfate (SO_4^{2-}), hidrogenphosphate $\left(HPO_4^-\right)$, carbonate $\left(CO_3^{2-}\right)$ and hydroxide (OH^-) repelled at a lower extent, whereas salts of calcium (Ca^{2+}) had no effect (Castro et al. 1990; Le Saux and Quénéhervé 2002; Qi et al. 2015).

Many of these salts are used as fertilizers and may have a disruptive effect on nematode orientation in soil (Qi et al. 2015). Besides repelling nematodes, some salts, such as the ones containing ammonium have a nematicidal activity (Oka and Pivonia 2002). It would be interesting to determine if these salts can increase the efficacy of chemical nematicides when they are combined in joint field applications.

3.6 Microorganisms Affecting *Meloidogyne* Chemotaxis

Plant roots are metabolically active organs that produce exudates and when these compounds are released, they attract microorganisms of different trophic levels, including saprophytes, symbionts and phytopathogens, such as plant-parasitic nematodes (Hol et al. 2013). The rhizosphere is one of the most complex ecosystems on earth, fostering millions of microbial cells that can affect the migration of nematodes (Korenblum et al. 2020). Surprisingly, despite the extensive number of reports demonstrating the influence of root exudates from host plants on the behaviour of plant-parasitic nematodes, there have been few studies on the behaviour of nematodes with respect to soil microorganisms. Several authors have demonstrated that bacteria, mainly in the genera *Bacillus* and *Pseudomonas*, are able to reduce *Meloidogyne* spp. penetration and reproduction (Leontopoulos et al. 2017; Cruz-Magalhães et al. 2021; Antil et al. 2022; Gowda et al. 2022). It is thought that microorganisms, in general, can alter the production of root exudates or modify their composition after secretion, thereby affecting nematode chemotaxis. One of the main effects of microorganisms is to decrease the attractiveness of the root exudates (Padgham and Sikora 2007; Hu et al. 2017; Zhao et al. 2022).

Bacteria such as *Pseudomonas oryzihabitans* were shown to inhibit the migration of *M. javanica* J2s by modifying the root exudates, making it less attractive to the nematode (Leontopoulos et al. 2017). The efficient colonization of roots by the biological control agent *Bacillus cereus* strain BCM2 was fundamental to repelling J2s of *M. incognita*, leading to 80% reduction in the number of galls (Hu et al. 2017). Based on these results, Li et al. (2019) studied the composition of root exudates released by tomato plants colonized by *B. cereus* BCM2 and showed that the bacterium changed the composition of the exudates, increasing the number of molecules produced, including 2,4-di-tert-butylphenol and 3,3-dimethiloctane, which reduced the number of galls and the number of nematodes in soil and plant tissue. The VOCs furfural acetone and decan-2-ol from the bacterium *Paenibacillus polymyxa* KM25021–1 attracted J2s of *M. incognita* in a strategy named "honey-trap" by the authors (Cheng et al. 2017). These J2s were subsequently killed either through fumigation or direct contact with the bacterium, which probably used the nematode as a food source.

In a screening of actinomycetes performed by Wang et al. (2019), 17% of the isolates attracted J2s of *M. incognita*, while 8% repelled them. The selected actinomycete *Streptomyces plicatus* strain G produced the VOC dibenzofuran (Fig. 3.2), that was a potent attractant to J2s, whereas benzothiazole (Fig. 3.2) was a repellent. The attractive effect prevailed when the mixture of purified VOCs or cultures of the bacterium were applied to tomato roots. This bacterium may attract the nematodes to the roots to use them for their nutrition.

Fungi were also shown to affect the chemotaxis of *Meloidogyne* species J2s. Common endophytic fungi such as *Fusarium* spp. were shown to alter the composition of root exudates (Hallmann and Sikora 2011) and thereby affect chemotaxis. *Purpureocillium lavendulum* produced the compound 5-methoxymethyl-1*H*-pyrrole-2-carboxaldehyde (Fig. 3.2), which attracted J2s of *M. incognita* at low concentrations and was toxic at high concentrations, causing up to 98% mortality and inhibiting egg hatching by 81% (Bao et al. 2022). The fungal species *Pochonia clamydosporia* has been widely studied for its antagonistic interaction with plant-parasitic nematodes. This fungal species produced several VOCs and among them, 1,4-dimethoxybenzene (Fig. 3.2), which attracted J2s of *M. incognita,* causing 89% mortality and reduced hatching by 86% (Pacheco et al. 2021). The nematophagous fungus *Arthrobotrys oligospora* perceives the presence of nematodes by detecting their ascarosides (Hsueh et al. 2013) and is then able to attract these nematodes with volatile furanones and at the same time increase the number of traps to capture nematodes by signaling with pyrones (Wang et al. 2018b).

Some of these rhizosphere microorganisms are active ingredients of commercial products because they reduce the reproduction of *Meloidogyne* spp. on plants. However, the mode of action of some of them is still unknown, but part of them is expected to act by disrupting chemoreception in J2s.

3.7 Prospects and Potential Uses of Chemotaxis to Manage *Meloidogyne* Species

Plants and microorganisms rely on chemical communication networks to determine the outcome of their interactions (Van Dam and Bouwmeester 2016). The composition and concentration of semiochemicals impact plant development and health as plants evolved strategies to interact with beneficial microorganisms and protect themselves against pathogens, such as nematodes (Siddique et al. 2022).

Several techniques were employed to study chemotaxis in vivo and in vitro (Dusenbery 1980, 1983; Castro et al. 1988; Haseeb and Fried 1988; Perry 1996; Rocha et al. 2016; Wang et al. 2009; Oka 2020, 2021; Pacheco et al. 2021). These techniques have advantages and disadvantages, but none of them is superior. The most used in vitro approach is agar plates with demarcated zones to calculate the chemotaxis index (Cheng et al. 2017; Zhai et al. 2018) and in vivo/*in planta* assays are pots connected with tubes filled with soil or sand (Wang et al. 2019; Oliveira et al. 2020). The most challenging task is extracting the nematodes from the soil (Oka 2021). Although assays in sand or soil may best simulate the natural environment, nematodes cannot be seen in these opaque substrates, instead, they must be extracted to monitor migration (Siddique et al. 2022). Nematode extraction techniques recover only around 10% of the total number of nematodes placed in soil (Oka 2020; Viglierchio and Schmitt 1983). Together these two factors may explain why most chemotaxis studies are conducted in vitro with Petri dishes. These in vitro assays are difficult to standardize because of the variation in set-ups. New apparatuses with microchannels filled with a gel appear to allow the quantitative and high-throughput efficient determination of chemotaxis in nematodes (Hida et al. 2015) or standardized chambers made by 3D printers could help standardize the chemotaxis tests (Laloum et al. 2020).

Many chemicals from plants and microorganisms that play a role in chemotaxis are being revealed. These chemicals may be used in nematode management in different ways, such as the development of synthetic nematicides by using them as lead structures. This may be necessary if the chemicals are not stable enough to be used in their natural form. Some chemicals such as carvacrol have dual effects as they attract and kill nematodes at the same time (Oka 2020) and can be used directly as a nematicide. Plants may not produce enough of these semiochemicals or may depend on specific conditions such as temperature and nutrition, and therefore the direct application of the purified product might be more efficient, especially when they can be produced at low costs. One of the difficulties with synthetic semiochemicals is that they appear to be highly specific. Finding compounds that would attract a broad range of parasitic nematodes seems to be impossible. Up to this moment, there is no universal attractant to all *Meloidogyne* species.

Interference with chemotaxis is one of the most promising management strategies for nematodes in general. Interference could be applied by using plants or/and microorganisms that produce or modify the semiochemicals in order to decrease or eliminate chemotaxis, produce repellents or increase the amount of attractive chemicals. The final outcome would be the impedance of host location by lack of attractants, presence of repellents and a confounding effect that would lead J2s overwhelmed and incapable of locating the host. Plants already naturally interfere with chemotaxis by perceiving nematode ascarosides, for example, and synthesizing chemicals that repel nematodes and induce systemic resistance (Manohar et al. 2020). Repellence may be selected in different plants, as shown for peppers, where resistant cultivars repelled *M. incognita* J2s whereas the susceptible ones attracted (Hu et al. 2017; Kihika et al. 2017). The selection of plants that host more microorganisms, such as bacteria and fungi that produce repellent semiochemicals, is a strategy that has not yet been exploited but holds promise. Another strategy of interest is the modification of plant root exudates by microorganisms. Exudates of lettuce are normally attractive to *M. incognita*, but the inoculation of roots with an isolate of *Bacillus subtilis* turned them repulsive to the nematode (VP Cavalcanti, unpublished data). Trap plants are regarded as attractive to *Meloidogyne* spp. and their use is considered effective, especially in small plots. For example, Dong et al. (2014) reported that five crown daisy plants can protect one tomato plant from *M. incognita*. Yet another way of interfering with chemotaxis is inserting a physical barrier between the nematode and plant roots, such as wrapping with banana tissue employed in Africa to control the potato cyst nematode (Ochola et al. 2022).

Transgenic plants, although not yet widely accepted, are interesting alternatives to manage nematodes through chemotaxis. Transgenic potato plants secreting peptides that interfere with chemoreception decreased *Globodera pallida* infection and development (Liu et al. 2005). This strategy, which aims to interfere with the invasion process rather than with the feeding process adopted in most transgenic plants (Atkinson et al. 2003), may be further explored to control *Meloidogyne* spp.

The number of studies with the olfactory genes in *Meloidogyne* spp. is still relatively small, but at least 14 genes were characterized in the genome of *M. incognita* (Dong et al. 2014; Shivakumara et al. 2019; Li et al. 2022). When these genes were interfered with iRNA by soaking, the J2s lost their attraction towards or repulsion away from different semiochemicals that were previously known to affect the chemotaxis of J2s of this species (Shivakumara et al. 2019; Li et al. 2022). These results indicate that these genes are targets for the development of new chemical nematicides that interfere with chemotaxis, new iRNA-based nematicides directed to these genes or the development of transgenic plants through host-induced gene silencing that would interfere with these genes and disrupt chemotaxis.

Nematode chemotaxis is tightly associated with microorganisms that colonize the rhizosphere and soil. Chemicals released by bacteria and fungi (Table 3.2) and other interactions that are not yet well understood influence chemotaxis. For example, most studies report that mycorrhized plants reduced the ability of nematodes to locate and penetrate plant roots by interfering with chemotaxis (Bacetty et al. 2009; Vos et al. 2012). Some studies show the contrary, increased infection in mycorrhized plants due to a decreased resistance induced by the symbiont (Borowicz 2001; Hol and Cook 2005; Frew et al. 2018). However, most studies showing increases in nematode populations were done with migratory nematodes, which appear to influence the outcome (Gough et al. 2020). Metataxonomic studies on the whole microbiome with NGS sequencing will shed more light on the complex interactions between nematodes and the other microorganisms with whom they share the infection court. In this context, the microbiome in nematode-suppressive soils may harbour the clues needed to build an unfavourable environment for these parasites (Topalovic et al. 2020). These types of studies showed changes in the bacterial and fungal communities (Wang et al. 2014; Toju and Tanaka 2019; Yergalieyev et al. 2020; Zhang et al. 2020; Liu et al. 2022) and nematode populations (Sikder et al. 2021) influenced by semiochemicals or by the presence of nematodes. However, in order to turn this knowledge into control measures, more field experiments with these anti-nematode microorganisms need to be pursued.

References

Antil S, Kumar R, Pathak DV, Kumar A, Panwar A, Kumari A (2022) Plant growth-promoting rhizobacteria – *Bacillus cereus* KMT-5 and *B. megaterium* KMT-8 effectively suppressed *Meloidogyne javanica* infection. Appl Soil Ecol 174:104419. https://doi.org/10.1016/j.apsoil.2022.104419

Atkinson HJ, Urwin PE, McPherson MJ (2003) Engineering plants for nematode resistance. Annu Rev Phytopathol 41:615–639. https://doi.org/10.1146/annurev.phyto.41.052002.095737

Bacetty AA, Snook ME, Glenn AE, Noe JP, Nagabhyru P, Bacon CH (2009) Chemotaxis disruption in *Pratylenchus scribneri* by tall fescue root extracts and alkaloids. J Chem Ecol 35:844–850. https://doi.org/10.1007/s10886-009-9657-x

Bao Z-X, Liu R, Li C-Q, Pan X-R, Zhao P-J (2022) Pathogenicity and metabolites of *Purpureocillium lavendulum* YMF1.00683 against *Meloidogyne incognita*. Pathogens 11:795. https://doi.org/10.3390/pathogens11070795

Bargmann CI (2006) Chemosensation in *C. elegans*. In: WormBook (ed) The C. elegans research community. WormBook. https://doi.org/10.1895/wormbook.1.123.1. http://www.wormbook.org

Bertin C, Yang XH, Weston LA (2003) The role of root exudates and allelochemicals in the rhizosphere. Plant Soil 256:67–83. https://doi.org/10.1023/A:1026290508166

Bhattarai KK, Xie Q-G, Mantelin S, Bishnoi U, Girke T, Navarre DA, Kaloshian I (2008) Tomato susceptibility to root-knot nematodes requires an intact jasmonic acid signaling pathway. Mol Plant-Microbe Int 21(9):1205–1214. https://doi.org/10.1094/MPMI-21-9-1205

Blossfeld S, Perriguey J, Sterckeman T, Morel J-L, Losch R (2010) Rhizosphere pH dynamics in trace-metal-contaminated soils, monitored with planar pH optodes. Plant Soil 330:173–184. https://doi.org/10.1007/s11104-009-0190-z

Borowicz VA (2001) Do arbuscular mycorrhizal fungi alter plant- pathogen relations? Ecology 82: 3057–3068. https://doi.org/10.1890/0012-9658(2001)082[3057:DAMFAP]2.0.CO;2

Campos HD, Campos VP, Silva JRC, Silva LHCP, Costa LSAS, Terra WC (2011) Atração e penetração de *Meloidogyne javanica* e *Heterodera glycines* em raízes excisadas de soja. Ciência Rural 41(9):1496–1502. https://doi.org/10.1590/S0103-84782011000900002

Castro CE, Belser NO, McKinney HE, Thomason IJ (1988) Quantitative bioassay for chemotaxis with plant parasitic nematodes. J Chem Ecol 15:1297–1309. https://doi.org/10.1007/BF01014831

Castro CE, Belser NO, Mckinney HE, Thomason IJ (1990) Strong repellency of the root knot nematode, *Meloidogyne incognita* by specific inorganic ions. J Chem Ecol 16:1297–1309. https://doi.org/10.1007/BF01021019

Čepulytė R, Danquah WB, Bruening G, Williamson VM (2018) Potent attractant for root-knot nematodes in exudates from seedling root tips of two host species. Sci Rep 8:1–10. https://doi.org/10.1038/s41598-018-29165-4

Cheng W, Yang J, Nie Q, Huang D, Yu C, Zheng L, Cai M, Thomashow LS, Weller DM, Yu Z, Zhang J (2017) Volatile organic compounds from *Paenibacillus polymyxa* KM2501-1 control *Meloidogyne incognita* by multiple strategies. Sci Rep 7(1):1–11. https://doi.org/10.1038/s41598-017-16631-8

Choe A, von Reuss SH, Kogan D, Gasser RB, Platzer EG, Schroeder FC, Sternberg PW (2012) Ascaroside signaling is widely conserved among nematodes. Curr Biol 22(9):772–780. https://doi.org/10.1016/j.cub.2012.03.024

Coyne DL, Cortada L, Dalzell JJ, Claudius-Cole AO, Haukeland S, Luambano N, Talwana H (2018) Plant-parasitic nematodes and food security in sub-Saharan Africa. Annu Rev Phytopathol 56:381–403. https://doi.org/10.1146/annurev-phyto-080417-045833

Cruz-Magalhães V, Guimarães RA, Silva JCP, Faria AF, Pedroso MP, Campos VP, Marbach PAS, Medeiros FHV, De Souza JT (2021) The combination of two Bacillus strains suppresses *Meloidogyne incognita* and fungal pathogens, but does not enhance plant growth. Pest Manag Sci 78(2):722–732. https://doi.org/10.1002/ps.6685

Curtis RHC (2008) Plant-nematode interactions: environmental signals detected by the nematode's chemosensory organs control changes in the surface cuticle and behaviour. Parasite 15:310–316. https://doi.org/10.1051/parasite/2008153310

Dong L, Li X, Huang L, Gao Y, Zhong L, Zheng Y, Zuo Y (2014) Lauric acid in crown daisy root exudate potently regulates root-knot nematode chemotaxis and disrupts Mi-flp-18 expression to block infection. J Exp Botany 65(1):131–141. https://doi.org/10.1093/jxb/ert356

Dong L, Li X, Huang C, Lu Q, Li B, Yao Y, Liu T, Zuo Y (2018) Reduced *Meloidogyne incognita* infection of tomato in the presence of castor and the involvement of fatty acids. Sci Hortic 237: 169–175. https://doi.org/10.1016/j.scienta.2018.03.066

Dowd CD, Chronis D, Radakovic ZS, Siddique S, Schmulling T, Werner T, Kakimoto T, Grundler FMW, Mitchum MG (2017) Divergent expression of cytokinin biosynthesis, signaling and catabolism genes underlying differences in feeding sites induced by cyst and root-knot nematodes. Plant J 92:211–228. https://doi.org/10.1111/tpj.13647

Dudareva N, Negre F, Nagegowda OI (2006) Plant volatile: recent advance and future perspectives. Crit Rev Plant Sci 25:417–440. https://doi.org/10.1080/07352680600899973

Dusenbery DB (1980) Responses of the nematode *Caenorhabditis elegans* to controlled chemical stimulation. J Comp Physiol 136:327–331. https://doi.org/10.1007/BF00657352

Dusenbery DB (1983) Chemotactic behavior in nematodes. J Nematol 15:168–173

Dutta TK, Powers SJ, Gaur HS, Birkett M, Curtis RHC (2012) Effect of small lipophilic molecules in tomato and rice root exudates on the behaviour of *Meloidogyne incognita* and *M. graminicola*. Nematology 14(3):309–320. https://doi.org/10.1163/156854111X612306

Eo J, Nakamoto TN, Otobe K, Mizukubo TM (2007) The role of pore size on the migration of *Meloidogyne incognita* juveniles under different tillage systems. Nematology 9:751–758. https://doi.org/10.1163/156854107782331252

Fleming TR, Maule AG, Fleming CC (2017) Chemosensory responses of plant parasitic nematodes to selected phytochemicals reveal long-term habituation traits. J Nematol 49(4):462–471

Frew A, Powell JR, Glauser G, Bennett AE, Johnson SN (2018) Mycorrhizal fungi enhance nutrient uptake but disarm defences in plant roots, promoting plant-parasitic nematode populations. Soil Biol Biochem 126:123–132. https://doi.org/10.1016/j.soilbio.2018.08.019

Fudali SL, Wang C, Williamson VM (2013) Ethylene signaling pathway modulates attractiveness of host roots to the root-knot nematode *Meloidogyne hapla*. Mol Plant-Microbe Inter 26(1): 75–86. https://doi.org/10.1094/MPMI-05-12-0107-R

Gough EC, Owen KJ, Zwart RS, Thompson JP (2020) A systematic review of the effects of arbuscular mycorrhizal fungi on root-lesion nematodes, *Pratylenchus* spp. Front Plant Sci 11: 923. https://doi.org/10.3389/fpls.2020.00923

Gowda MT, Meena BR, Krishnan N, Manjunath M, Sellaperumal C, Rai AB, Singh A, Manimurugan C, Patil J, Pandey KK, Singh J (2022) Antimicrobial peptides producing native *Bacillus* spp. for the management of root-knot nematode *Meloidogyne incognita* infecting okra (*Abelmoschus esculentus* L. Moench). Biol Control 171:104951. https://doi.org/10.1016/j.biocontrol.2022.104951

Hallmann J, Sikora RA (2011) Endophytic fungi. In: Davies KG, Spiegel K (eds) Biological control of plant-parasitic nematodes. Springer, Dordrecht, pp 227–258. https://doi.org/10.1007/978-1-4020-9648-8

Hamada N, Yimer HZ, Williamson VM, Siddique S (2020) Chemical hide and seek: nematode's journey to its plant host. Mol Plant 13:541–543. https://doi.org/10.1016/j.molp.2020.03.005

Haseeb MA, Fried B (1988) Chemical communication in helminths. In: Baker JR, Muller R (eds) Advances in parasitology. Academic Press

Hida H, Nishiyama H, Sawa S, Higashiyama T, Arata H (2015) Chemotaxis assay of plant-parasitic nematodes on a gel-filled microchannel device. Sensors Actuat B 221:1483–1491. https://doi.org/10.1016/j.snb.2015.07.081

Hol WHG, Cook R (2005) An overview of arbuscular mycorrhizal fungi–nematode interactions. Basic Appl Ecol 6:489–503. https://doi.org/10.1016/j.baae.2005.04.001

Hol WHG, Bezemer TM, Biere A (2013) Getting the ecology into interactions between plants and the plant growth-promoting bacterium *Pseudomonas fluorescens*. Front Plant Sci 4:81. https://doi.org/10.3389/fpls.2013.00081

Holz M, Zarebanadkouki M, Kaestner A, Kuzyakov Y, Carminati A (2018) Rhizodeposition under drought is controlled by root growth rate and rhizosphere water content. Plant Soil 423:429–442. https://doi.org/10.1007/s11104-017-3522-4

Holz M, Becker JN, Daudin G, Oburger E (2020) Application of planar optodes to measure CO2 gradients in the rhizosphere of unsaturated soils. Rhizosphere 16:100266. https://doi.org/10.1016/j.rhisph.2020.100266

Hsueh Y-P, Mahanti P, Schroeder FC, Sternberg PW (2013) Nematode-trapping fungi eavesdrop on nematode pheromones. Curr Biol 23:83–86. https://doi.org/10.1016/j.cub.2012.11.035

Hu HJ, Chen YL, Wang YF, Tang YY, Chen SL, Yan SZ (2017) Endophytic *Bacillus cereus* effectively controls *Meloidogyne incognita* on tomato plants through rapid rhizosphere occupation and repellent action. Plant Dis 101(3):448–455. https://doi.org/10.1094/PDIS-06-16-0871-RE

Kihika R, Murungi LK, Coyne D, Ng'ang'a M, Hassanali A, Teal PEA, Torto B (2017) Parasitic nematode *Meloidogyne incognita* interactions with different *Capsicum annum* cultivars reveal the chemical constituents modulating root herbivory. Sci Rep 7:2903. https://doi.org/10.1038/s41598-017-02379-8

Kirwa HK, Murungi LK, Beck JJ, Torto B (2018) Elicitation of differential responses in the root-knot nematode *Meloidogyne incognita* to tomato root exudate cytokinin, flavonoids and alkaloids. J Agric Food Chem 66:11291–11300. https://doi.org/10.1021/acs.jafc.8b05101

Klingler J (1965) On the orientation of plant nematodes and of some other soil animals. Nematologica 11:4–18

Korenblum E, Dong Y, Szymanski J, Panda S, Jozwiak A, Massalha H, Meir S, Rogachev I, Aharoni A (2020) Rhizosphere microbiome mediates systemic root metabolite exudation by root-to-root signaling. Proc Natl Acad Sci U S A 117(7):3874–3883. https://doi.org/10.1073/pnas.1912130117

Kuzyakov Y, Razavi BS (2019) Rhizosphere size and shape: temporal dynamics and spatial stationarity. Soil Biol Biochem 135:343–360. https://doi.org/10.1016/j.soilbio.2019.05.011

Laloum Y, Ngala B, Ianszen M, Boulogne I, Plasson C, Fournet S, Gotté M, Nguema-Ona E, Le Roux A-C, Gobert V, Driouich A, Vicré M (2020) A novel *in vitro* tool to study cyst nematode chemotaxis. Front Plant Sci 11:1024. https://doi.org/10.3389/fpls.2020.01024

Le Saux R, Quénéhervé P (2002) Differential chemotactic responses of two plant-parasitic nematodes, *Meloidogyne incognita* and *Rotylenchulus reniformis*, to some inorganic ions. Nematology 4(1):99–105

Leitão DAH, Pedrosa EMR, Dickson DW, Brito JA, Oliveira AKS, Rolim MM (2021a) Upward migration of second-stage juveniles of *Meloidogyne floridensis* and *M. incognita* under different plant stimuli. Eur J Plant Pathol 161:301–311. https://doi.org/10.1007/s10658-021-02322-8

Leitão DAH, Pedrosa EMR, Dickson DW, Oliveira AKS, Rolim MM (2021b) Temperature: a driving factor for *Meloidogyne floridensis* migration toward different hosts. J Nematol 53: e2021–e2074. https://doi.org/10.21307/jofnem-2021-074

Leontopoulos S, Petrotos K, Anatolioti V, Skenderidis P (2017) Chemotactic responses of Pseudomonas oryzihabitans and second stage juveniles of *Meloidogyne javanica* on tomato root tip exudates. Int J Food Biosyst Eng 5(1):75–100

Li X, Hu HJ, Li JY, Wang C, Chen SL, Yan SZ (2019) Effects of the endophytic bacteria *Bacillus cereus* BCM2 on tomato root exudates and *Meloidogyne incognita* infection. Plant Dis 103(7): 1551–1558. https://doi.org/10.1094/PDIS-11-18-2016-RE

Li Y, Ren Q, Bo T, Mo M, Liu Y (2022) AWA and ASH homologous sensing genes of *Meloidogyne incognita* contribute to the tomato infection process. Pathogens 11:1322. https://doi.org/10.3390/pathogens11111322

Lindford MB (1939) Attractiveness of roots and excised shoot tissues to certain nematodes. Proc Helminthol Soc Wash 6:11–18

Liu B, Hibbard JK, Urwin PE, Atkinson HJ (2005) The production of synthetic chemodisruptive peptides in planta disrupts the establishment of cyst nematodes. Plant Biotechnol J 3:487–496. https://doi.org/10.1111/j.1467-7652.2005.00139.x

Liu W, Jones AL, Gosse HN, Lawrence KS, Park S-W (2019) Validation of the chemotaxis of plant parasitic nematodes toward host root exudates. J Nematol 51(1):1–10. https://doi.org/10.21307/jofnem-2019-063

Liu M, Philp J, Wang Y, Hu J, Wei Y, Li J, Ryder M, Toh R, Zhou Y, Denton MD, Wu Y, Yang H (2022) Plant growth-promoting rhizobacteria *Burkholderia vietnamiensis* B418 inhibits root-knot nematode on watermelon by modifying the rhizosphere microbial community. Sci Rep 12(1):1–13. https://doi.org/10.1038/s41598-022-12472-2

Maheshwari DK, Anwar M (1990) Nematocidal activity of some phenolics on root-knot, growth and yield of *Capsicum frutescens* cv California Wonder. J Phytopathol 129:159–164. https://doi.org/10.1111/j.1439-0434.1990.tb04299.x

Manohar M, Tenjo-Castano F, Chen S, Zhang YK, Kumari A, Williamson VM, Wang X, Klessig DF, Schroeder FC (2020) Plant metabolism of nematode pheromones mediates plant-nematode interactions. Nat Commun 11:208. https://doi.org/10.1038/s41467-019-14104-2

Manosalva P, Manohar M, von Reuss SH, Chen S, Koch A, Kaplan F, Choe A, Micikas RJ, Wang X, Kogel K-H, Sternberg PW, Williamson VW, Schroeder FC, Klessig DF (2015) Conserved nematode signalling molecules elicit plant defenses and pathogen resistance. Nat Comm 6:7795. https://doi.org/10.1038/ncomms8795

Murungi LK, Kirwa H, Coyne D, Teal PEA, Beck JJ, Torto B (2018) Identification of key root volatiles signaling preference of tomato over spinach by the root knot nematode *Meloidogyne incognita*. J Agric Food Chem 66:7328–7336. https://doi.org/10.1021/acs.jafc.8b03257

Nandi B, Kundu K, Banerjee N, Babu SPS (2003) Salicylic acid-induced suppression of *Meloidogyne incognita* infestation of okra and cowpea. Nematology 5:747–752. https://doi.org/10.1163/156854103322746922

Ochola J, Cortada L, Mwaura O, Tariku M, Christensen SA, Ng'ang'a M, Hassanali A, Pirzada T, Khan S, Pal L, Mathew R, Guenther D, Davis E, Sit T, Coyne D, Opperman C, Torto B (2022) Wrap-and-plant technology to manage sustainably potato cyst nematodes in East Africa. Nat Sustain 5:425–433. https://doi.org/10.1038/s41893-022-00852-5

Oka Y (2020) Screening of chemical attractants for second-stage juveniles of *Meloidogyne* species on agar plates. Plant Pathol 70:912–921. https://doi.org/10.1111/ppa.13336

Oka Y (2021) Aromatic compounds that attract *Meloidogyne* species second-stage juveniles in soil. Pest Manag Sci 77:4288–4297. https://doi.org/10.1002/ps.6506

Oka Y, Pivonia S (2002) Use of ammonia-releasing compounds for control of the root-knot nematode *Meloidogyne javanica*. Nematology 4:65–71. https://doi.org/10.1163/156854102760082212

Oliveira AKS, Pedrosa EMR, Dickson DW, Vau SJSSO, Leitão DAH, Silva EFF (2020) Migration and penetration of *Meloidogyne enterolobii* and *M. incognita* in soil columns with tomato and marigold. Eur J Plant Pathol 158:591–598. https://doi.org/10.1007/s10658-019-01889-7

Oota M, Tsai AY-L, Aoki D, Matsushita Y, Toyoda S, Fukushima K, Saeki K, Toda K, Perfus-Barbeoch L, Favery B, Ishikawa H, Sawa S (2019) Identification of naturally occurring polyamines as root-knot nematode attractants. Mol Plant 13:658–665. https://doi.org/10.1016/j.molp.2019.12.010

Pacheco PVM, Campos VP, Terra WC, Pedroso MP, Paula LL, Silva MSG, Monteiro TSA, Freitas LG (2021) Attraction and toxicity: ways volatile organic compounds released by *Pochonia chlamydosporia* affect *Meloidogyne incognita*. Microbiol Res 255:126925. https://doi.org/10.1016/j.micres.2021.126925

Padgham JL, Sikora RA (2007) Biological control potential and modes of action of *Bacillus megaterium* against *Meloidogyne graminicola* on rice. Crop Prot 26:971–977. https://doi.org/10.1016/j.cropro.2006.09.004

Pausch J, Kuzyakov Y (2018) Carbon input by roots into the soil: quantification of rhizodeposition from root to ecosystem scale. Glob Chang Biol 24:1–12. https://doi.org/10.1111/gcb.13850

Perry RN (1996) Chemoreception in plant parasitic nematodes. Annu Rev Phytopathol 34:181–199

Pinkerton JN, Mojtahedi H, Santo GS, O'Bannon JH (1987) Vertical migration of *Meloidogyne chitwoodi* and *M. hapla* under controlled temperature. J Nematol 19(2):152–157

Pline M, Dusenbery DB (1987) Responses of plant-parasitic nematode *Meloidogyne incognita* to carbon dioxide determined by video camera-computer tracking. J Chem Ecol 13:873–888

Prot J (1976) Amplitude et cinétique des migrations du nématode *Meloidogyne javanica* sous l'influence d'um plant de tomate. Cahiers – ORSTOM. Série biologie 6(3):157–166

Prot J, van Gundy SD (1981) Effect of soil texture and the clay component on migration of *Meloidogyne incognita* second-stage juveniles. J Nematol 13(2):213–217

Qi Y, Meng L, Cao S, Li M, Chen S, Ye D (2015) Chemotaxis of *Meloidogyne incognita* in response to different salts. Agric Sci 06(09):900–907. https://doi.org/10.4236/as.2015.69086

Reynolds AM, Dutta TK, Curtis RHC, Powers SJ, Gaur HS, Kerry BR (2011) Chemotaxis can take plant-parasitic nematodes to the source of a chemoattractant via the shortest possible routes. J R Soc Interface 57:568–577. https://doi.org/10.1098/rsif.2010.0417

Robinson AF, Perry RN (2006) Behaviour and sensory perception. In: Perry RN, Moens M (eds) Plant nematology. CABI, Wallingford

Rocha FS, Campos VP, De Souza JT (2010) Variation in lipid reserves of second-stage juveniles of *Meloidogyne exigua* in a coffee field and its relationship with infectivity. Nematology 12:365–371. https://doi.org/10.1163/138855409X12548945788367

Rocha FS, Campos VP, Fernandes MFG, Muniz MFS (2016) Migration and reproduction of *Meloidogyne incognita* in two soil textures. Nematropica 46:162–171

Schroeder FC (2015) Modular assembly of primary metabolic building blocks: a chemical language in *C. elegans*. Chem Biol 22(1):7–16. https://doi.org/10.1016/j.chembiol.2014.10.012

Shivakumara TN, Dutta TK, Rao U (2018) A novel in vitro chemotaxis bioassay to assess the response of *Meloidogyne incognita* towards various test compounds. J Nematol 50:487–494. https://doi.org/10.21307/jofnem-2018-047

Shivakumara TN, Dutta TK, Chaudhary S, von Reuss SH, Williamson VM, Rao U (2019) Homologs of *Caenorhabditis elegans* chemosensory genes have roles in behavior and chemotaxis in the root-knot nematode *Meloidogyne incognita*. Mol Plant-Microbe Interact 32:876–887. https://doi.org/10.1094/MPMI-08-18-0226-R

Siddique S, Coomer A, Baum T, Williamson VM (2022) Recognition and response in plant-nematode interactions. Annu Rev Phytopathol 60:7.1–7-20. https://doi.org/10.1146/annurev-phyto-020620-102355

Sikder MM, Vestergård M (2020) Impacts of root metabolites on soil nematodes. Front Plant Sci 10:1792. https://doi.org/10.3389/fpls.2019.01792

Sikder MM, Vestergård M, Kyndt T, Fomsgaard IS, Kudjordjie EN, Nicolaisen M (2021) Benzoxazinoids selectively affect maize root- associated nematode taxa. J Exp Botany 72: 3835–3845. https://doi.org/10.1093/jxb/erab104

Tian C, Ni J, Chang F, Liu S, Xu N, Sun W, Xie Y, Guo Y, Ma Y, Yang Z, Dang C, Huang Y, Tian Z, Wang Y (2016) Bio-source of di-n-butyl phthalate production by filamentous fungi. Sci Rep 6:19791. https://doi.org/10.1038/srep19791

Toju H, Tanaka Y (2019) Consortia of anti-nematode fungi and bacteria in the rhizosphere of soybean plants attacked by root-knot nematodes. R Soc Open Sci 6:181693. https://doi.org/10.1098/rsos.181693

Topalovic O, Hussain M, Heuer H (2020) Plants and associated soil microbiota cooperatively suppress plant-parasitic nematodes. Front Microbiol 11:313. https://doi.org/10.3389/fmicb.2020.00313

Torto B, Cortada L, Murungi LK, Haukeland S, Coyne DL (2018) Management of cyst and root knot nematodes: a chemical ecology perspective. J Agric Food Chem 66(33):8672–8678. https://doi.org/10.1021/acs.jafc.8b01940

Tsai AY-L, Higaki T, Nguyen C-N, Perfus-Barbeoch L, Favery B, Sawa S (2019) Regulation of root-knot nematode behavior by seed-coat mucilage-derived attractants. Mol Plant 12:99–112. https://doi.org/10.1016/j.molp.2018.11.008

Tsai AY-L, Iwamoto Y, Tsumuraya Y, Oota M, Konishi T, Ito S, Kotake T, Ishikawa H, Sawa S (2021) Root-knot nematode chemotaxis is positively regulated by L-galactose sidechains of mucilage carbohydrate rhamnogalacturonan-I. Sci Adv 7:eabh4182. https://doi.org/10.1126/sciadv.abh4182

Van Dam NM, Bouwmeester HJ (2016) Metabolomics in the rhizosphere: tapping into below-ground chemical communication. Trends Plant Sci 21:256–265. https://doi.org/10.1016/j.tplants.2016.01.008

Viglierchio DR, Schmitt RV (1983) On the methodology of nematode extraction from field samples: Baermann funnel modifications. J Nematol 15(3):438–44, 27, 169–207. https://doi.org/10.1016/S0065-308X(08)60355-3

Vos C, Van Den Broucke D, Lombic FM, De Waele D, Elsen A (2012) Mycorrhiza-induced resistance in banana acts on nematode host location and penetration. Soil Biol Biochem 47:60–66. https://doi.org/10.1016/j.soilbio.2011.12.027

Wallace HR (1968) The dynamics of nematode movement. Annu Rev Phytopathol 6:91–114. https://doi.org/10.1146/annurev.py.06.090168.000515

Wang C, Bruening G, Williamson VM (2009) Determination of preferred pH for root-knot nematode aggregation using pluronic F-127 gel. J Chem Ecol 35:1242–1251. https://doi.org/10.1007/s10886-009-9703-8

Wang X, Li G-H, Zou C-G, Ji X-L, Liu T, Zhao P-J, Liang L-M, Xu J-P, An Z-Q, Zheng X, Qin Y-K, Tian M-Q, Xu Y-Y, Ma Y-C, Yu Z-F, Huang X-W, Liu S-Q, Niu X-M, Yang J-K, Huang Y, Zhang K-Q (2014) Bacteria can mobilize nematode-trapping fungi to kill nematodes. Nat Commun 5:5776. https://doi.org/10.1038/ncomms6776

Wang C, Masler EP, Rogers ST (2018a) Responses of *Heterodera glycines* and *Meloidogyne incognita* infective juveniles to root tissues, root exudates, and root extracts from three plant species. Plant Dis 102(9):1733–1740. https://doi.org/10.1094/PDIS-09-17-1445-RE

Wang B-L, Chen Y-H, He J-N, Xue H-X, Yan N, Zeng Z-J, Bennett JW, Zhang K-Q, Niu X-M (2018b) Integrated metabolomics and morphogenesis reveal volatile signaling of the nematode-trapping fungus *Arthrobotrys oligospora*. Appl Environ Microbiol 84:e02749–e02717. https://doi.org/10.1128/AEM.02749-17

Wang P, Sun Y, Yang L, Hu Y, Li J, Wang J, Zhang F, Liu Y (2019) Chemotactic responses of the root-knot nematode *Meloidogyne incognita* to *Streptomyces plicatus*. FEMS Microbiol Let 366:1–7. https://doi.org/10.1093/femsle/fnz234

Wang J, Ding Z, Bian J, Bo T, Liu Y (2021) Chemotaxis response of *Meloidogyne incognita* to volatiles and organic acids from root exudates. Rhizosphere 17:1–5. https://doi.org/10.1016/j.rhisph.2021.100320

Williamson V, Wang C, Lower S (2009) Application of pluronic gel to the study of root-knot nematode behaviour. Nematology 11:453–464. https://doi.org/10.1163/156854109X447024

Wuyts N, Swennen R, De Waele D (2006) Effects of plant phenylpropanoid pathway products and selected terpenoids and alkaloids on the behaviour of the plant-parasitic nematodes *Radopholus similis*, *Pratylenchus penetrans* and *Meloidogyne incognita*. Nematology 8:89–101

Yang G, Zhou B, Zhang X, Zhang Z, Wu Y, Zhang Y, Lu S, Zou Q, Gao Y, Teng L (2016) Effects of tomato root exudates on *Meloidogyne incognita*. PLoS One 11(4):e0154675. https://doi.org/10.1371/journal.pone.0154675

Yergaliyev TM, Alexander-Shani R, Dimerets H, Pivonia S, Bird DM, Rachmilevitch S, Szitenberg A (2020) Bacterial community structure dynamics in *Meloidogyne incognita*-infected roots and its role in worm- microbiome interactions. mSphere 5:e00306–20. https://doi.org/10.1128/mSphere.00306-20

Zhai Y, Shao Z, Cai M, Zheng L, Li G, Huang D, Cheng W, Thomashow LS, Welle DM, Yu Z, Zhang J (2018) Multiple modes of nematode control by volatiles of *Pseudomonas putida* 1A00316 from Antarctic soil against *Meloidogyne incognita*. Front Microbiol 9:253. https://doi.org/10.3389/fmicb.2018.00253

Zhang Y, Li S, Li H, Wang R, Zhang K-Q, Xu J (2020) Fungi–nematode interactions: diversity, ecology, and biocontrol prospects in agriculture. J Fungi 6:206. https://doi.org/10.3390/jof6040206

Zhao X, Schmitt M, Hawes MC (2000) Species-dependent effects of border cell and root tip exudates on nematode behavior. Phytopathology 90(11):1239–1245. https://doi.org/10.1094/PHYTO.2000.90.11.1239

Zhao Y, Zhou Q, Zou C, Zhang K, Huang X (2022) Repulsive response of *Meloidogyne incognita* induced by biocontrol bacteria and its effect on interspecific interactions. Front Microbiol 13: 994941. https://doi.org/10.3389/fmicb.2022.994941

Zinovieva SV, Udalova ZV, Seiml-Buchinger VV, Khasanov FK (2021) Gene expression of protease inhibitors in tomato plants with invasion by root-knot nematode *Meloidogyne incognita* and modulation of their activity with salicylic and jasmonic acids. Biol Bull 48(2):130–139. https://doi.org/10.1134/S1062359021020175

Zuckerman BM, Jansson H-B (1984) Nematode chemotaxis and possible mechanisms of host/prey recognition. Annu Rev Phytopathol 22:95–113. https://doi.org/10.1146/annurev.py.22.090184.000523

Phytohormone-Mediated Feeding Site Development

4

Sagnik Nag, Prachi Pandey, Souvik Samanta, Oishi Mitra, Anwesha Chatterjee, and Monalisa Chakraborty

Abstract

Infection caused by nematodes leads to the abnormal, localized swelling or outgrowth of plant tissue forming root-gall disease. In horticulture and agriculture, root-knot nematodes (RKNs) are among the most economically damaging parasitic nematodes. Approximately 14.5% of yearly vegetable crop losses are attributed to plant-pathogenic nematodes. In the root-knot nematode-feeding site, *Meloidogyne* sp. accentuates a fundamental change in the cells around the plant's root through molecules secreted by the three esophageal gland cells of the nematode. RKNs induce the inception and shaping of nematode-feeding sites (NFS) in the root tissue by implementing chemicals produced by three esophageal gland cells to coordinate a fundamental alteration in the plant root cells. Phytohormones modulate nematode-plant relationships and coordinate and accentuate cellular and metabolic responses linked with the development of nematodes. The key regulators for manipulating plant tissues that allow galls to promulgate are presumed to be phytohormones genes. The allocation of crucial clues to root-gall disease treatment may be mediated by genes. The intricate pattern between growth and defense processes makes understanding of exact

S. Nag (✉) · S. Samanta · O. Mitra
Department of Bio-Sciences, School of Bio-Sciences and Technology (SBST), Vellore Institute of Technology (VIT), Vellore, Tamil Nadu, India

P. Pandey
Drs Kiran and Pallavi Patel Global University, Vadodara, Gujarat, India

A. Chatterjee
Department of Microbiology, St. Xavier's College (Autonomous), Kolkata, West Bengal, India

M. Chakraborty
Department of Biotechnology, Mount Carmel College (Autonomous), Karnataka, Bengaluru, India

F. Ahmad, G. Nombela (eds.), *Root-Galling Disease of Vegetable Plants*,
https://doi.org/10.1007/978-981-99-3892-6_4

roles of different phytohormones difficult, and thus to combat this pest, new diverse research strategies are underway. Thus, this chapter provides insight into phytohormone-mediated feeding site development and focuses on different regulatory mechanisms to elevate vegetable crop production globally.

Keywords

Meloidogyne species · Root galls · Compounds secretion · Crop production · Agriculture

4.1 Introduction

The most economically significant plant parasitic nematode, which is a root-knot nematode (*Meloidogyne* spp.) severely damages a range of plants, including tomato (*Lycopersicon esculentum*), eggplant (*Solanum melongena*), potato (*Solanum tuberosum*), pepper (*Capsicum frutescens*), cucumber (*Cucumis satis*), and carrot (*Daucus carota*). The disease typically reduces yield by 15–25%, resulting in an estimated $100 billion loss per year globally, but it can sometimes be as much as 75% (Cabrera et al. 2015). The feeding cells, known as giant cells (GCs), initiated by root-knot nematodes (RKNs), develop within a new organ called a gall in the root of the plant. Through the elongated area, they enter the root to move intracellularly toward the root apical meristem (RAM). This helps in establishing themselves in a vasculature. Around 5–8 vascular cells are transformed into specialized cells for transfer called GCs. In addition, the endodermis and cortical enlargement caused by the nematode result in cell proliferation, affecting the vasculature and forming galls (Singh et al. 2021). Recent years have seen a large and focused change in the expression of syncytia and GCs, as shown by transcriptome investigations in conjunction with molecular cell biology. The genes involved in hormone-regulated pathways in roots, that are connected to auxins and cytokines, are among those whose expressions are altered to feeding sites (Ibrahim et al. 2019; Zhang et al. 2017). Rich in auxin, callus-induced media (CIM-C), and callus with wounding-induced are the key mechanisms by which a pluripotent callus forms. With the RAM marker genes of molecules like CIM-C development entails the distinction of pericycle-like cells in a procedure that stimulates root-tip variation. LBD genes (like LBD16) are required for CIM-C growth and lateral root (LR) primordia production. Ectopic activation of the LR developmental program is a standard method in callus generation from various organs. It is interesting to note that following cutting, RAM regeneration starts a new distal RAM sequence that resembles an embryo. The identity of pluripotent meristematic cells within the galls is enforced by nematodes, which take developmental pathways of any new organogenesis and root renewal (Cabrera et al. 2015; Deveshwar et al. 2020).

4.2 Physiological Implications of Root-Gall Disease

Plants infected with nematodes result in certain symptoms or diseases on their roots as well as the portions present above the ground. In cases where the nematode infections have been associated with saprophytic bacteria or bacteria which are pathogenic to plants or fungi, it may result in conditions such as root lesions, root galls, root knots, excessive branching in the roots, wounded root tips or even root rots. The symptoms of the root are frequently associated with non-specific symptoms in the plant's aboveground portions, such as stunted growth and deficiency indications in the nutrition, which may, in turn, include yellowing of foliage, increased wilting in both hot and dry conditions (Agrios 2005). Galls are tiny in size, usually less than 5 mm in terms of length. They may be located on the root apex or down the root axis, and affected plants may have many LRs. Because of those modifications, the central cylinder becomes asymmetrical, resulting in an aberrant root function and decreased plant development (Palomares-Rius et al. 2017; Siddique and Grundler 2015).

A nematode injects secretory proteins into a plant cell when it first uses its stylet to pierce it, causing the parasitized cells to undergo alterations. As nuclear division occurs when the cell wall formation is absent, parasitized cells quickly become multinucleate. Cell division is believed to be uncoupled from this mechanism. Cells get larger and carry more nuclear material; they do not divide to form new cells. This enables the giant cell to generate significant quantities of proteins, which the nematode will later consume. Giant cells also serve as nutrient sinks, channeling the plant's nutrients to the nematode feeding on them. The RKNs do not consume cells straight away. It creates a feeding tube secreted into the plant cell's cytoplasm by the stylet and functions similarly to a sieve to filter the cytosol the nematode consumes. Giant cells can become highly enormous, as the name suggests. It has been shown that this increase in cell size and division is caused by a rise in plant growth regulator production, which is stimulated by the secretions of the nematode esophageal gland cells. Plant growth regulator diffusion is likely the cause of the nearby root cells of the giant cells' rapid expansion and division, which leads to gall formation (Mitkowski and Abawi 2003).

4.3 Root-Gall Nematodes (RKNs)

The RKNs belong to the genus *Meloidogyne* and have described 100 different *Meloidogyne* species. The species *M. incognita, M. javanica, M. hapla, M. chitwoodi, and M. graminicola* are some of the most common and economically significant (Mitkowski and Abawi 2003). Phylogenetic analyses are generally performed using Bayesian inference and maximum likelihood techniques. These findings deduced that the examined species can be divided into 11 clades, including *M. nataliei* and *M. indica* making up the basal lineage. Seven clades of the *Meloidogyne* superclade comprise 75% of these species (Álvarez-Ortega et al. 2019). RKNs do not have an internal skeletal system; thus, their cuticle acts as a barrier from internal turgor pressure to keep their bodies in form and facilitate movement. Once they locate a feeding place, they adhere to the root permanently (Mitkowski and Abawi 2003). By causing an impact on the surface cuticle of *Meloidogyne* spp., Auxin can function as a signaling chemical, which is necessary for infection. These worms may sense an auxin gradient and follow it via amphidial or phasmidial receptors while penetrating and migrating within roots. Auxin interacts with the chemosensory organs, amphids, and phasmids, as well as with numerous tail neurons of *Meloidogyne* spp. (Curtis 2007). The root-knot nematodes also require specific concentrations of cytokinin to form the feeding site within the roots of the plants (Gheysen and Mitchum 2019). These studies have implied that gall formers may cause species-specific and temporally variable alterations in the chemical composition of gall tissue. The levels of nutrients as well as secondary compounds in gall tissue have also been indicated to be typically noticeably different from those of the surrounding plant tissue (Hartley 1998).

4.4 Feeding Site Development and Associated Factors

The vascular system of plants behaves like a network for the transportation of nutrients, water, and vital photosynthates from their organ of origin in the direction of the site of requirement. This presents the vascular system as the ultimate target for access and accumulation of host resources by pests, especially nematodes that induce root-gall diseases (Bartlem et al. 2014). As mentioned in the previous section, root-knot nematodes infect the host plant's root by converting their vascular parenchyma cells into specialized structures called nematode-feeding sites (NFS). Such feeding sites act as the sole source of nutrients required for the reproduction and growth of the nematode that develops within the root tissues. At the initial stages of infection, these juveniles migrate through the roots during elongation, and giant cells multinucleate cells are observed by manipulation of normal root physiology. The feeding site is initiated via the creation of binucleate cells. Finally, it progresses through repetitive nuclear divisions (hyperplasia) and cellular growth (hypertrophy) devoid of cytokinesis resulting in root swelling and vascular deformity seen (Favery et al. 2020). The RKNs extract nutrition from the giant cells and develop into males or females, releasing the eggs directly into the gall surroundings or rhizosphere. Each feeding site contains a minimum of 4–10 giant cells. The physiological characteristics of the giant cells involve a dense cytoplasm, small vacuoles, ingrowths in cell walls adjacent to surrounding vascular tissue regions, expanded nuclei, and visible procreation of xyloglucan endo-transglycosylase/XTH (SER; organized into swirls), mitochondria, ribosomes, and plastids (Kyndt et al. 2013). The other tissues that entirely or partially surround the feeding sites develop to protect the inducer from natural enemies and/or abiotic stress. Events involved in the development of a nematode-feeding site include the following:

4.4.1 Vascularization

The giant cells are enclosed within a xylem network, and the phloem forms de novo. To preserve vascular continuity, a network of thick cell-walled xylem cells with lignified secondary deposits are found either enveloping the giant cells or confining within regions skirting the border of giant cells (Bartlem et al. 2014). Unlike normal xylem cells, these cells are not elongated; instead, these are asymmetrical: similar to wound-type xylem elements, and form irregular networks that are interconnected (Jones and Goto 2011). Giant cells may be deficient of plasmodesmata on the ingrowths of their cell wall and isolated, removed from the surrounding tissue but connected through plasmodesmata or not only connected to one another by the help of plasmodesmata but also reshuffles the cytoskeleton to include plasmodesmata on the cell walls present toward the neighboring cells. The type of attachment among giant cells and their adjacent cells varies depending on the stage of development (Hofmann et al. 2010). In case the giant cells are simplistically isolated, nutrients and assimilate are loaded with the assistance of transporters. Thus, vast-scale phloem formation is induced in the periphery of giant cells. The phloem in which feeding sites are virtually embedded exclusively consists of sieve elements and lacks companion cells. These sieve elements are capable of performing transcriptional responses and are routinely nucleated (Absmanner et al. 2013). The absence of companion cells may be justified in that they, if present, would interfere with the reuptake of solutes like sucrose by the phloem from the apoplast and, thereby, hinder the direct flow of solutes into the giant cells. The lack of companion cells may be due to two reasons: consumption of these cells during the process of vascularization or in-expression of the gene encoding for the identity of these cells (SUC2). In addition to nutrient delivery, vascularization also functions as a management system for waste procured from the feeding sites by virtue of the parasite.

4.4.2 Cell Expansion and Cell Wall Modification

Cell expansion occurs due to the loosening of the cell wall brought about by the upregulation of genes-encoding proteins, for example, expansins and pectinases. The plant cell wall comprises hemicellulose, a component of which xyloglucans (impart rigidity and elasticity to cell walls) are a constituent (Scheller and Ulvskov 2010). Either non-hydrolytic cleavage (XTH) or chain shortening (xyloglucan endo-hydrolase) of xyloglucan chains aids in the cell wall relaxation (Eklöf and Brumer 2010). Once cell expansion is fulfilled, the cell wall of the giant cells undergoes certain modifications because of the increased turgor pressure from metabolite re-allocation. Thickening the cell walls takes place as a major modification to withstand the excruciating pressure. The cells induce the same by reinforcing mechanisms involving XTHs, depositions of lignin or callose, and peroxidase activity. Feeding sites formed by RKNs upregulate cellulose synthesizing genes such as cellulose synthase A (*CesA*), a gene in charge of both primary and secondary cell wall synthesis, and either upregulate or downregulate XTH depending on the kind of feeding site and its respective stage of development. Increased expression of *CesA* is primarily observed in members of the plant genus *Arabidopsis* (Kyndt et al. 2013).

4.4.3 Host Cell Cycle Sustenance

The parasite ensures coordinated cell cycles between the host and itself to lead to the formation of multinucleate cells. In the case of RKNs, the appearance of NFS consists of a DNA synthesis period linked to endoreduplication and an acytokinetic mitotic phase, causing nuclear enlargement. The cell cycle machinery in eukaryotes, as known, is regulated by transcriptional or posttranscriptional mechanisms that ensure the activation of certain cyclin-dependent kinases (CDKs) (Kyndt et al. 2013). Formation of feeding sites induces not only heightened transcription of CDKs and cyclins but also participates in the activity of genes positively associated with G2-M transition. Thus, CDK inhibitors play a pivotal role in regulating giant cell formation. If overexpressed, CDK inhibitors like KRP1, KRP2, and KRP4 may drastically reduce the root gall's size. Apart from this, they may also obstruct multiple nuclear divisions in the giant cells. APC (anaphase-promoting complex) genes are essential components of endocycle machinery which is an integral part of NFS generation (Koltai et al. 2001; Vieira et al. 2012).

4.5 Overview of Phytohormones

The most disadvantageous physiological characteristic of plants is their sessility, which makes them vulnerable. To overcome this, plants have now evolved to master mechanisms capable of detecting changes in their surroundings. Once they sense the presence of a suitable environment for vegetative growth, they employ certain chemicals to initiate root and shoot growth. Plants also use the same chemicals for surviving stress factor-induced detrimental situations (such as biotic and/or abiotic stress). These chemicals are often referred to as phytohormones and are specifically synthesized in defined organs of plants. In other words, phytohormones are chemicals that transcribe signals received from the environment into an observable phenotypic action within the plant. Such hormones thus regulate plant growth and development on the basis of their concentration gradient and distribution pattern; consequently, they also dictate plant metabolism (Zhao 2010). The majorly significant phytohormones are listed below (Table 4.1).

4.5.1 Auxin

Identification of auxin as a potential phytohormone occurred through studies executed to seek the source for differentiation in plants corresponding to light stimulus. Naturally occurring auxin (indole-3-acetic acid/IAA) is mainly involved

Table 4.1 Tabulated representation of phytohormones and their overall functioning

Phytohormone	Site of action	Biosynthesis precursor	Regulatory genes	Function	References
Auxin	Young leaves, apical meristem of shoots, and seeds	Tryptophan (Trp)	YUC2, YUC6, SPL (sporocyteless), SHI (short intenotes), STY1, NGA, and TAA	Induces proteolysis via ubiquitin in indole-3-acetic acid and modulates the plant's response toward light	Zhao (2010)
Gibberellin	Shoot, flower, fruit, and seed	Acetate/GA intermediate	AtGA3ox1, AtGA3ox2, AtCPS, GID1, DELLA, and SCR3	Induces shoot elongation, flower and fruit maturation, and seed germination	Binenbaum et al. (2018)
Cytokinin	Bud, leaf, root, and shoot	Furfural	LOG3, LOG4, cyclin D3, ANT (aintegumenta), and cyclin 1	Induces differentiation and division of cells in bud, leaf senescence and development of chloroplast	Wybouw and de Rybel (2019)
Abscisic acid	Depending on the type of stress, mostly the site of ABA action is bud, shoot, guard cells, and seed	Violaxanthin	CYP707A1/3, NCED3, HAT1, ABA3, NCED3, and SnRK2.3	Induces reduced seeding growth and germination, modulates metabolism across guard cells and ion homeostasis	Chen et al. (2020)
Ethylene	Leaf and shoot	1-Aminocyclopropane-1- carboxylic acid (ACC)	ERS1, ERS2 (ethylene response sensor), EIN4 (ethylene insensitive), ETR1, ETR2 (ethylene resistance), ARGOS proteins, and RTE (reversion of ethylene sensitivity)	Regulates senescence and growth	Dubois et al. (2018)

in developmental procedures, including division, differentiation, and elongation of the plant cells. This hormone is related to female gametophyte development as studies performed on *Arabidopsis* affirm the presence of auxin within the developing embryo sac. When supplied with exogenous auxin, plants have been recognized to suffer reduced primary root elongation and a surge in shoot development. In addition to plants, auxin is also generated by several plant pathogens to disrupt the balance of auxin concentration in the host system, thereby interfering with the physiological development of the host (Zhao 2010).

4.5.2 Gibberellin

Gibberellin supports several developmental phenomena such as seed development, flower, and fruit, the transformation from the vegetative to the reproductive phase of the plant, cell growth resulting in organ expansion (hypocotyl xylem expansion), and elongation. Gibberellin localization within the plant varies based on developmental stage, tissue type, and organ. It is capable of movement within the plant biological system in basipetal and acro-petal directions, as predicted in reports published approximately 50 years ago. Even though it moves in both directions, the acro-petal movement of gibberellin is more significant (Binenbaum et al. 2018).

4.5.3 Cytokinin

First identified in the late 1950s as a class of structurally diverse phytohormones, cytokinin plays a major role in shoot development in association with auxin. This leads to dedifferentiation and proliferation of the cells, resulting in callus formation and shoot regeneration. Cytokinin influences the fate of shoot cells, the development of the flower, female gametophyte, vasculature, and root nodules (Wybouw and de Rybel 2019).

4.5.4 Abscisic Acid (ABA)

Primarily recognized as an ether and water-soluble substance that inhibits growth and enhances bud dormancy (previously called dormin), abscisic acid (ABA) is the phytohormone of interest in stress conditions. It regulates biotic and/or abiotic stress environments via either transcriptional or posttranscriptional mechanisms. ABA is also intertwined with physiological processes like regulation of osmosis, senescence of leaf, stomatal closure, seed germination, and so on (Chen et al. 2020).

4.5.5 Ethylene (ET)

Ethylene (ET) is the smallest known gaseous phytohormone with a very simple structure, renowned for its significance in fruit ripening, leaf development, and senescence. Acts as a response against biotic and abiotic stress, including heat, alkalinity, salinity, metal ion concentration, and shade. Thus, it bridges the gap between changes in the territory of the plants and their corresponding developmental adaptation. Ethylene acts as a growth inhibitor, as the presence of ethylene in high concentrations inhibits leaf development. However, ethylene's effect on cellular division depends on the organ of concern. For example, in the case of the formation of the apical hook or vascular development, ethylene is found to be an active participant in the stimulation of cell division. Under these specific contexts, ethylene may have a positive impact in terms of plant growth (Dubois et al. 2018).

4.6 Role of Phytohormones in Plant-Nematode Interaction

Phytohormones are chemical compounds that are not abundantly found in plants. They are integral to the plant's growth and development and enhance its ability to endure and adapt to the changing environment. Phytohormones influence plant-nematode interactions by functioning as chemoattractants or chemorepellents or inadvertently influencing the root-associated microbiota or the host's defensive mechanisms (Sikder et al. 2021). Nematodes foster the feeding sites by employing certain plant developmental pathways, including hormonal cross talk. They must simultaneously inhibit the plant defense and corresponding hormone pathways, thus making it challenging to unravel the precise functions of the phytohormones in these intricate interactions (Gheysen and Mitchum 2019).

Auxin plays a vital role in organogenesis, making the local auxin accumulation for the onset and development of the nematode-feeding site evident. It also influences hypertrophy, cell cycle activation, and cell wall ingrowth. It functions as a chemical messenger and helps the nematodes for root invasion by modifying the nematode cuticles and their behavior. Auxin moves to the elongation zone through the epidermal cells from the root tip. However, auxin is a crucial part but inadequate for gall formation. It majorly contributed to the cell division and enlargement of the surrounding cells. The aggregation of auxin at the feeding site results from auxin efflux inhibition by the nematodes. Flavonoids are a type of polyphenolic compound produced by plants that can regulate auxin levels directly by utilizing the auxin-degrading enzymes or indirectly by functioning as a transport inhibitor. The root-gall nematodes stimulate this flavonoid pathway to alter the plant's auxin levels. The auxin transport inhibition is due to the up-regulation of the *WRKY23* gene, which induces flavonoid biosynthesis. Auxin initiates from the root epidermis of the plant after it attaches to the juveniles of *Meloidogyne* spp. and passes through the root tip and meristematic tissue to finally reach the vascular cylinder to come across a suitable feeding site. Thus, auxin is a plant-specific excitatory messenger for communication (Curtis 2007).

Cytokinin is a signaling molecule essential for the cell cycle, cell division, and nutrient metabolization. These vital cellular processes are modified for the feeding site development during the plant-nematode interaction. Cytokinin signaling is necessary not only for initial nematode infection but also for effective gall formation. With reference to an experiment on *Lotus japonicus,* having enzyme cytokinin oxidase (cytokinin degrading enzyme), it was observed that a lesser number of nematodes-induced root galls were identified, thus making the utility of cytokinin evident. The overexpression of *AtCKX3* and *ZmCKX1* cytokinin oxidase is the major cause for the same. The level of cytokinin in the plant is majorly regulated by a family of seven cytokinin oxidases (*CKX1–7*). The expression of these genes is dependent on the tissues, for example, *CKX1* and *CKX2* are primarily present in the early floral tissues and the shoot apex; *CKX4* is found in the leaf stipules, stomata, and root cap; *CKX5* is mostly found in the growing leaves, pollen, stamen, and apical meristem; *CKX6* is abundant in the vascular cylinder of young roots and shoot tissues and *CKX7* is primarily found in the early developmental roots and the mature embryo sac. Cytokinin biosynthesis and signaling might not be the same for all nematodes; specifically, zeatin and benzyl adenine are two cytokinins secreted by root-knot nematodes. After the infection, there is an aberrant increase in cell division which eventually results in the development of root galls (Dowd et al. 2017).

Ethylene is a gaseous phytohormone that contributes to the opening of flowers, fruit ripening, and shedding of leaves. In a study conducted using tomato plants, ethephon, an ethylene-releasing agent, was introduced in tomato plants already affected by the root-knot nematodes. There was a rise in the galling weight of the infected individuals due to the parenchymatous cell proliferation compared to the uninfected cultures. Ethylene is also known to promote protein and RNA synthesis and incorporation of glucose and protein into the cell wall cellulose resulting in abundant growth in this area (Giazer et al. 1983).

Certain defense hormones like jasmonic acid (JA) and strigolactone (SA) interact antagonistically or synergistically. SA or other similar chemicals majorly decrease the extent of nematode infection. Lower SA levels or signaling in mutants and transgenics generally make them more vulnerable to nematodes, whereas higher SA levels or signaling make them less sensitive to nematode infections. JA increases the expression of pathways that create secondary metabolites with antiherbivore action and protease inhibitors. Protease inhibitors inhibit insect development and reproduction by limiting the proteolytic activity of the digestive enzymes. Although introducing JA to plants increases their ability to fight against nematodes, it is its effects on the development of metabolites and proteins (such as proteinase inhibitors) that are to be blamed (such as terpenes and oxylipins). The degree to which JA-related gene alterations impact these antiherbivore chemical determines how sensitive (or not) the plant is to nematode infection (Gheysen and Mitchum 2019).

Other hormones like gibberellic acid (GA), ABA, and brassinosteroids regulate gall formation. GA has an antagonistic effect on JA action and induces SA signaling when studied in *Arabidopsis,* whereas GA has antagonistic actions on both JA and SA signaling in rice. ABA is incorporated to enhance the susceptibility of rice and tomato to root-knot nematode infection. Brassinosteroids function by repressing the rice defense, interacting antagonistically with the JA pathway (Gheysen and Mitchum 2019).

4.7 Mechanism of Phytohormone-Mediated Feeding Site Development

4.7.1 Auxin-Mediated Regulatory Networks in Nematode Feeding Cells

IAA, also known as *auxin*, is a crucial regulator of organogenesis in plants. Therefore, this is not astonishing local auxin accumulation is linked to the onset and maturation of NFS. Auxin mutants are hence far less vulnerable to RKN. Numerous alterations that occur during the growth of feeding sites, including cell wall growths, hypertrophy, and activation of the cell cycle, may be mediated by auxin. The upregulation of plasma membrane proton pumps and cell wall-modifying proteins, which control acid growth, is how auxin is known to contribute to cell expansion. Auxin and ET work together to create cell wall ingrowths during the creation of transfer cells. Additionally, auxin plays a role in numerous other cell cycle phases and is a crucial trigger for cell cycle entry (Siddique and Grundler 2015). Auxin biosynthesis, signaling, and gene-related genes are up- and downregulated in a complicated temporal and geographic manner in NFS, according to studies of the transcriptome and promoter-reporter data. Auxin production and auxin-response genes are primarily upregulated shortly after nematode infection, whereas genes producing repositories are generally downregulated, supporting an early involvement for auxin during infection. Auxin buildup at the starting NFS may result from nematode secretion, locally stimulated plant biosynthesis, or modifications in auxin transport. Auxin has been found in the secretions of RKN primarily in its conjugated form, but how this NFS production affects that is unclear (Gheysen and Mitchum 2019).

4.7.2 Cytokinin-Mediated Regulatory Networks in Feeding Cell Development

In conjunction with auxin, cytokinin is an N6-substituted adenine derivative that regulates cell division and differentiation in plants. The timing and amplitude of the oscillating levels of cytokinin, which are essential for controlling the cell cycle, may influence whether cells enter mitosis or DNA reduplication. Cytokinin modifies nutrient translocation to postpone senescence and transform tissues into sinks (Siddique and Grundler 2015; Zhou et al. 2020). Cytokinin has long been speculated to play a vital role in developing NFS because they are engrossed in cell cycle regulation and nutrient mobilization. Scientists have found cytokinin secrete RKN *M. incognita*. In addition to this, in *H. schachtii*, it was confirmed by the identification of nematode cytokinin synthesis genes that are produced at the beginning of infection. Silencing this gene reduces infectivity by nematodes.

On the other hand, *Arabidopsis thaliana* mutants that biosynthesize cytokinins exhibit small syncytia than wild-type plant species. A detailed analysis has not properly been performed for RKN, but a similar scenario will likely occur. Signaling mutants of cytokinin and plants with lowered cytokinin range are less susceptible to both nematodes. Nevertheless, cytokinin biosynthesis, signaling, and catabolic gene expression differ in syncytia and gall, which may underlie different cell cycle progression modes. The hypothesis has been confirmed by analysis of cytokinin-cognitive mutants, showing *Ahk4* is one of the major *Ahk* genes (encoding Arabidopsis His kinase) involved in the development of syncytia and that *Ahk2* and *Ahk3* are significant for gall formation. Comparing the expression of genes in young syncytia and gall to callus revealed that syncytia resembled sprout-forming callus and gall resembled solid callus due to their high cytokinin/auxin ratio. However, it remains unclear how cytokinin signaling is involved in various cell cycle abnormalities in giant cells and syncytia (Cabrera et al. 2015; Zhao 2010).

4.7.3 Ethylene (ET)-Mediated Regulatory Networks in Feeding Cell Development

Ethylene ($H_2C{=}CH_2$), is a gaseous type hormone that is involved in multiple plant processes and is known for senescence and ripening of fruit, which includes the activation of the cell wall to degrade. ET can produce various outcomes in other plant processes through positive interactions with the auxin pathway or the JA pathway. Although the available information on the role of ETs in *Caenorhabditis elegans* infection appears contradictory, several important features can be distinguished. ET consistently suppresses RKN infection. Early reports showed that ET positively affected bile weight and giant cell hypertrophy, but this effect is not always consistent with increased nematode infection. Indeed, all subsequent studies on several plant species convincingly show that ET inhibits RKN infection, presumably by reducing nematode attraction to roots. Tolerant plants show greater upregulation of ET biosynthesis and response genes than susceptible plants, consistent with their role in plant defense (Gheysen and Mitchum 2019).

4.8 Remedial Measures for Curbing Feeding Site Development

Cultivating vegetable crops in controlled environments is a recent development that is highly preferred by farmers all over the country. RKNs are regarded as the most mutilating species of feeding site nematodes, even under controlled environments. Severe chlorosis and stunting decrease the plant's yield by introducing multiple root galls. Numerous commonly found vegetables, fruits, trees, ornamental, medicinal plants, cereals, and weeds are targets of RKNs. These nematodes can spread widely and are challenging as they can spread from garden to garden via tools and boots very conveniently. The development of these galls harms the ability of the roots to conduct water and nutrients. Galls can break, particularly on the roots of vegetable plants, enabling pathogenic microorganisms from the soil to enter. Moisture and temperature are the two most effective contributing factors to the multiplication of root-knot nematodes (Patil and Yadav 2021). It is essential to discuss the different remedial measures that can be taken to prevent root gall formation and its detrimental consequences. Nematicides can prove problematic for the environment, human health, and nematode resistance if they are extensively used. Applying efficient, affordable, and secure alternative control mechanisms to producers, consumers, and the environment is crucial. Some of the remedies worth mentioning are silver nanoparticles, plant growth-promoting rhizobacteria (PGPR), and phenolic compounds. Silver-based nanoparticles (AgNPs) are one of root gall's most effective preventive measures. It was experimentally observed that high doses of AgNPs (around 90.4 mg/m^2) could reduce the number of *Meloidogyne* sp. Based on the assessment of Ag-nano formulations against root-knot nematodes, it was established that the Ag nanoparticles and petroleum ether extract effect can be effective and ecologically safe for reducing *Meloidogyne incognita*. Following the application of AgNPs (silver nanoparticles), root galls caused by root-knot nematodes were significantly reduced. Various findings demonstrated that *Conyza dioscoridis* leaf extracts produced as silver nanoparticles (AgNPs) had significant nematicidal action against *Meloidogyne* eggs and juveniles in the second stage (J2). Laboratory tests revealed that the minimal dose for 100% irreversible nematode mortality after 12 h was 0.1 g mL 1 in the water screening test. Furthermore, findings from the sand screening test after 24 h of incubation revealed that AgNP at 2 g mL 1 had a 100% nematicidal effect. In glasshouse experiments using the soilless rice culture method, applying 1 g mL of AgNP straight to the trays effectively suppressed the growth of root gall (Mohamed et al. 2021).

PGPR can decrease the nematode population through antagonistic actions. Through the suppression of plant disease-causing organisms, competition for resources and ecological niches, production of antimicrobial compounds, or production of phytohormones and peptides that act as biostimulants without harming the user, consumer, or the environment, PGPR appears to promote the growth of the plants. In an experiment on tomato plant growth and root-knot nematode, the beneficial effect of six PGPR isolates *Pseudomonas putida*, *Pseudomonas fluorescens*, *Serratia marcescens*, *Bacillus amyloliquefaciens*, *Bacillus subtilis*, and *Bacillus cereus* were studied after 45 days of nematode infection. After 45 days of growth, the plants were harvested to measure the plant growth parameters, such as shoot dry weight, plant height, number of fruits per plant, and yield weight per plant. The number of J2 in the soil, the number of galls per root system, and the number of egg masses per root system were also counted to observe positive changes as an effect of the PGPR isolates, thus making PGPR a potential remedy against RKNs (Umar and Safiya 2021). There are several phenolic compounds that have toxic effects on *Meloidogyne incognita*. Forty-nine different phenolic compounds were to check their toxic behaviors against the nematode and found that having the ability to reduce gall formation. It was observed that 7 out of the 49 compounds were capable of increasing J_2 mortality at 500 μg/mL. P-anisaldehyde was the most active compound, whose LC_{50} value was half that of the synthetic nematicide carbofuran. Hydroquinone at 500 μg/mL acted against the tomato plant's nematodes, although most of the other competent phenols failed to show its effect in tomato plants. It was also concluded that in vivo assay is crucial to assess the potential of phenols as nematicides (Oliveira et al. 2019).

4.9 Recent Advancements and Future Prospects

Meloidogyne spp. is responsible for an annual worldwide loss of $157 billion (Cabrera et al. 2015). *Meloidogyne* spp. can drastically diminish yields after harvest, raise the manufacturing cost through greater fertilizer treatment, and increase crop loss levels during growth, depending on the degree of nematode populations (Onkendi et al. 2014). Therefore, due to the increase in the economic losses caused by RKN, it is essential to establish new environmental-friendly and well-organized strategies. Currently, targeted sequencing techniques like 16S and 18S rDNA sequencing have a great deal of potential to be used to detect novel biological agents for managing RKNs (Forghani and Hajihassani 2020). The biocontrol investigations will become quicker, more affordable, and more valuable. Future research on the RKN suppressive soils' microbiomes would also be beneficial to investigate the potential for creating more comprehensive management plans with many targets for action. Compared to conventional chemical techniques, environment friendly methods are now insufficient to protect plants fully against RKN. As a result, it is essential to think about the creation and enhancement of multidisciplinary management techniques for RKN, such as integrating microbial tactics which involve the use of bacterial and fungal agents with other cultural control procedures or host resistance (Yadav 2017). Although both biocontrol and the application of soil amendments have been partially investigated against nematodes, there is still an opportunity for more research on how these two techniques can work together. For instance, research on how specific amendments may affect the soil microbiota in connection to nematode inhibition.

Additionally, more technologies are becoming accessible, such as O3wat, which may be included in multi-aspect tactics created (Ahmad and Ullah, n.d.; Anwar et al. 2021). Utilizing biological, cultural, and chemical techniques in accordance with integrated pest management (IPM) protocols is the most efficient way to handle harmful nematodes, including *Meloidogyne* spp. For precise identification, a combination of conventional and molecular-based diagnostic techniques should be applied (Gowda et al. 2019; Onkendi et al. 2014). This will eventually decrease the high amounts of damage caused to diverse crops by *Meloidogyne* spp. Growers will subsequently profit from this tactic, and exorbitant production expenses will be avoided. To prevent the entry of *Meloidogyne* spp. onto their farms, growers should also receive comprehensive phytosanitary training (Ansari et al. 2020).

4.10 Conclusion

RKNs contribute and assist in the genesis of the feeding site, including the initiation of a multinuclear cell which acts as the source of the nutrients crucial for reproduction, multiplication, growth, and development of the nematode. Many critical phenomena in plants, such as differentiation, maturation, and development of cells and their responses to abiotic and biotic stimuli, are administered and regulated by phytohormones such as auxin and ET, which play a significant role in plant-pathogen interaction. Phytohormones mediate nematode-induced feeding sites. These plant hormones trigger a remarkable change in the morphological characteristics of the host cell. These changes are due to overexpression or underexpression of phytohormone-responsible genes. Various plant hormones interact, modify, and influence these main hormones; hence, they showcase contrasting and distinct outcomes whose dependency is highly specific to the interaction between the host and the nematode. The participation of plant hormones in the construction of nematode-feeding sites and their role in plant responses involving the defense system is complicated. In conclusion, future and upcoming research should focus on ecologically friendly techniques built on interdisciplinary methods and strategies that may cover the gaps left by single-sided management systems.

References

Absmanner B, Stadler R, Hammes UZ (2013) Phloem development in nematode-induced feeding sites: the implications of auxin and cytokinin. Front Plant Sci 4:241. https://doi.org/10.3389/fpls.2013.00241

Agrios GN (2005) Plant diseases caused by nematodes. Plant Pathol 5. https://doi.org/10.1016/B978-0-08-047378-9.50021-X

Ahmad S, Ullah I (n.d.) Prospective chapter: integrated root-knot nematodes (Meloidogyne) management approaches. https://doi.org/10.5772/intechopen.102882

Álvarez-Ortega S, Brito JA, Subbotin S (2019) Multigene phylogeny of root-knot nematodes and molecular characterization of Meloidogyne nataliei Golden, Rose & Bird, 1981 (Nematoda: Tylenchida). Sci Rep 9(1):1–11. https://doi.org/10.1038/s41598-019-48195-0

Ansari RA, Rizvi R, Mahmood I (2020) Management of phytonematodes: recent advances and future challenges. Springer. https://doi.org/10.1007/978-981-15-4087-5

Anwar A, Mughal NM, Shahnaz E, Banday S, Bashir T, Nisa Q, Jeelani G (2021) Management of root-knot nematode, Meloidogyne Incognita dreaded invading in pointed gourd (Trichosanthes dioica Roxb.) crop prone to eastern UP of India. In: Nematodes-recent advances, management and new perspectives. IntechOpen. https://doi.org/10.5772/intechopen.98923

Bartlem DG, Jones MGK, Hammes UZ (2014) Vascularization and nutrient delivery at root-knot nematode feeding sites in host roots. J Exp Bot 65(7):1789–1798. https://doi.org/10.1093/jxb/ert415

Binenbaum J, Weinstain R, Shani E (2018) Gibberellin localization and transport in plants. Trends Plant Sci 23(5):410–421. https://doi.org/10.1016/j.tplants.2018.02.005

Cabrera J, Díaz-Manzano FE, Fenoll C, Escobar C (2015) Developmental pathways mediated by hormones in nematode feeding sites. In: Adv bot res. Elsevier, vol 73, pp 167–188. https://doi.org/10.1016/bs.abr.2014.12.005

Chen K, Li G, Bressan RA, Song C, Zhu J, Zhao Y (2020) Abscisic acid dynamics, signaling, and functions in plants. J Integr Plant Biol 62(1):25–54. https://doi.org/10.1111/jipb.12899

Curtis R (2007) Do phytohormones influence nematode invasion and feeding site establishment? Nematology 9(2):155–160. https://doi.org/10.1163/156854107780739072

Deveshwar P, Prusty A, Sharma S, Tyagi AK (2020) Phytohormone-mediated molecular mechanisms involving multiple genes and QTL govern grain number in rice. Front Genet 11: 586462. https://doi.org/10.3389/fgene.2020.586462

Dowd CD, Chronis D, Radakovic ZS, Siddique S, Schmülling T, Werner T, Kakimoto T, Grundler FMW, Mitchum MG (2017) Divergent expression of cytokinin biosynthesis, signaling and catabolism genes underlying differences in feeding sites induced by cyst and root-knot nematodes. Plant J 92(2):211–228. https://doi.org/10.1111/tpj.13647

Dubois M, van den Broeck L, Inzé D (2018) The pivotal role of ethylene in plant growth. Trends Plant Sci 23(4):311–323. https://doi.org/10.1016/j.tplants.2018.01.003

Eklöf JM, Brumer H (2010) The XTH gene family: an update on enzyme structure, function, and phylogeny in xyloglucan remodeling. Plant Physiol 153(2):456–466. https://doi.org/10.1104/pp.110.156844

Favery B, Dubreuil G, Chen M-S, Giron D, Abad P (2020) Gall-inducing parasites: convergent and conserved strategies of plant manipulation by insects and nematodes. Annu Rev Phytopathol 58: 1–22. https://doi.org/10.1146/annurev-phyto-010820-012722

Forghani F, Hajihassani A (2020) Recent advances in the development of environmentally benign treatments to control root-knot nematodes. Front Plant Sci 11:1125. https://doi.org/10.3389/fpls.2020.01125

Gheysen G, Mitchum MG (2019) Phytoparasitic nematode control of plant hormone pathways. Plant Physiol 179(4):1212–1226. https://doi.org/10.1104/pp.18.01067

Giazer I, Orion D, Apelbaum A (1983) Interrelationships between ethylene production, gall formation, and root-knot nematode development in tomato plants infected with Meloidogyne javanica. J Nematol 15(4):539

Gowda MT, Rai AB, Singh B (2019) Root knot nematodes menace in vegetable crops and their management in India: a review. Vegetable. Science 46(1 and 2):1–16

Hartley SE (1998) The chemical composition of plant galls: are levels of nutrients and secondary compounds controlled by the gall-former? Oecologia 113(4):492–501. https://doi.org/10.1007/s004420050401

Hofmann J, el Ashry AEN, Anwar S, Erban A, Kopka J, Grundler F (2010) Metabolic profiling reveals local and systemic responses of host plants to nematode parasitism. Plant J 62(6): 1058–1071. https://doi.org/10.1111/j.1365-313X.2010.04217.x

Ibrahim HMM, Ahmad EM, Martínez-Medina A, Aly MAM (2019) Effective approaches to study the plant-root knot nematode interaction. Plant Physiol Biochem 141:332–342. https://doi.org/10.1016/j.plaphy.2019.06.009

Jones MGK, Goto DB (2011) Root-knot nematodes and giant cells. In: Genomics and molecular genetics of plant-nematode interactions. Springer, pp 83–100. https://doi.org/10.1007/978-94-007-0434-3_5

Koltai H, Dhandaydham M, Opperman C, Thomas J, Bird D (2001) Overlapping plant signal transduction pathways induced by a parasitic nematode and a rhizobial endosymbiont. Mol Plant-Microbe Interact 14(10):1168–1177. https://doi.org/10.1094/MPMI.2001.14.10.1168

Kyndt T, Vieira P, Gheysen G, de Almeida-Engler J (2013) Nematode feeding sites: unique organs in plant roots. Planta 238(5):807–818. https://doi.org/10.1007/s00425-013-1923-z

Mitkowski NA, Abawi GS (2003) Root-knot nematodes. Plant Health Instructor 10. https://doi.org/10.1094/PHI-I-2003-0917-01

Mohamed HI, Abd-Elsalam KA, Tmam AMM, Sofy MR (2021) Silver-based nanomaterials for plant diseases management: today and future perspectives. In: Abd-Elsalam KA (ed) Silver nanomaterials for agri-food applications. Elsevier, pp 495–526. https://doi.org/10.1016/B978-0-12-823528-7.00031-7

Oliveira DF, Costa VA, Terra WC, Campos VP, Paula PM, Martins SJ (2019) Impact of phenolic compounds on Meloidogyne incognita in vitro and in tomato plants. Exp Parasitol 199:17–23. https://doi.org/10.1016/j.exppara.2019.02.009

Onkendi EM, Kariuki GM, Marais M, Moleleki LN (2014) The threat of root-knot nematodes (M eloidogyne spp.) in Africa: a review. Plant Pathol 63(4):727–737. https://doi.org/10.1111/ppa.12202

Palomares-Rius JE, Escobar C, Cabrera J, Vovlas A, Castillo P (2017) Anatomical alterations in plant tissues induced by plant-parasitic nematodes. Front Plant Sci 8:1987. https://doi.org/10.3389/fpls.2017.01987

Patil JA, Yadav S (2021) Root-knot nematodes a major peril to protected cultivation system in India: current status and its management. In: Nematodes-recent advances, management and new perspectives. IntechOpen. https://doi.org/10.5772/intechopen.100541

Scheller HV, Ulvskov P (2010) Hemicelluloses. Annu Rev Plant Biol 61:263–289. https://doi.org/10.1146/annurev-arplant-042809-112315

Siddique S, Grundler FMW (2015) Metabolism in nematode feeding sites. In: Adv Bot Res. Elsevier, vol 73, pp 119–138. https://doi.org/10.1016/bs.abr.2015.02.001

Sikder MM, Vestergård M, Kyndt T, Kudjordjie EN, Nicolaisen M (2021) Phytohormones selectively affect plant parasitic nematodes associated with Arabidopsis roots. New Phytol 232(3):1272–1285. https://doi.org/10.1111/nph.17549

Singh PR, Karssen G, Couvreur M, Subbotin SA, Bert W (2021) Integrative taxonomy and molecular phylogeny of the plant-parasitic nematode genus Paratylenchus (Nematoda: Paratylenchinae): linking species with molecular barcodes. Plan Theory 10(2):408. https://doi.org/10.3390/plants10020408

Umar B, Safiya M (2021) Nematicidal efficacy of soil fungal isolates against the root-knot nematode (Meloidogyne incognita) of tomato. PhD Thesis, pp 1–70. http://repository.futminna.edu.ng:8080/jspui/handle/123456789/13614

Vieira P, Engler G, de Almeida Engler J (2012) Whole-mount confocal imaging of nuclei in giant feeding cells induced by root-knot nematodes in Arabidopsis. New Phytol 195(2):488–496. https://doi.org/10.1111/j.1469-8137.2012.04175.x

Wybouw B, de Rybel B (2019) Cytokinin–a developing story. Trends Plant Sci 24(2):177–185. https://doi.org/10.1016/j.tplants.2018.10.012

Yadav U (2017) Recent trends in nematode management practices: the Indian context. Int J Eng Technol 12:482–489. https://www.irjet.net/archives/V4/i12/IRJET-V4I1290.pdf

Zhang W, Chen Y, Wang Z, Yang J (2017) Polyamines and ethylene in rice young panicles in response to soil drought during panicle differentiation. Plant Growth Regul 82(3):491–503. https://doi.org/10.1007/s10725-017-0275-2

Zhao Y (2010) Auxin biosynthesis and its role in plant development. Annu Rev Plant Biol 61:49. https://doi.org/10.1146/annurev-arplant-042809-112308

Zhou J, Li Z, Xiao G, Zhai M, Pan X, Huang R, Zhang H (2020) CYP71D8L is a key regulator involved in growth and stress responses by mediating gibberellin homeostasis in rice. J Exp Bot 71(3):1160–1170. https://doi.org/10.1093/jxb/erz491

Current and Future Studies on the Genes for Parasitism in *Meloidogyne*

5

A. Mounika, P. V. Phanindra, Uday Kumar Thera, Sparsh Tiwari, Ashmita Timsina, Mandla Rajashekar, and Lalith Pandey

Abstract

Root-knot nematodes (RKNs; *Meloidogyne* spp.) are among the major pests of economic importance causing disease in over 5500 plant species across the world incurring national agricultural yield losses up to 14.1%. RKNs circumvent the plant immune system and hijack the cell cycle and metabolism of plants abetted by various effector molecules to successfully establish feeding sites, that is, giant cells. The efficacious management of these parasites necessitates a better understanding of their genetic adaptations underlying their successful evolution of parasitism and the knowledge of associated parasitism genes. Tracing back the origin of this parasitism gene led to the proposition of many theories like horizontal gene transfer, neofunctionalization, and gene duplication. The extensive parasitism of some of the species of *Meloidogyne* might result from either macroevolutionary events like whole genome duplications and massive HGT or microevolutionary changes like gene family expansions and intragenomic

A. Mounika
Department of Plant Pathology, Indira Gandhi Krishi Viswavidyalaya, Raipur, Chattisgarh, India

P. V. Phanindra · S. Tiwari · A. Timsina
Department of Mycology and Plant Pathology, Institute of Agricultural Sciences, Banaras Hindu University, Varanasi, India

U. K. Thera (✉)
Department of Plant Sciences and Landscape Architecture, University of Maryland, College Park, USA

M. Rajashekar
Krishi Vigyan Kendra, Professor Jayashankar Telangana State Agricultural University, Hyderabad, India

L. Pandey
Department of Soil Science and Agricultural Chemistry, Banaras Hindu University, Varanasi, India

F. Ahmad, G. Nombela (eds.), *Root-Galling Disease of Vegetable Plants*,
https://doi.org/10.1007/978-981-99-3892-6_5

duplications. However, the ancestors of root-knot nematodes are still unknown, and their worldwide occurrence is far from clear. Rapidly developing omic technologies and bioinformatic tools are standing upfront in characterizing parasitism genes, their functions, and associated molecular targets in host plants. Genome sequences of highly parasitic species, effector profiling, and plant susceptible gene studies will increase our understanding in this respect. A better understanding of the functions of these parasitism genes is hindered by the absence of homologous protein databases, insufficient information on deciphered functions of these homologous proteins, or the non-amenability of these microscopic biotrophs for molecular transformation. This chapter is an attempt to put forth a detailing of parasitism genes of *Meloidogyne* spp., their origin, different signature events for adaptation of parasitism, genetic maps as well sequencing of these genes, and various techniques under use to understand parasitism genes.

Keywords

Genes · Genome · *Meloidogyne* · Parasitism · Root-knot nematode · Sequencing

5.1 Introduction

Plant parasitic nematodes (PPNs) have evolved to be the most destructive plant pathogens threatening food security. They cause considerable amount of crop losses of up to 21.3% globally, costing 1.58 billion USD annually (Kumar et al. 2020). In India, the yield losses in vegetable crops resulted from these obligate biotrophs extend up to 19.6%, worth 242.1 billion annually and an overall annual yield loss of up to 60% in horticultural crops (Gowda et al. 2017). Root-knot nematodes (*Meloidogyne* spp.) form a major group among PPNs with their ability to parasitize nearly 5500 species of different crop plants (Feyisa 2022), causing national agricultural yield loss of up to 14.1% (Jain et al. 2007).

Degenkolb and Vilcinskas (2016) estimated there to be approximately 97 different species of *Meloidogyne*, with *M. incognita, M. arenaria, M. javanica,* and *M. hapla* being the most harmful (Moens et al. 2009; Sereno. 2002). The efficacious management of these parasites necessitates a better understanding of their genetic adaptations underlying their successful evolution of parasitism. *Meloidogyne* spp. are sedentary endoparasites that, in addition to causing physical damage by stylet injection, establish specialized feeding sites in host root parenchymal cells by bringing about sophisticated cellular modifications with substantial demand for nutrients. The initiation and maintenance of these feeding sites are governed by effector molecules encoded by parasitism genes of the nematode that are prime determinants of a successful interaction. Plants are naturally resistant to nematodes and are made susceptible through complex interactions in which *Meloidogyne* spp. are pioneers by altering host gene expression.

An apparent comprehension of the origin of RKN's parasitism genes is yet to be deciphered. However, multiple theories have been put forth, showing how they would have acquired parasitism to this greater extent with substantial evidence. Events like horizontal gene transfer, neofunctionalization, and gene duplications might have played a role in this respect (Castagnone-Sereno et al. 2013). Interestingly, the highly evolved sophisticated parasitism is restricted to a few species of *Meloidogyne* genera, especially the ones with the asexual mitotic parthenogenetic mode of reproduction. This intricacy is dictated by interspecific hybridization and polyploidization events during evolution. With the rapidly advancing molecular biology, especially sequencing technology, new comprehensive studies emerged, and one such milestone in the field is the whole genome sequencing of *Caenorhabditis elegans*. High-quality genome sequences for different RKNs result from efficient sequencing technologies like Illumina, Pac-Bio, etc. Various techniques are used to identify and characterize parasitism genes. With the advent of omic technology, the physiological assessment of nematode genes involved in interaction with plants and the associated plant genes liable for alteration is in progress. Comparative genomics has much to offer in this respect, where sequence comparisons between parasitic and nonparasitic species and life stages explain the mutational events or adaptational changes in the nematode genome. The practical difficulty in amenability of these sedentary endoparasites for genetic transformation experiments owing to their obligatory biotrophy and microscopic nature is a significant limitation for their genetic studies. The importance of parasitism genes secretome profiling and their effect on the host system cannot go unnoticed, especially the proteomic studies. Studies on effector transgenes that target plant processes to interfere with giant cell formation and site-directed mutagenesis techniques like RNA interference (RNAi) and CRISPR are being used to identify the parasitism genes by linking their phenotypic expression.

Considering the importance of all these aspects of nematode parasitism, this chapter is an attempt to provide comprehension of *Meloidogyne* parasitism genes in terms of their origin, signatures of adaptation in nematode genome for plant parasitic nature, sequencing, and gene maps, as well as different adapted techniques in understanding their parasitism.

5.2 Origin of Parasitism Genes

The phylogenetic studies of the phylum Nematoda based on SSU rDNA data places *root-knot* nematodes as the highly evolved lineage among all the plant parasitic genera. RKN parasitism genes may have shared a common ancestor with those of other parasitic nematodes that feed on plants. Parasitism genes are involved in the successful infection and establishment of the pathogen on its host. These genes may govern morphological and behavioral changes or reproductive abilities in nematodes. The root-knot nematode produces cell wall-degrading enzymes, expansins, pectate lyases, cellulases, and endoglucanases that degrade and loosen the plant cell wall (Caillaud et al. 2008). However, evidence suggests that the plant parasitic ability has evolved multiple times in nematodes. The phylogenetic analyses of GHF5 cellulases present among the Tylenchida members revealed the presence of two types of domains and their coding sequences, suggesting that RKNs do not appear to have acquired these genes via a lateral gene transfer mechanism (Rybarczyk-Mydłowska et al. 2012). Moreover, these genes might have passed on from common ancestors of the family Pratylenchidae.

The other and more convincing mechanism of origin of the parasitism genes is horizontal gene transfer. It is an asexual mechanism in which genetic material movement occurs between different species irrespective of their phylogenetic relationship. For example, the polygalacturonase of the GH 28 family was often found in bacteria, fungi, and oomycetes. This enzyme is unique to *M. incognita* and has been isolated and biochemically described. The phylogenetic analysis of these enzymes shows a close relationship with that of the bacteria *Ralstonia solanacearum* (Danchin et al. 2010). It can be hypothesized that the presence of both these organisms in the same ecological niche for extended periods has resulted in the transfer of genes. In addition, pectate lyase PL3, an enzyme that causes the breakdown of β- 1,4 galactouronan, present in *M. incognita* and *M. hapla* showed a close association with that of *Clavibacter michiganensis,* pointing to the possibility that the bacterium's gene was acquired from a common origin or a close relationship. The establishment of the feeding site by the RKNs has certain similarities with that of nodules of *Rhizobia*. The gene *NodL* encoding a signalling peptide is responsible for nodule formation in the roots of leguminous plants. Lateral gene transfer (LGT) is also a source of peptide mimic IDA-like effectors. These are unique to RKNs, acting like the plant signalling peptides that cause flower abscission and lateral root emergence (Kim et al. 2018). The exact identification of the donor and the underlying mechanisms are still far from clear. However, the phylogenetic studies provide information about potential donors at the phylum and kingdom levels. The bacteria, either plant pathogens or symbionts sharing a common niche, are the most recognized potential donors. Similarly, several plant-pathogenic fungi have been recognized in donor clades. The genes acquired from these organisms through LGT in RKNs mostly have a role in parasitic interactions with the plant. The RKNs pan-genomic analysis reveals the presence of 3.34% of protein-coding genes with known and predicted functions acquired through LGT from non-metazoan animals (Paganini et al. 2012).

Gene duplications of these acquired genes or other genes already present in the genome of nematodes have promoted to gain new or more specialized functions through neo- or sub-functionalization (Castagnone-Sereno et al. 2013). The occurrence of multigene families indicates that multiple copies of genes might have favored the individuals upon positive selective pressure. For instance, *Meloidogyne incognita* and other allopolyploid species have duplicated genomes from hybridization (Schoonmaker et al. 2020). The extensive duplication and mutations within the housekeeping glutathione synthetase gene have led to the development of glutathione synthase-like effectors (Lilley et al. 2018). The expansin-like effector gene family MiMAP1 is restricted only to the *Meloidogyne* genus and includes at least seven members (Tomalova et al. 2012). These genes help soften the cell walls of plants and are secreted by either migratory or sedentary stages that indicate a possible role in establishing a feeding site (Vieira et al. 2011; Rosso et al. 2011). Variation in gene organization and number of internal repeats correlates with (a)-virulence in near-isogenic strains of *M. incognita*. It was recently shown that map-1 is part of a small gene family (Castagnone-Sereno et al. 2009). Taken together, these facts suggest that dynamic variations in repeats, genome loss, and gene duplication have been the primary drivers of the expansion of the map-1 gene family. Similarly, 30 MiMSP32 gene variants specifically restricted to root-knot nematodes of the *Meloidogyne* genus are identified (Verhoeven et al. 2022).

5.3 Signatures of Adaptation to Plant Parasitism in the Genome of Root-Knot Nematodes

The parasitic ability of *Meloidogyne*, especially *M. incognita, M. javanica,* and *M. arenaria* (collectively known as *Meloidogyne incognita* group, MIG), is considered an evolutionary paradox about the idea that asexually reproducing species are dead ends of evolution. This wide adaptability is not only attributed to the single nucleotide variations acquired by point and short-scale mutations but to other mechanisms such as epigenetics, copy number variations, and transfer of large-scale variations in genome structure (Koutsovoulos et al. 2019). The adaptability to plant parasitism in *Meloidogyne* species can be viewed at three stages, that is, the genus, species, and intraspecific levels. The genus *Meloidogyne* has some important plant parasitic genes in the genome that are acquired through horizontal gene transfers. However, an evolutionary homogenization process was required for acquired genes to function in their new genetic environment. The presence of cell wall-degrading enzymes, plant hormone mimic peptides that affect root primordia, xylanases, and arabinases, along with pectate lyases and cellulases, has an essential role in plant parasitism (Bird et al. 2015). However, the genome size of RKNs is reduced compared to the free-living nematode *C. elegans,* with a reduced gene pool that contributes to defense, detoxification, and immunization against fungus and bacteria (Abad et al. 2008). For example, simplified glutathione and a condensed set of chitinases exist in *M. hapla* and other RKNs (Abad et al. 2008; Opperman et al. 2008). Conversely, these nematodes have several novel effector-producing genes that might have originated from modification of some housekeeping genes or repeated gain of some gene portions.

In addition, epigenetic changes in the genome are governed by RNA-associated gene silencing, DNA methylation, and posttranslational histone modifications. Root-knot nematodes have conserved histone (de)acetylation and (de)methylation types of machinery, and some histone methyl transferases (HMT) known as HMT-PPN are known only in cyst and RKNs (Pratx et al. 2018). However, HMT SET domains are only present in RKNs, indicating their possible roles in plant parasitism. Of the 54 species, 32 were confined to single plant species or subclass. For example, *M. megatyla* only feeds on *Pinus* spp., *M. spartinae* only feeds on cordgrass *Spartina* spp., and *M. ichinoei* only feeds on *Iris laevigata* (Reviewed by Castagnone-sereno et al. 2013). In contrast, the MIG has characteristic hosts in each subclass and is truly polyphagous. These differences in the host preferences at the species level are attributed to the genomic differences in interspecific hybridization between two different RKN strains and the successive loss of meiosis. Hybridization allows the genomes to diverge in species so that any kind of recombination has not acted to homogenize the alleles. In addition to understanding how the *M. floridensis* genome is related to the published *M. incognita* genome, Lunt et al. (2014) proposed using a complicated double-hybridization process.

Root-knot nematodes have extreme divergence in terms of chromosome numbers (Triantaphyllou 1985). The species *M. spartianae* and *M. kikuyensis* have the smallest chromosome number of 9. However, the characteristic haploid number of the *Meloidogyne* genus is $n = 18$. Cytological evidence suggests that the genome of the species of the MIG group is polyploid, majorly triploid or hypo triploid, and several loci are present in three divergent copies. However, the careful examination of *M. incognita* and *M. arenaria* genomes has revealed the presence of only two divergent copies, and none contained a third divergent homolog. In contrast, the second copy of one of the two divergent genes has been found in all the species of MIG (Szitenberg et al. 2017). Gene duplications of specific genomic regions have a crucial role in promoting functional differences between the resulting gene copies following selection. The genome size differences between *M. incognita* and *M. hapla* and the proportion of repetitive elements between them are most probably linked with the mode of reproduction, that is, asexual in the former and sexual reproduction in the latter. *Meloidogyne incognita* has a within genomic nucleotide divergence of 7–8% that could result in functional divergence among the protein products (Abad et al. 2008). Such divergence in the genome brings plasticity in adaptation to different hosts through neofunctionalizsation. A more comprehensive study on genome assembly would accurately estimate the triploid genome proportion and the extent to which these loci differ.

The abundance of transposable elements (TE), nevertheless of their origin, in the genome improves the genomic plasticity through their active movements across the genomes. It passively promotes the shuffling of chromosomes between these regions. Compared to the 29% of the genome that TE occupied in *M. hapla,* Blanc-Mathieu et al. (2017) observed that TE made up almost 50% of the genomes of the three mitotic species. This suggests that in the absence of sexual reproduction, these regions have proliferated in the genomes due to their hybrid origin contributing to genetic diversity. Interestingly, a *Tm*1 transposon is associated with the phenotype changes of *M. javanica* and plays a role in the species' genetic variability (Gross and Williamson 2011).

The intraspecific variation occurred due to microevolutionary forces like gene family expansions and intragenic domain duplications in different *Meloidogyne* species has not only played a role in adopting to various hosts but also in the ability of some isolates of the species to multiply on selected hosts. Expression patterns of these genes vary with individuals of a species, geographical locations, and susceptibility of hosts. The resistance breakdown in the hosts has recently been reported to be associated with convergent gene copy variations (Castagnone-Sereno et al. 2019).

It is also noteworthy to mention that the differential expression patterns of parasitic genes have contributed to adaptation to host susceptibility as revealed by the prolonged expression of genes encoding cell wall-degrading enzymes, neuropeptides, and peptidases in resistant cultivars of rice as noticed in *M. graminicola* (Petitot et al. 2020).

5.4 Sequence and Genetic Map of RKNs

The genetic makeup of root-knot nematodes (RKNs) includes the following:

Whole genome sequencing, genetic mapping, protein-encoding genes, and multigene phylogeny of RKNs. These studies can provide insights into the biology, ecology, and evolution of RKNs, which are essential plant parasites causing significant damage to crop worldwide.

5.4.1 Whole Genome Sequencing of RKN

Genome sequences for 10 RKNs, including *M. incognita* (Abad et al. 2008), *M. hapla* (Opperman et al. 2008), etc., were publicly available nearly 20 years after the sequencing of the *C. elegans* genome, providing a unique chance to compare their genomes. Projects for enhancing the assemblies and annotation of all RKN genomes are in progress. Concerning *M. hapla*, genome reannotation based on mapping 2 billion RNA-Seq reads was in concurrence with most of the already available gene models with minor editing in a few (Guo et al. 2014). The *M. graminicola* genome assembly is among the smallest of any root-knot nematode identified to date (41.5 Mb). In contrast, the genome assembly of *M. arenaria* is roughly six times larger than that of *M. graminicola*, which may indicate gene duplications in this hypotriploid RKN. Many distinct nematode genome studies are now underway, and as a result, entire genome sequences for many different nematode genera and species are already available (Table 5.1). Across the whole mtDNA sequence of *M. graminicola*, the nucleotide distribution skewed toward A and T, and codon use reflects this. The proportion of A + T bases in the *M. graminicola* genome is 83.51%, and the mitochondrial genome consists of 36 genes (excluding atp8) that are transcribed in the same orientation (Sun et al. 2014). Two main features make it challenging to reconstruct the *M. graminicola* genome. First, the *M. graminicola* genome is fragile due to its low GC content (GC content = 23.5%). Second, the heterozygous nature of the genome (heterozygosity = ca. 2%) makes it challenging to assemble because of the prevalence of divergent haplotypes, particularly while using short reads (Besnard et al. 2019). Assemblies may be performed in various ways, with some homologous regions being assembled independently while others are combined to form a single consensus sequence (Besnard et al. 2019). Among the MIG species studied, *M. arenaria* and *M. floridensis* showed the highest levels of genetic diversity (Adam et al. 2014; Carneiro et al. 2008). Based on classical and molecular characterization techniques,

Table 5.1 Summarized data from *Meloidogyne* spp. genomic assemblies in DDBJ/ENA/GenBank

	Species	Isolate, strain, pathotype, genotype	Sequencing platform	Assembly approach	Number of scaffolds	Number of contigs	Total assembly length (Mb)	% Complete CEGMA (C) (copies) %Partial (P)	GC %	Assembly level	References
1	*Meloidogyne arenaria*	A2-O	PacBio RSII	CANU v. 1.3	–	2224	281.68	C: 94.76 (3.57) P: 96.77	30.1	Contig	Sato et al. (2018)
2	*M. chitwoodi*	PNW-race 1	PacBio Sequel RSII	SMRT v. 2.3.0	38	30	47.48	C: 94.29 (2.23) P: 98.78%	24.8	Contig	Humphreys-Pereira & Elling (2014)
3	*M. enterolobii*	Swiss pop	Illumina HiSeq and PacBio RS II	Platanus v. 1.2.4	4437	4451	240	C: 94.76 (3.30) P:96.77	30	Scaffolds	Koutsovoulos et al. (2020)
4	*M. exigua*	Mex1	Oxford Nanopore technology	CANU v. 1.8	206	206	42.1	C: 95.97% (C + P: 97.18%)	25.6	Contig	Phan et al. (2021)
5	*M. floridensis*	SJF1	Illumina HiSeq SJF1	De-novo, Platanus	8887	13,362	74.84	C: 77.42 (1.71) P: 76.61	30	Scaffolds	Szitenberg et al. (2017))
6	*M. graminicola*	IARI strain	Illumina HiSeq; Oxford Nanopore technology	Canu v. 1.8	283	286	41.55	C:84.27 (1.34) P: 90.73	23.2	Scaffolds	Somvanshi et al. (2018)
7	*M. hapla*	VW9	ABI3730, megabase sequence analyzer	De-novo, Arachne v2.0.1	3450	3450	53.01	C: 93.55 (1.19) P: 95.56	27.4	Contig	Opperman et al. (2008)

8	*M. incognita*	Kmmt_Gs004	PacBio Sequel	Falcon-kit v. 1.4.2	374	374	193.2	C: 94.76 (2.93) P: 96.77	29.5	Contig	Asamizu et al. (2020)
9	*M. javanica*	Avignon	Illumina VW4	De-novo, Platanus	34,316	38,690	150.35	C: 89.52 (2.71) P: 95.16	29.9	Scaffolds	Szitenberg et al. (2017)
10	*M. luci*	SI-Smartno V13	Illumina HiSeqX		327	327	209.16	C: 95.56 (2.92) P: 96.77	30.2	Contig	Susič et al. (2020)

many more RKN species are expected to be included within the MIG phylogenetic cluster (Pagan et al. 2015; Holterman et al. 2009).

Genome sequences often feature gaps, sometimes thousands of them, with the significant exception of *Caenorhabditis elegans*, whose number of scaffolds is equal to its number of chromosomes, which is equal to its number of linkage groups. Because of the cloning and sequencing methods that have been applied, it is feasible to make a reliable estimate of the gap sizes, particularly for *M. hapla*, which is usually as little as 1 nt. It is vital to include specific measures to estimate the quantity of the genome in the assembly and coverage until technological advancements allow gapless assembly (which is anticipated to take place shortly). A value is known as the "scaffold N50," and to a limited extent, the contig numbers are frequently used to assess the assembly (Table 5.1). Contigs are ranked by size, with N50 corresponding to the position in the list, where the total among all larger contigs on the list is equal to half of the estimated size of the genome. The N50 estimates that higher tend to indicate more robust assemblies. The Core Eukaryotic Genes Mapping Approach (CEGMA) is an approved standard used to evaluate the assembly's completeness and accuracy by mapping the genes in the genome (Parra et al. 2007, 2009). The probability of detecting a gene can be inferred from the proportion of the complete CEGMA complement observed in a particular genome assembly, which is a proxy for assembly quality that considers the number of gaps. This supports the idea that a genome coverage estimate may be obtained using CEGMA (Parra et al. 2007, 2009). The recently completed *M. chitwoodi* genome assembly and its CEGMA score of 99% indicates nearly complete genome coverage.

5.4.2 Genetic Mapping of RKNs

Physical mapping of nematode genomes and various cloning procedures are used in conjunction with forwarding genetic approach investigations of RKN interactions with resistant plant genotypes to isolate parasitism genes (Bird et al. 1999). There are gaps in genetic and physical maps corresponding to low recombination regions or the genome regions that are challenging to clone and sequence or with less-frequent polymorphic markers. It is often difficult to confirm and establish how the genetic linkage groups correspond to physical chromosomes. Genetic map building of *M. hapla* with 15 linkage groups was made possible by analyzing polymorphism in 293 AFLP markers based on segregation in 183 F2 lines. The sum of the genetic distances of markers in resulted linkage groups is 771 cM. From this, Opperman et al. (2008) estimated a total genetic distance of ~1000 cM, corresponding to an average of ~50 kb/cM. Therefore, integrating genetic analysis with physical maps of RKN genomes will help to extract nematode (a)virulence genes. However, it must be shown whether these genes reflect an altered portion of parasitism genes or serve some other purpose unrelated to plant parasitism.

5.4.3 Protein-Encoding Genes

Estimates of the total number of genes encoding proteins are affected not only by the actual biological sources of variation but also by the annotation technique and genomic contiguity. Fragmentation of anticipated coding sequences due to low-quality genomes can lead to an inflated gene count, whereas improper assembly may cause a decreased gene count (Koutsovoulos et al. 2020). Some gene families are significantly more abundant in the *C. elegans* genome than in the RKN genomes (Opperman et al. 2008; Abad et al. 2008), which may indicate a higher necessity for these capabilities in the niche of *C. elegans*. The G protein-coupled receptor family (GPCR: 1011 genes) is the most prominent gene family of *C. elegans* (Robertson and Thomas 2006; Bargmann 2006), although this family is severely reduced in the *M. hapla* gene repository (147 genes). Compared to *C. elegans*, *M. hapla* has a smaller number of genes, suggesting that *C. elegans* has expanded many large families into much larger ones and many specific genes into small families. On the note is that the GPCR constitutes the most prominent gene family in *C. elegans*, with 1280 genes. Only 147 GPCRs are encoded by *M. hapla*. In *C. elegans*, many GPCRs are smell sensory receptors, reflecting the requirement to forage for nourishment in its complex soil habitat. Similarly, In *C. elegans*, around 180 cuticle collagens gene family members are divided into six subfamilies (Page and Johnstone 2007). There are only 81 collagens encoded by *M. hapla*. Thus, in total, *M. incognita* encodes 122 collagens and 108 GPCRs. Protein-coding sequences ranged from 14,144 in *M. floridensis* to 30,308 in *M. arenaria* among the apomicts (Table 5.1). *M. hapla* possesses 14,700 protein-encoding genes in its homozygous genome (Opperman et al. 2008), which is quite close to the number expected in the largely homozygous *M. floridensis*. In addition, MiMsp40 is a novel *Meloidogyne* immunomodulatory effector released by early parasitic stages of the nematode into plant cells that suppresses PTI and ETI signals to facilitate RKN parasitism (Niu et al. 2016). The novel effector, MgGPP, is specifically expressed in the nematode sub ventral esophageal gland cells and upregulated in the early parasitic stage of *M. graminicola* (Chen et al. 2017).

5.4.4 Multigene Phylogeny of RKNs

A strongly supported superior clade (Álvarez-Ortega et al. 2019) (PP and BS = 100%) was revealed in the phylogenetic tree generated (Fig. 5.1) from the D2-D3 of 28S rRNA gene sequence analysis (42 *Meloidogyne* species, 791 bp), of four major *Meloidogyne* clades (*M. spartelensis-M. hapla; M. arabicida-M. inornata; M. trifoliophila-M. minor; M. graminis* with *M. maylandi*) and the clade with *M. christiei*. The remaining nematodes were classified into six clades (*M. artiellia-M. oleae; M. mali; M. daklakensis-M. aberrans; M. camelliae; M. indica* with *M. nataliei; M. africana*). *M. nataliei*, a sister species to *M. indica*, was close to the bottom of the genus and displayed similar, possibly ancestral traits. Molecular evidence contradicts Goldstein and Triantaphyllou (1986) hypothesis that the grape

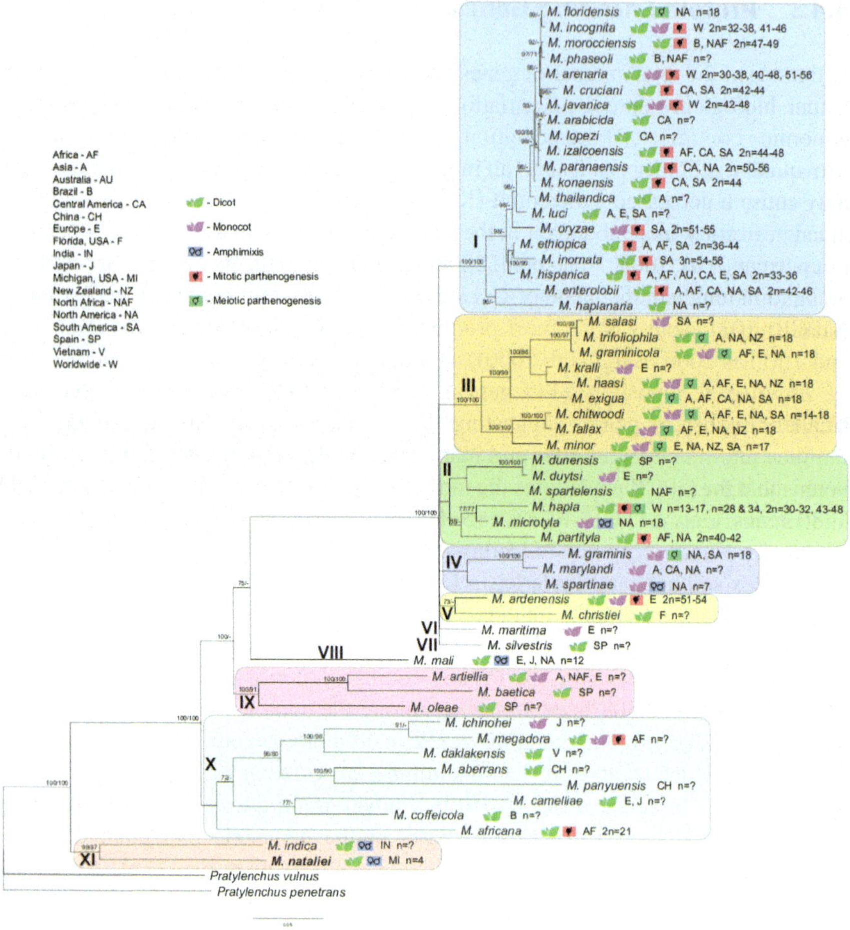

Fig. 5.1 The phylogenetic tree generated from the D2-D3 of 28S rRNA gene sequence analysis (42 *Meloidogyne* species, 791 bp) (n =?—chromosome number information unknown) (Álvarez-Ortega et al. 2019)

root-knot nematode collected from Michigan is not a *Meloidogyne* species. Based on molecular findings, Phani et al. (2018) molecularly characterized *M. indica* and suggested that this RKN species should be regarded as the most primitive taxon of the genus.

5.5 Adapted Techniques for Understanding the Genes in Parasitism

Understanding nematode parasitism necessitates a better understanding of the genes governing it and the associated molecular triggers. The genes governing a nematode's life cycle processes, although they do not involve parasitism directly, have a more significant indirect role concerning genes associated with esophageal gland secretions that drastically increase during parasitism governing them. So, the knowledge of parasitism genes and the associated effectors are prime necessities that go hand in hand and are required to decipher the mechanisms behind the successful establishment of *Meloidogyne* species as plant parasites.

5.5.1 Molecular Genetic Techniques

With the advent of molecular techniques, a massive redirection in approaches to understanding parasitic nematode interaction with host plants has resulted from conventional microscopic techniques. To date, an indispensable technique in this field of study is q-PCR, a gold standard for specific detection and quantification of target nucleic acids. This technique has been used to study several parasitism genes of different *Meloidogyne* spp. for their developmental expressions. For instance, a parasitism gene of *M. incognita, Mi8D05,* and the corresponding target protein-encoding gene, tonoplast intrinsic protein 2 (tip2; AY731066) in tomato plant was studied by RT-PCR technique using the gene-specific primers designed (Xue et al. 2013).

Nevertheless, identification of such specific parasites and associated reproductive differentials is also of prime importance for understanding species-specific parasitism. Molecular markers best serve this purpose by enabling deciphering of the parasitic species complexes and are also widely used in resistance breeding programs by Marker Assisted Selection (MAS). A PCR-based method called amplified fragment length polymorphism (AFLP) fingerprinting enables a comparative examination of the *M. incognita* population, including near-isogenic lines with reproducible differential ability on *Mi*-resistant tomatoes. Few DNA fragments were documented as reproductive differentials between avirulent and virulent lines (Semblat et al. 2001). The differential expression of one of those fragments, designated as map-1, specifically restricted to avirulent lines, was further confirmed by RT-PCR studies (Semblat et al. 2001). The *Mj-1* locus conditioning resistance to *M. javanica* in an inbred carrot line, Brasilia-1252, was analyzed for linked randomly amplified polymorphic DNA (RAPD) markers to put forth a linkage map encompassing the locus (Boiteux et al. 2000). Alongside the benefits offered, related limitations are inevitable regarding the failure of qPCR to detect species beyond the used primers and high precision, standardized protocols, and complex detection systems of molecular markers if considered.

5.5.2 Omic Approaches

Omic studies based on genes and proteins stand upfront in gathering information on the genes and proteins associated with nematode parasitism and subsequent host resistance. Rapidly becoming available molecular databases of different *Meloidogyne* spp., novel genomic tools, sequencing technologies, bioinformatic tools, and nematode secretome analytical techniques play critical roles in understanding the nematode plant interactions. Modern genome-editing techniques can precisely relate the parasitism genes to their phenotypic performances.

5.5.2.1 DNA Level

Gene mapping is usually the preliminary study of genes that directs toward the downstream understanding of an organism's genome. The mapping of genes to know their relative locations on the genome and underlying biological functions has evolved from conventional cloning to genome sequencing and computational analysis. Linking genetic mapping with long read sequencing enables detecting and characterizing parasitism genes in nematodes. Mapping of *Meloidogyne* genes explicitly expressed in secretory gland cells provided information on the evolutionary conservation of effectors. The alignment of retrieved gene coding sequences (CDS) to *M. incognita* genome sequence and four other *Meloidogyne* species, viz., *M. enterobolii, M. arenaria, M. hapla,* and *M. javanica* using the splice aware aligner SPALN presented the effector lineage. Clade I *Meloidogyne*, to which *M. arenaria, M. javanica*, and *M. enterobolii* also belong, is responsible for the inheritance of almost 87% and 82% of encoded proteins from the sub-ventral and dorsal glands of *M. incognita*, respectively. (Da Rocha et al. 2021). Identifying and characterizing target protein-encoded genes of nematode parasitism genes in host plants is equally important to understand their interactions better. Numerous wild plant species harbor natural host resistance that does not suppress the development and reproduction of *Meloidogyne* spp. (Roberts 1995). Three dominant resistance genes, designated *Mi-1*, *Mi-3*, and *Mi-9*, against *M. incognita* have been mapped on tomato chromosomes (Kaloshian et al. 1998; Ammiraju et al. 2003; Yaghoobi et al. 1995).

5.5.2.2 DNA Sequencing

Sequencing of DNA presents genetic information of a specific DNA segment or even the whole genome. The gene sequence information can screen for parasitism genes, their characteristic features, and regulatory elements present in that target DNA. More importantly, sequencing forms the basis for comparative studies between different stages, that is, parasitic, and pre-parasitic, and different organisms. Remarkably, sequencing discloses the changes in genes that may determine nematode parasitism. The successful whole genome sequence of *Caenorhabditis elegans* revolutionized the field of genome sequencing, after which many PPN genomes have been sequenced. Different sequencing techniques, viz., Sanger sequencing, Illumina Hi-sequencing, PacBio sequencing, Oxford Nanopore technology, ABI3730 megabase sequence, etc., are used for DNA sequencing of various

Meloidogyne spp. as presented in Table. 5.1. As a result, high-quality genome sequences are available for major *Meloidogyne* spp. (Blanc-Mathieu et al. 2017; Szitenberg et al. 2017; Sato et al. 2018; Mani et al. 2021), viz., *M. arenaria, M. enterolobii, M. incognita, M. javanica*, and *M. floridensis,* as well as for the less-distributed ones, viz., *M. luci* (Susič et al. 2020), *M. enterolobii* (Koutsovoulos et al. 2020), *M. exigua* (Phan et al. 2021), *M. chitwoodi* (Bali et al. 2021), and *M. graminicola* (Somvanshi et al. 2018). In this respect, a whole genome shotgun technique of *M. arenaria* presented the long read-based assembly directing the identification of parasitism-related genes that are frequently encountered in highly variable and repeat-rich regions (Sato et al. 2018).

5.5.2.3 Comparative Genomics

Comparative genomics involves comparing the genetic material of two organisms to understand the evolutionary changes between them, which enables the identification of conserved and novel genes. Such comparative studies between parasitic and nonparasitic nematodes or between stages of a nematode offer a better understanding of the associated genes governing parasitism. Using *C. elegans*, a free-living worm, to study the genomes of two root-knot nematodes, *M. hapla,* and *M. incognita*, researchers could deduce that the parasitic species contain complexes of enzymes that mainly target the host plant, explaining their parasitic success (Bird et al. 2015). Phylogenetics is an efficient comparative genomic tool to investigate evolutionary changes in genes.

Phylogenetics has been used to decipher the evolution of nematode parasitism by making informative genomic comparisons between free-living and parasitic species. A phylogenomic comparison of different nematodes, including parasitic and free-living, identified more than 24,000 families of proteins explicit to the parasites, with *M. incognita* constituting 10,000 proteins orthologous to those of phytoparasitic species. Of these, 1000 proteins were found to be like the prior identified secreted effectors with an indispensable role in nematode parasitism (Grynberg et al. 2020). The expansin-like proteins determined by the *map-1* gene family aid in successfully establishing RKN in plant roots. In contrast to *M. floridensis*, phylogenetic analyses of the distribution of *Meloidogyne*-specific genes (i.e., *map-1* genes) show that they are only present in species that reproduce through mitotic parthenogenesis, an evolutionary deviation between meiotic and mitotic RKN species (Tomalova et al. 2012).

5.5.3 RNA Level: Transcriptomics

The studies based on RNA, offer much reliable information about the gene expression patterns, and enable us to better understand the parasitism of RKNs and the underlying molecular events. Several techniques have been devised with subsequent increase in efficiency like in situ hybridization and microarrays. The RNA sequencing and site-directed mutagenesis-based techniques have added a wealth of information on parasitism genes of RKNs and have become indispensable in the present-day research.

5.5.3.1 ISH

In situ hybridization (ISH) technique allows the localization of a nucleic acid segment in the histological sections. The target nucleic acid is detected utilizing complementary probes tagged with a reporter molecule, thereby its localization. The probes could either be DNA or RNA, but RNA probes (riboprobes) are more common owing to their strong binding to the targets and offer an advantage in assessing the gene expression levels. The in situ hybridization experiments localized the expression of a candidate effector protein encoded by *M. graminicola* in sub-ventral glands, which are very active during migratory and pre-parasitic stages of the nematode. This candidate effector protein was thus identified to play a vital role during the early parasitic stages of *M. graminicola* (Naalden et al. 2018). Fluorescent in situ hybridization (FISH) is an ISH technique that uses fluorescently labelled probes for complementary nucleic acid localization that can be visualized under a fluorescent microscope. The TAG lipase in *M. javanica* was localized in dorsal and sub-ventral glands using FISH employing a Cy5-probe that precisely unveils the spatiotemporal expression of many candidate-effector encoding genes carrying a signal peptide (Fitoussi et al. 2021).

5.5.3.2 Microarrays/cDNA Microarray

Microarrays use an array of nucleic acid molecules fixed onto a surface bathed with a test sample allowing complementary base pairing. The chip-immobilized nucleic acid molecules are fluorescently labelled and specifically bind to the corresponding complementary molecules producing detectable light through fluorescence. This technique thus can be used in comparative genomic hybridization, analysis of quantification of parasitism genes, and their differential expression patterns. RKNs alter host gene expression to establish a feeding site successfully. To examine soybean (*Glycine max*) gene expression in RKN-induced gall tissues, an Affymetrix Soybean GeneChip constituting 37,500 *Glycine max* probe sets was used by integrating the gene expression patterns with biochemical pathways. It was observed that genes expressing enzymes associated with the cell wall and carbohydrate metabolism, genes monitoring cell cycle, and those related to plant defense were differentially expressed (Ibrahim et al. 2011). However, the microarrays also tend to result in cross-hybridization patterns, inefficient quantification of over- and underexpressed genes, and the prior sequence information.

5.5.3.3 RNA-Seq Based

RNA-sequencing (RNA-seq) technique uses high-throughput sequencing methods like next-generation sequencing (NGS) to provide information on transcriptome regarding its presence and quantity. It facilitates analysis of the posttranscriptional modifications, alternative gene spliced transcripts, mutations, gene fusions, and continuous changes in gene expression. It ensures excellent coverage and resolution of transcriptome nature compared to microarray-based techniques. This technique has dramatically upgraded the knowledge of *Meloidogyne* parasitism in various aspects, a few exemplified here. A spatiotemporal RKN gene expression analysis was conducted using RNA sequencing to identify biological signatures at different transitionary developmental stages. A motif by name, Mel-DOG, was identified which is noncoding and explicitly abundant in the effector genes promoter regions with specific functions related to pathogenicity such as CAZymes. Mel-DOG is suggested to transcriptionally regulate degrading or modifying enzymes involved in tissue maceration during nematode penetration (Da Rocha et al. 2021). Similarly, in an RNA-sequencing study of RKN, *M. graminicola*-induced large cells in rice roots revealed a systemic upregulation of primary metabolism. Significant downregulation of defense-related genes and overexpression of genes involved in photosynthesis, tetrapyrrole synthesis, and chloroplast biogenesis were seen in large cells (Ji et al. 2013).

Differentially expressed transcripts may be found and cloned with RNA fingerprinting. To investigate the metamorphosis from the nematode's pre-parasitic to its parasitic condition, differential gene expression analysis was performed (Ding et al. 1998, 2000). With the use of RNA fingerprinting, Ding et al. (2000) were able to determine that *M. incognita* produces a cDNA-encoded protein (MI-MSP-1) that is structurally like the allergen AG5. Micro-aspiration was initially used to obtain extraneous tissue contamination-free contents of nematode esophageal gland cells (Shields et al. 1998), where a new process is being developed to extract cDNA from single cells using RT-PCR (Karrer et al. 1995). More than 40 parasitism genes are identified in *M. incognita* by the esophageal gland cell micro-aspiration technique along with transcript mining assays (Huang et al. 2003, 2004).

The dual RNA-seq technique simultaneously enables transcriptome analysis by sequencing nematodes and root tissues. The parasitism genes and encoded secretory proteins of *M. chitwoodi* were transcriptionally analyzed by dual RNA-seq, substantially reducing the list of genes to be studied to encode secretum (Roze et al. 2005). Few genes expressed during the early parasitic stages of *M. chitwoodi* were analyzed by this technique (Zhang and Gleason 2021).

5.5.3.4 Subtractive Hybridization

Subtractive hybridization is a technique to specifically study the expression of genes in particular cell types or tissues or even at a definite stage of development of an organism. This technique removes the common nucleotide sequences between comparative organisms, thereby discerning the different sequences. Huang et al. (2004) used this solid-phase subtractive hybridization technique to identify candidate parasitism genes expressed in esophageal gland cells of *M. incognita*. Subtraction of gland cell cDNA library constituting 1000 clones with already cloned genes of parasitism removed 89 cDNA clones enabling effective identification of new candidate parasitism genes attributed to having a role in *M. incognita* parasitism.

5.5.3.5 cDNA Libraries and ESTs

The cDNA libraries constitute the active transcribing regions synthesized from mRNAs by cloning them into a suitable vector that is then transferred to the host.

Their construction can also be done effectively by a new technique, suppression subtractive hybridization (SSH; Diatchenko et al. 1996, 1999), which significantly boosts the low quantity of cDNAs produced from variably expressed mRNAs. Utilizing this method, Liang et al. (2004) identified differentially expressed genes and genes controlled by symbiosis by constructing a cDNA library containing a gene whose expression was shown to be different between two cell lines (Voiblet et al. 2001; Morales and Thurston 2003). The ESTs are immediate information on transcriptomes being used in gene discovery. These are single-shot sequence reads at 3′ or 5′ ends of cDNAs that are individual clones from a cDNA library and represent the portions of expressed genes. The ESTs generated from cDNA libraries are assembled into clusters and contigs. The sequences are submitted to dbEST-database for "expressed sequence tags as individual reads or to GenBank if assembled through the Transcriptome Shotgun Assembly."

Dautova et al. 2001 investigated the expression of genes in *M. incognita* at the onset of parasitism by generating ESTs from a cDNA library of fresh pre-parasitic J2s and produced sequence for candidate parasitism genes along with generating ESTs for all parasitism genes reported till date. The clustering and sequence analysis resulted in 5832 ESTs with protein lengths ranging from 150 to 299 amino acids. They showed transmembrane regions and their orientation for 4024 clusters that could be the novel target genes for nematode control (Kang et al. 2010).

5.5.3.6 Site-Directed Mutagenesis: RNAi and CRISPR

Gene functional studies by knockout experiments and ectopic expression mostly rely upon the genetic transformation of the organisms, which has been difficult in nematodes owing to their less-conducive biological nature (Eves-van den Akker et al. 2021). These microscopic and obligately biotrophic nematodes with very few approachable immature germlines render genetic transformation techniques difficult (Kranse et al. 2021). However, with RNA interference (RNAi) and CRISPR that bring about site-directed mutagenesis, a wealth of information on functional analysis of nematode parasitism genes is being put forth.

5.5.3.6.1 RNAi

RNA interference is a posttranscriptional gene silencing technique directed by dsRNA molecules and an argonaut, a catalytic component of RNA-induced silencing complex (RISC). The discovery of this technique in *C. elegans* revolutionized the gene function analysis in PPNs (Fire et al. 1998). This could be in vitro where nematode parasitism genes under target can be silenced or *in planta* where host plants are genetically transformed to encode dsRNA molecules having sequences of the target gene to correlate the phenotypic effect to specific silencing directly. The secretory product of the RKN parasitism gene, *16D10,* promotes root growth and acts as a ligand for a putative transcription factor. The silencing of this gene by RNAi technique by ingestion of encoded dsRNA into RKN declined the infectivity of nematode. The in vivo expression of dsRNA encoded by the *16D10* gene in the Arabidopsis plant resulted in resistance against four major RKN species, viz., *M. arenaria, M. incognita, M. javanica, or M. hapla* (Huang et al. 2006). When expressed in tomato hairy roots, the RNAi assay of a hairpin molecule of the *M. javanica* gene, mj-far-1, reduced nematode infection levels. Due to defects in female development, there was a dramatic decrease in the number of giant cells (Iberkleid et al. 2013). Knocking down the *Mi-Rpn7* gene of *M. incognita* by RNAi technique resulted in specific transcript absence, subsequently causing episodic locomotion of juveniles in the pluronic gel medium used for attraction assay Niu et al. (2016).

5.5.3.6.2 CRISPR

CRISPR-Cas9 is an efficient alternative to RNAi for studying genetically interacting nematode-host systems. This gene-editing system includes precise excision of genes by cas9 enzyme guided by clustered repeat interspaced palindromic repeat (CRISPR) sequences and allows take over by natural repair process resulting the changes. The CRISPR technique helps to identify the targets for nematode effectors and associated functions by altering the expression of nematode resistance genes in plants. SlWRKY45 is a transcription factor that interacts with a critical repressor of jasmonic acid signalling, that is, jasmonic acid-ZIM domain family proteins (JAZ). Mutants of slwrky45 generated by CRISPR/Cas9 technology in tomatoes showed increased resistance to *M. incognita*. The slwrky45 mutants showed a decrease in gall numbers and number of eggs per gram of roots compared to the wild type (Huang et al. 2022). Several software has been created specifically for designing CRISPR experiments for plant parasitic nematodes. CRISPR as an emerging tool significantly adds to the knowledge about parasitism of plant nematodes and is yet to be exploited for various *Meloidogyne* spp.

5.5.3.7 Bioinformatics

Bioinformatics includes the computational tools that collect, store, analyze, and disseminate the biological data of nucleic acid sequences, amino acid sequences, or annotations about them. The use of a bioinformatic pipeline of five tools to predict the excretory/secretory proteins in the genome of *M. incognita,* viz., Phobius (Käll et al. 2004), SignalP (Petersen et al. 2011), TMHMM (Kahsay et al. 2005), SecretomeP (Bendtsen et al. 2004), and TargetP (Emanuelsson et al. 2000). Mani et al. (2021) used Blast2GO tool to obtain functional annotations of transcriptomes sequenced at various stages of *M. incognita* development and their comparison to identify potential regulatory networks.

5.5.4 Protein Level: Proteomics

The knowledge of the effector proteins encoded by parasitism genes, their corresponding targets, and interactions between them is equally essential to understand their function in the establishment of a successful parasitic relationship with the host. This has become possible with the advent of different techniques like western blotting, immunological techniques, yeast two-hybrid screen system,

LC-MS, and others. The spatiotemporal analysis of parasitism gene expression patterns is being accounted with techniques like GUS reporter system.

5.5.4.1 Immunological Techniques

Immunological techniques are also widely used to localize and study the expression of known target proteins using specific antibodies. The putative function of the map-1 gene was understood using antibodies generated against the protein's representative of amphidal secretions of J2s of *M. incognita* aided by immunofluorescence microscopy experiments. The study suggested the possible role of map-1 in recognition events during plant-nematode interaction (Abad et al. 2003). A similar study by Huang et al. (2006) used polyclonal antiserum produced by immunized rabbits against Mi8D05 encoded product to localize expression of 8D05 in *M. incognita*. The co-immunoprecipitation assay is quite a popular technique that identifies the physiologically relevant protein-protein interactions using antibodies specific to the target protein molecules to capture the proteins bound to the target protein indirectly. Analyzing such protein complexes in nematode-host interaction systems ensures the identification of novel binding partners, their affinities, and their associated functions. Plant immune responses are dictated by several factors, of which transmembrane receptors play a more significant role. FERONIA is a receptor-like kinase playing a role in stress-related responses and cell growth in plants and has peptide ligands called rapid alkalinization factors (RALFs). A mutation of FERONIA resulted in reduced susceptibility of *Arabidopsis thaliana* to RKN, *M. incognita*. To better understand the underlying mechanism, several assays of which co-immunoprecipitation assay confirmed the interaction of FER with RALF-like peptides, viz., MiRALF1, AtRALF1, and MiRALF3. An anti-FLAG antibody (DYKDDDDK Tag [D6W5B] rabbit mAb) and His-Tag (2A8) Mouse mAb were used to detect FER-FLAG and RALF-HIS proteins (Zhang et al. 2020).

5.5.4.2 Yeast Two-Hybrid Screens

Nematode effectors influence various plant cellular processes, and these interactions have been isolated using yeast two-hybrid screens (Gheysen and Mitchum 2011; Rosso et al. 2011). The host targets of a specific effector, MiEF1, restricted to feeding cells of phytopathogenic nematodes, were investigated in Arabidopsis by a yeast two-hybrid approach. The cytosolic glyceraldehyde-3-phosphate dehydrogenases (GAPCs) and universal stress protein (USP) were identified to be targets of MiEFF1 (Truong et al. 2021). Yeast two-hybrid screens, in combination with co-immunoprecipitation assays, exposed the interaction of a novel effector encoded by *M. graminicola* conspicuously during the third or fourth stages of its parasitic life cycle with three endogenous proteins in rice plants. The proteins characterized were cysteine-rich repeat secretory protein 55 (OsCRRSP55), 1,3-β-glucan synthase component (OsGSC), and pathogenesis-related BetvI family protein (OsBetvI) suggested to have a role in host defense, in turn, implying the role of MgM0237 in nematode parasitism (Chen et al. 2018).

5.5.4.3 Techniques for Effector Studies

Nematodes are not incredibly conducive for transformation experiments owing to their obligate biotrophic nature, so plant transformation with effector encoding genes could be a better option to study the transient expression of effectors that elucidates the biology of nematode interaction with plants. Techniques discussed above are widely used to identify and characterize the nematode effectors interfering with plant defense systems, viz., immunolocalization techniques and yeast two-hybrid screening. Few other techniques, like LC-MS, reporter genes/proteins, western blotting, APEX, FRET-based techniques, etc., add a wealth of information to decipher the plant-nematode interactions as discussed below.

Gus reporter system studies the activity of a gene transcription promoter either quantitively or qualitatively, localizes the intracellular gene product, aids in detecting protein-protein or protein- DNA interactions, and efficiently determines gene delivery systems. Gus-promoter fusion constructs allow spatiotemporal analysis of gene expression changes in nematode-induced giant cells. One such study by Fitoussi et al. (2021), based on the GUS system, demonstrated the induction of oxylipin biosynthesis genes, *OPR2*, *α-DOX1*, *AOS1,* and *LOX1.2* in two-week-old hairy root lines of tomato plants on *M. javanica* infection (Fitoussi et al. 2021).

The liquid chromatography-mass spectroscopy (LC-MS) technique is a synergistic combination of the physical separation ability of liquid chromatography and the mass-analysis aspect of mass spectroscopy, enabling the identification of each separated compound. LC-MS technique, in combination with two-dimensional electrophoresis, identified 222 differently abundant proteins in wild Arachis as a response to RKN, *M. arenaria* infection that might have a role in the synthesis, folding, degradation, and posttranscriptional modifications necessary for cell physiological function maintenance and redox homeostasis (Martins et al. 2020).

The yeast signal sequence trap system is another powerful technique to study the dynamics of nematode-host interactions. The nematode host system involves the fusion of cDNAs of nematode effector molecules with the invertase-reported yeast gene. Thus, a vector containing the *SUC2* gene without the signal sequence and the start codon might be used to turn the resulting fusion library into an invertase-deficient yeast strain. When these transformants are plated onto sucrose solution, the transformants constituting cDNA of a secreted protein can rescue the mutants. Then the plasmid DNA can be sequenced for secreted protein identification. This system found application in validating the association of a signal peptide with *M. incognita* effector, *MiISE6* (Shi et al. 2018).

5.6 Conclusion and Further Prospective

The knowledge of plant nematode interactions is essential to devise novel control strategies. The genes governing RKNs parasitism are not understood to govern any functions apart from parasitism or are representative subsets of modified parasitism genes (Davis et al. 2000). A striking similarity between cellulase encoding genes in plant parasitic nematodes and a few microbial genes questioned their origin, leading to the proposition of many theories like HGT, followed by gene duplications and neofunctionalization. The mechanism underlying the extensive parasitism of some of the species of *Meloidogyne* spp. might be either macroevolutionary events like whole genome duplications and massive HGT or microevolutionary changes like gene family expansions and intragenomic duplications. However, the ancestors of root-knot nematodes are still unknown, and their worldwide occurrence is far from clear. Progress in omic technologies and bioinformatic tools are boosting the information on secretome complexes, genetic changes, and pathways associated with RKN parasitism altering host gene expression and response. Genome sequences of highly parasitic species, along with the effector profiling and plant susceptible gene studies, will increase our understanding in this respect. Despite the more significant application of novel techniques like RNAi and CRISPR in other organisms, the in vivo studies in PPN, especially the sedentary endoparasites like RKNs, are very much limited attributed to the practical difficulties in handling these microscopic obligate biotrophs. Though distantly related, insights into parallel defense evolution mechanisms in plants and animals against pathogens might also answer a few questions related to parasitic nematode functions (Dubreuil et al. 2007). Overall, the knowledge of the genetic basis of RKN-plant interaction is gradually adding up with the improving technologies; however, it demands an increased research focus to develop efficient means to understand better the mechanisms underlying their parasitism.

References

Abad P, Favery B, Rosso MN et al (2003) Root-knot nematode parasitism and host response: molecular basis of a sophisticated interaction. Mol Plant Pathol 4:217–224. https://doi.org/10.1046/j.1364-3703.2003.00170.x

Abad P, Gouzy J, Aury JM et al (2008) Genome sequence of the metazoan plant parasitic nematode *Meloidogyne incognita*. Nat Biotehnol 26(8):909–915. https://doi.org/10.1038/nbt.1482

Adam M, Heuer H, Hallmann J (2014) Bacterial antagonists of fungal pathogens also control root-knot nematodes by induced systemic resistance of tomato plants. PLoS One 9(2):e90402. https://doi.org/10.1371/journal.pone.0090402

Álvarez-Ortega S, Brito JA, Subbotin SA (2019) Multigene phylogeny of root-knot nematodes and molecular characterization of *Meloidogyne nataliei* Golden, Rose & Bird, 1981 (Nematoda: Tylenchida). Sci Rep 9:11788. https://doi.org/10.1038/s41598-019-48195-0

Ammiraju S, Veremis C, Huang X et al (2003) The heat-stable root-knot nematode resistance gene mi-9 from *Lycopersicon peruvianum* is on the short arm of chromosome 6. Theor Appl Genet 106:478–484. https://doi.org/10.1007/s00122-002-1106-y

Asamizu E, Shirasawa K, Hirakawa H, Iwahori H (2020) Root-knot nematode genetic diversity associated with host compatibility to sweetpotato cultivars. Molecular Plant Pathology 21(8):1088–1098. https://doi.org/10.1111/mpp.12961

Bali S, Hu S, Vining K et al (2021) Nematode genome announcement: draft genome of *Meloidogyne chitwoodi*, an economically important pest of potato in the Pacific northwest. Mol Plant-Microbe Interact 34(8):981–986. https://doi.org/10.1094/MPMI-12-20-0337-A

Bargmann CI. (2006) Chemosensation in *Caenorhabditis elegans*. Worm Book: the online review of *Caenorhabditis elegans* biology [Internet]

Bendtsen JD, Jensen LJ, Blom N et al (2004) Feature-based prediction of non-classical and leaderless protein secretion. Protein Eng Des Sel 17(4):349–356. https://doi.org/10.1093/protein/gzh037

Besnard G, Phan NT, Ho-Bich et al (2019) On the close relatedness of two rice-parasitic root-knot nematode species and the recent expansion of *Meloidogyne graminicola* in Southeast Asia. Genes 10(2):175. https://doi.org/10.3390/genes10020175

Bird DM, Jones JT, Opperman CH et al (2015) Signatures of adaptation to plant parasitism in nematode genomes. Parasitology 142(S1):S71–S84. https://doi.org/10.1017/S0031182013002163

Bird DM, Opperman CH, Jones SJM et al (1999) The *Caenorhabditis elegans* genome: a guide in the post-genomics age. Annu Rev Phytopathol 37:247–265. https://doi.org/10.1146/annurev.phyto.37.1.247

Blanc-Mathieu R, Perfus-Barbeoch L, Aury JM et al (2017) Hybridization and polyploidy enable genomic plasticity without sex in the most devastating plant-parasitic nematodes. PLoS Genet 13(6):e1006777. https://doi.org/10.1371/journal.pgen.1006777

Boiteux LS, Belter JG, Roberts PA, Simon PW (2000) RAPD linkage map of the genomic region encompassing the root-knot nematode (*Meloidogyne javanica*) resistance locus in carrot. Theor Appl Genet 100(3):439–446. https://doi.org/10.1007/s001220050057

Caillaud MC, Dubreuil G, Quentin M et al (2008) Root-knot nematodes manipulate plant cell functions during a compatible interaction. J Plant Phy 165(1):104–113. https://doi.org/10.1016/j.jplph.2007.05.007

Carneiro RMDG, Cofcewicz ET (2008) The taxonomy of Meloidogyne spp. from coffee. In: RM Souza (ed) Plant Parasitic Nematodes of Coffee. APS Press & Springer, New York, pp 87–122

Castagnone-Sereno P, Danchin EG, Perfus-Barbeoch L et al (2013) Diversity and evolution of root-knot nematodes, genus *Meloidogyne*: new insights from the genomic era. Annu Rev Phytopathol 51:203–220. https://doi.org/10.1146/annurev-phyto-082712-102300

Castagnone-Sereno P, Mulet K, Danchin EG et al (2019) Gene copy number variations as signatures of adaptive evolution in the parthenogenetic, plant-parasitic nematode *Meloidogyne incognita*. Mol Ecol 28(10):2559–2572. https://doi.org/10.1111/mec.15095

Castagnone-Sereno P, Semblat JP, Castagnone C (2009) Modular architecture and evolution of the map-1 gene family in the root-knot nematode *Meloidogyne incognita*. Mol Genet Genomics 282(5):547–554. https://doi.org/10.1007/s00438-009-0487-x

Chen J, Hu L, Sun L et al (2018) A novel *Meloidogyne graminicola* effector, MgMO237, interacts with multiple host defence-related proteins to manipulate plant basal immunity and promote parasitism. Mol Plant Pathol 19(8):1942–1955. https://doi.org/10.1111/mpp.12671

Chen J, Lin B, Huang Q et al (2017) A novel *Meloidogyne graminicola* effector, MgGPP, is secreted into host cells and undergoes glycosylation in concert with proteolysis to suppress plant defenses and promote parasitism. PLoS Pathog 13(4):e1006301. https://doi.org/10.1371/journal.ppat.1006301

Da Rocha M, Bournaud C, Dazenière J et al (2021) Genome expression dynamics reveal the parasitism regulatory landscape of the root-knot nematode *Meloidogyne incognita* and a promoter motif associated with effector genes. Genes. 12(5):771. https://doi.org/10.3390/genes12050771

Danchin EG, Rosso MN, Vieira P et al (2010) Multiple lateral gene transfers and duplications have promoted plant parasitism ability in nematodes. PNAS 107(41):17651–17656. https://doi.org/10.1073/pnas.1008486107

Dautova M, Gommers F, Abad P et al (2001) Single-pass cDNA sequencing–a powerful tool to analyse gene expression in preparasitic juveniles of the southern root-knot nematode *Meloidogyne incognita*. Nematol 3:129–139. https://doi.org/10.1163/156854101750236259

Davis EL, Hussey RS, Baum TJ et al (2000) Nematode parasitism genes. Annu Rev Phytopathol 38(1):365–396. https://doi.org/10.1146/annurev.phyto.38.1.365

Diatchenko L, Lau YF, Campbell AP et al (1996) Suppression subtractive hybridization: a method for generating differentially regulated or tissue-specific cDNA probes and libraries. PNAS 93(12):6025–6030. https://doi.org/10.1073/pnas.93.12.6025

Diatchenko L, Lukyanov S, Lau YF et al (1999) Suppression subtractive hybridization: a versatile method for identifying differentially expressed genes. Methods Enzy 303:349–380. https://doi.org/10.1016/s0076-6879(99)03022-0

Ding X, Shields J, Allen R, Hussey RS (1998) A secretory cellulose-binding protein cDNA cloned from the root-knot nematode (*Meloidogyne incognita*). Mol Plant-Microbe Interact 11:952–959. https://doi.org/10.1094/MPMI.1998.11.10.952

Ding X, Shields J, Allen R, Hussey RS (2000) Molecular cloning and characterisation of a venom allergin AG5-like cDNA from *Meloidogyne incognita*. Int J Parasitol 30:77–81. https://doi.org/10.1016/s0020-7519(99)00165-4

Dubreuil G, Magliano M, Deleury E et al (2007) Transcriptome analysis of root-knot nematode functions induced in the early stages of parasitism. New Phytol 176(2):426–436. https://doi.org/10.1111/j.1469-8137.2007.02181.x

Emanuelsson O, Nielsen H, Brunak S, von Heijne G (2000) Predicting subcellular localization of proteins based on their N-terminal amino acid sequence. J Mol Biol 300(4):1005–1016. https://doi.org/10.1006/jmbi.2000.3903

Eves-van den Akker S, Stojilković B, Gheysen G (2021) Recent applications of biotechnological approaches to elucidate the biology of plant–nematode interactions. Curr Opin Biotechnol 70: 122–130. https://doi.org/10.1016/j.copbio.2021.03.008

Feyisa B (2022) Factors associated with plant parasitic nematode (PPN) population: a review. Anim Vet Sci 10(2):41–45. https://doi.org/10.11648/j.avs.20221002.15

Fire A, Xu S, Montgomery MK et al (1998) Potent and specific genetic interference by double-stranded RNA in *Caenorhabditis elegans*. Nature 391(6669):806–811. https://doi.org/10.1038/35888

Fitoussi N, Borrego E, Kolomiets MV et al (2021) Oxylipins are implicated as communication signals in tomato–root-knot nematode (*Meloidogyne javanica*) interaction. Sci Rep 11(1):1–6. https://doi.org/10.1038/s41598-020-79432-6

Gheysen G, Mitchum MG (2011) How nematodes manipulate plant development pathways for infection. Curr Opin Plant Biol 14:1–7. https://doi.org/10.1016/j.pbi.2011.03.012

Goldstein P, Triantaphyllou AC (1986) The synaptonemal complex of *Meloidogyne* nataliei and its relationship to that of other *Meloidogyne* species. Chromosoma 93(3):261–266. https://doi.org/10.1007/BF00292747

Gowda MT, Rai AB, Singh A (2017) Root knot nematode: a threat to Vegetable Production & Management, technical bulletin no. 76, ICAR–Indian Institute of Vegetable Research

Gross SM, Williamson VM (2011) Tm1: a mutator/foldback transposable element family in root-knot nematodes. PLoS One 6(9):e24534. https://doi.org/10.1371/journal.pone.0024534

Grynberg P, Coiti Togawa R, Dias de Freitas L et al (2020) Comparative genomics reveals novel target genes towards specific control of plant-parasitic nematodes. Genes 13(11):1347. https://doi.org/10.3390/genes11111347

Guo Y, Bird DM, Nielsen DM (2014) Improved structural annotation of protein-coding genes in the Meloidogyne hapla genome using RNA-Seq. Worm 3:e29158. https://doi.org/10.4161/worm.29158

Holterman M, Karssen G, van den Elsen S et al (2009) Small subunit rDNA-based phylogeny of the Tylenchida sheds light on relationships among some high-impact plant-parasitic nematodes and the evolution of plant feeding. Phytopathology 99:227–235. https://doi.org/10.1094/PHYTO-99-3-0227

Huang G, Allen R, Davis EL et al (2006) Engineering broad root-knot resistance in transgenic plants by RNAi silencing of a conserved and essential root-knot nematode parasitism gene. PNAS 103(39):14302–14306. https://doi.org/10.1073/pnas.0604698103

Huang G, Dong R, Maier T et al (2004) Use of solid-phase subtractive hybridization for the identification of parasitism gene candidates from the root-knot nematode *Meloidogyne incognita*. Mol Plant Pathol 5(217–222):30. https://doi.org/10.1111/j.1364-3703.2004.00220.x

Huang G, Gao B, Maier T et al (2003) A profile of putative parasitism genes expressed in the oesophageal gland cells of the root-knot nematode *Meloidogyne incognita*. Mol Plant-Microbe Interact 16:376–381. https://doi.org/10.1094/MPMI.2003.16.5.376

Huang H, Zhao W, Qiao H et al (2022) SlWRKY45 interacts with jasmonate-ZIM domain proteins to negatively regulate defense against the root-knot nematode *Meloidogyne incognita* in tomato. Horti Res 9:uhac197. https://doi.org/10.1093/hr/uhac197

Iberkleid I, Vieira P, de Almeida EJ et al (2013) Fatty acid-and retinol-binding protein, Mj-FAR-1 induces tomato host susceptibility to root-knot nematodes. PLoS One 8(5):e64586. https://doi.org/10.1371/journal.pone.0064586

Ibrahim HM, Hosseini P, Alkharouf NW et al (2011) Analysis of gene expression in soybean (Glycine max) roots in response to the root-knot nematode *Meloidogyne incognita* using microarrays and KEGG pathways. BMC genom 12(1):1–6. https://doi.org/10.1186/1471-2164-12-220

Jain RK, Mathur KN, Singh RV (2007) Estimation of losses due to plant parasitic nematodes on different crops in India. Indian J Nematol 37(2):219–221

Ji H, Gheysen G, Denil S et al (2013) Transcriptional analysis through RNA sequencing of giant cells induced by *Meloidogyne graminicola* in rice roots. J Exp Bot 64(12):3885–3898. https://doi.org/10.1093/jxb/ert219

Kahsay RY, Gao G, Liao L (2005) An improved hidden Markov model for transmembrane protein detection and topology prediction and its applications to complete genomes. Bioinformatics 21(9):1853–1858. https://doi.org/10.1093/bioinformatics/bti303

Käll L, Krogh A, Sonnhammer EL (2004) A combined transmembrane topology and signal peptide prediction method. J Mol Biol 338(5):1027–1036. https://doi.org/10.1016/j.jmb.2004.03.016

Kaloshian I, Yaghoobi J, Liharska T et al (1998) Genetic and physical localization of the root-knot nematode resistance locus mi in tomato. Mol Gen Genet 257:376–385. https://doi.org/10.1007/s004380050660

Kang MJ, Kim YH, Hahn BS (2010) Expressed sequence tag analysis generated from a normalized full-length cDNA library of the root-knot nematode (Meloidogyne incognita). Genes Genom 32:553–562

Karrer EE, Lincoln JE, Hogenhout S et al (1995) In situ isolation of mRNA from individual plant cells: creation of cell-specific cDNA libraries. PNAS 92:3814–3818. https://doi.org/10.1073/pnas.92.9.3814

Kim J, Yang R, Chang C et al (2018) The root-knot nematode *Meloidogyne incognita* produces a functional mimic of the Arabidopsis inflorescence deficient in abscission signaling peptide. J Exp Bot 69(12):3009–3021. https://doi.org/10.1093/jxb/ery135

Koutsovoulos GD, Poullet M, Elashry A et al (2020) Author correction: genome assembly and annotation of *Meloidogyne enterolobii*, an emerging parthenogenetic root-knot nematode. Sci Data 7(1):413. https://doi.org/10.1038/s41597-020-00747-0

Koutsovoulos GD, Marques E, Arguel MJ et al (2019) Population genomics supports clonal reproduction and multiple independent gains and losses of parasitic abilities in the most devastating nematode pest. Evolutionary Applications 13(2):442–457. https://doi.org/10.1111/eva.12881

Kranse O, Beasley H, Adams S et al (2021) Towards genetic modification of plant-parasitic nematodes: delivery of macromolecules to adults and expression of exogenous mRNA in second stage juveniles. G3 Genes Genomes Genet 11(2):jkaa058. https://doi.org/10.1093/g3journal/jkaa058

Kumar V, Khan MR, Walia RK (2020) Crop loss estimations due to plant-parasitic nematodes in major crops in India. Nat Acad Sci Letters 43:409–412. https://doi.org/10.1007/s40009-020-00895-2

Liang L, Ding YQ, Li X et al (2004) Construction of a metastasis-associated gene subtracted cDNA library of human colorectal carcinoma by suppression subtraction hybridization. World J Gastroenterol 10(9):1301–1305. https://doi.org/10.3748/wjg.v10.i9.1301

Lilley CJ, Maqbool A, Wu D et al (2018) Effector gene birth in plant parasitic nematodes: Neofunctionalization of a housekeeping glutathione synthetase gene. PLoS Genet 14(4): e1007310. https://doi.org/10.1371/journal.pgen.1007310

Lunt DH, Kumar S, Koutsovoulas G et al (2014) The complex hybrid origins of the root-knot nematodes revealed through comparative genomics. PeerJ 2:e356. https://doi.org/10.7717/peerj.356

Mani V, Assefa AD, Hahn BS (2021) Transcriptome analysis and miRNA target profiling at various stages of root-knot nematode *Meloidogyne incognita* development for identification of potential regulatory networks. Int J Mol Sci 22(14):7442. https://doi.org/10.3390/ijms22147442

Martins AC, Mehta A, Murad AM et al (2020) Proteomics unravels new candidate genes for *Meloidogyne* resistance in wild Arachis. J Proteome 217:103690. https://doi.org/10.1016/j.jprot.2020.103690

Moens M, Perry RN, Starr JL (2009) Meloidogyne species—A diverse group of novel and important plant parasites. Root-Knot Nematodes:1–17. https://doi.org/10.1079/9781845934927.0001

Morales P, Thurston FC (2003) Efficient isolation of genes differentially expressed on cellulose by suppression subtractive hybridization in Agaricus bisporus. Mycol Res 107:401–407. https://doi.org/10.1017/s0953756203007366

Naalden D, Haegeman A, de Almeida-Engler J et al (2018) The *Meloidogyne graminicola* effector Mg16820 is secreted in the apoplast and cytoplasm to suppress plant host defense responses. Mol Plant Pathol 19:2416–2430. https://doi.org/10.1111/mpp.12719

Niu J, Liu P, Liu Q et al (2016) Msp40 effector of root-knot nematode manipulates plant immunity to facilitate parasitism. Sci Rep 6(1):1–13. https://doi.org/10.1038/srep19443

Opperman CH, Bird DM, Williamson VM et al (2008) Sequence and genetic map of *Meloidogyne hapla*: a compact nematode genome for plant parasitism. PNAS 105(39):14802–14807. https://doi.org/10.1073/pnas.0805946105

Pagan C, Coyne D, Carneiro R (2015) Mitochondrial haplotype-based identification of ethanol-preserved root-knot nematodes from Africa. Phytopathology 105:350–357. https://doi.org/10.1094/PHYTO-08-14-0225-R

Paganini J, Campan-Fournier A, Da Rocha M et al (2012) Contribution of lateral gene transfers to the genome composition and parasitic ability of root-knot nematodes. PLoS One 7(11):e50875. https://doi.org/10.1371/journal.pone.0050875

Page AP, Johnstone IJ. (2007) The cuticle. The *C. elegans* Research Community, Worm book

Parra G, Bradnam K, Korf I (2007) CEGMA: a pipeline to accurately annotate core genes in eukaryotic genomes. Bioinformatics 23(9):1061–1067. https://doi.org/10.1093/bioinformatics/btm071

Parra G, Bradnam K, Ning Z (2009) Assessing the gene space in draft genomes. Nucleic Acids Res 37(1):289–297. https://doi.org/10.1093/nar/gkn916

Perry RNM, Moens JL, Starr Root-knot nematodes CABI UK

Petersen TN, Brunak S, von Heijne G et al (2011) SignalP 4.0: discriminating signal peptides from transmembrane regions. Nat Methods 8(10):785–786. https://doi.org/10.1038/nmeth.1701

Petitot AS, Dereeper A, Da Silva C, Guy J, Fernandez D (2020) Analyses of the root-knot nematode (*Meloidogyne graminicola*) transcriptome during host infection highlight specific gene expression profiling in resistant rice plants. Pathogens 9(8):644. https://doi.org/10.3390/pathogens9080644. PMID: 32784493; PMCID: PMC7460394

Phan NT, Besnard G, Ouazahrou R et al (2021) Genome sequence of the coffee root-knot nematode. J Nematol 53(1):1–6. https://doi.org/10.21307/jofnem-2021-065

Phani V, Bishnoi S, Sharma A et al (2018) Characterization of *Meloidogyne* indica (Nematoda: Meloidogynidae) parasitizing neem in India, with a molecular phylogeny of the species. J Nematol 50(3):387–398. https://doi.org/10.21307/jofnem-2018-015

Pratx L, Rancurel C, Da Rocha M et al (2018) Genome-wide expert annotation of the epigenetic machinery of the plant-parasitic nematodes *Meloidogyne* spp., with a focus on the asexually reproducing species. BMC Genomics 19(1):1–21. https://doi.org/10.1186/s12864-018-4686-x

Roberts PA (1995) Conceptual and practical aspects of variability in root-knot nematodes related to host plant resistance. Annu Rev Phytopathol 33:199–221. https://doi.org/10.1146/annurev.py.33.090195.001215

Robertson HM, Thomas JH (2006) *Caenorhabditis elegans* Worm Book, ed The *C. elegans* research community, 10.1895/wormbook.1.66.1. Available at: http://www.wormbook.org. Accessed June 10, 2008

Rosso MN, Hussey RS, Davis EL, et al (2011). Nematode effector proteins: targets and functions in plant parasitism. Effect Plant-Microbe Interact 327-354. https://doi.org/10.1002/9781119949138.ch13

Roze E, Hanse B, Mitreva M et al (2005) Mining the secretome of the root-knot nematode *Meloidogyne chitwoodi* for candidate parasitism genes. Mol Plant Pathol 9:1–10. https://doi.org/10.1111/j.1364-3703.2007.00435.x

Rybarczyk-Mydłowska K, Maboreke HR, van Megen H et al (2012) Rather than by direct acquisition via lateral gene transfer, GHF5 cellulases were passed on from early Pratylenchidae to root-knot and cyst nematodes. BMC Evol Biol 12(1):1–10. https://doi.org/10.1186/1471-2148-12-221

Sato K, Kadota Y, Gan P et al (2018) High-quality genome sequence of the root-knot nematode *Meloidogyne arenaria* genotype A2-O. Genome Announce 6:e00519–e00518. https://doi.org/10.1128/genomeA.00519-18

Schoonmaker A, Hao Y, Bird DMK, Conant GC (2020) A single, shared Triploidy in three species of parasitic nematodes. G3 Genes|Genomes|Genetics 10(1):225–233. https://doi.org/10.1534/g3.119.400650

Semblat JP, Rosso MN, Hussey RS et al (2001) Molecular cloning of a cDNA encoding an amphid-secreted putative avirulence protein from the root-knot nematode *Meloidogyne incognita*. Mol Plant-Microbe Interact 14(1):72–79. https://doi.org/10.1094/MPMI.2001.14.1.72

Sereno PC (2002) Genetic Variability in parthenogenic root-knot nematodes, *Meloidogyne* Sp. and their ability to overcome plant resistance genes. Nematologica 4:605–608

Shi Q, Mao Z, Zhang X et al (2018) The novel secreted *Meloidogyne incognita* effector MiISE6 targets the host nucleus and facilitates parasitism in Arabidopsis. Front Plant Sci 9:252. https://doi.org/10.3389/fpls.2018.00252

Shields JP, Ding X, Hussey RS (1998) Microaspiration of esophageal gland contents from plant-parasitic nematodes. J Nematol 30:515. https://doi.org/10.1007/978-1-61737-998-7_9

Somvanshi VS, Tathode M, Shukla RN et al (2018) Nematode genome announcement: a draft genome for Rice root-knot nematode, Meloidogyne graminicola. J Nematol 50(2):111–116. https://doi.org/10.21307/jofnem-2018-018

Sun L, Zhuo K, Lin B et al (2014) The complete mitochondrial genome of *Meloidogyne graminicola* (Tylenchina): a unique gene arrangement and its phylogenetic implications. PLoS One 9(6):e98558. https://doi.org/10.1371/journal.pone.0098558

Susič N, Koutsovoulos GD, Riccio C et al (2020) Genome sequence of the root-knot nematode. J Nematol 52:1–5. https://doi.org/10.21307/jofnem-2020-025

Szitenberg A, Salazar-Jaramillo L, Blok VC et al (2017) Comparative genomics of apomictic root-knot nematodes: hybridization, ploidy, and dynamic genome change. Genome Biol Evol 9(10):2844–2861. https://doi.org/10.1093/gbe/evx201

Tomalova I, Iachia C, Mulet K, Castagnone-Sereno P (2012) The map-1 gene family in root-knot nematodes, *Meloidogyne* spp.: a set of taxonomically restricted genes specific to clonal species. PLoS One 7(6):e38656. https://doi.org/10.1371/journal.pone.0038656

Triantaphyllou AC (1985) Cytological methods for the study of oogenesis and an advanced treatise on *Meloidogyne*. Methodology 2:107

Truong NM, Chen Y, Mejias J, Soulé S, Mulet K, Jaouannet M, Jaubert-Possamai S, Sawa S, Abad P, Favery B, Quentin M (2021) The Meloidogyne incognita Nuclear Effector MiEFF1

Interacts With Arabidopsis Cytosolic Glyceraldehyde-3-Phosphate Dehydrogenases to Promote Parasitism. Front Plant Sci 12:641480. https://doi.org/10.3389/fpls.2021.641480

Verhoeven A, Finkers-Tomczak A, Prins P et al (2022) The root-knot nematode effector MiMSP32 targets host 12-oxophytodienoate reductase 2 to regulate plant susceptibility. New Phytol 237(6):2360–2374. https://doi.org/10.1111/nph.18653

Vieira P, Danchin EG, Neveu C et al (2011) The plant apoplasm is an important recipient compartment for nematode-secreted proteins. J of Exp Bot 62(3):1241–1253. https://doi.org/10.1093/jxb/erq352

Voiblet C, Duplessies S, Encelot W, Martin F (2001) Identification of symbiosis regulated genes in Eucalyptus globulus-Pisolithus tinctorius ectomycorrhiza by differential hybridization of arrayed cDNAs. The Plant Journal 25:181–191

Xue B, Hamamouch N, Li C et al (2013) The 8D05 parasitism gene of *Meloidogyne incognita* is required for successful infection of host roots. Phytopathology 3(2):175–181. https://doi.org/10.1094/PHYTO-07-12-0173-R

Yaghoobi J, Kaloshian I, Williamson VM (1995) Mapping a new nematode resistance locus in *Lycopersicon peruvianum*. Theor Appl Genet 91:457–464. https://doi.org/10.1007/BF00222973

Zhang L, Gleason C (2021) Transcriptome analyses of pre-parasitic and parasitic *Meloidogyne chitwoodi* race 1 to identify putative effector genes. J Nematol 53:e2021–e2084. https://doi.org/10.21307/jofnem-2021-084

Zhang X, Peng H, Zhu S et al (2020) Nematode-encoded RALF peptide mimics facilitate parasitism of plants through the FERONIA receptor kinase. Mol Plant 13(10):1434–1454. https://doi.org/10.1016/j.molp.2020.08.014

Natural Product Repertoire for Suppressing the Immune Response of *Meloidogyne* Species

6

Jihane Kenfaoui, Khadija Goura, Ikram Legrifi, Najwa Seddiqi Khalil, Hajar El Hamss, Fouad Mokrini, Said Amiri, Zineb Belabess, and Rachid Lahlali

Abstract

Root-knot nematodes, *Meloidogyne* spp., are obligate endoparasites with major global economic significance. Their reproductive techniques display a broad variety continuum, ranging from amphimixis to obligatory mitotic parthenogenesis. Root-knot nematodes (RKN) developed invasion and colonization techniques, including the expression of immune suppressors to get beyond the host plant's defences. Natural products are used to combat these potentially dangerous microorganisms as part of sustainable agriculture, which strives to regulate soil and plant health while using fewer chemical inputs. Most of these natural products are biodegradable, and their investigation is subject to less-stringent regulatory approval procedures. In this study, we provide an extensive overview of the biological control of *Meloidogyne* and the main mechanisms of action of plant products against RKN. We discussed the different nematicidal

J. Kenfaoui · K. Goura · I. Legrifi · N. Seddiqi Khalil · H. El Hamss · S. Amiri
Phytopathology Unit, Department of Plant Protection, Ecole Nationale d'Agriculture de Meknès, Meknès, Morocco

F. Mokrini
Biotechnology Unit, Regional Center of Agricultural Research of Rabat, National Institute of Agricultural Research, Rabat, Morocco

Z. Belabess
Plant Protection Laboratory, Regional Center of Agricultural Research of Meknès, National Institute of Agricultural Research, Meknès, Morocco

R. Lahlali (✉)
Phytopathology Unit, Department of Plant Protection, Ecole Nationale d'Agriculture de Meknès, Meknès, Morocco

Plant Pathology Laboratory, AgroBioSciences, College of Sustainable Agriculture and Environmental Sciences, Mohammed VI Polytechnic University, Ben Guerir, Morocco
e-mail: rlahlali@enameknes.ac.ma

F. Ahmad, G. Nombela (eds.), *Root-Galling Disease of Vegetable Plants*,
https://doi.org/10.1007/978-981-99-3892-6_6

activities imposed by natural products alongside their large-scale efficiency in controlling RKN. Finally, we provide an overview of the major factors affecting the success or failure of using natural products as a reliable strategy to control RKN.

Keywords

Natural Product · Root-knot nematodes · Biological control · Immune response

6.1 Introduction

The majority of soil microorganisms are nematodes, which are by far the most prevalent animals on Earth (van den Hoogen et al. 2019; Bardgett and van der Putten 2014). Plant-parasitic nematodes, also known as PPNs, are thought of as hidden enemies because they have been connected to heavily infested farms and can drastically affect crop production by up to 80% (McCarter 2008). With an estimated annual output loss of more than $100 billion in a variety of plants and agricultural goods worldwide, they constitute a danger to agriculture (Moens et al. 2009).

The PPNs include a wide variety of species; approximately 4100 species have been identified so far (Geng et al. 2016). Root-knot nematodes (RKN; *Meloidogyne* spp.) are the main class of PPNs that affect yield. RKN are obligate stationary endoparasites that are simple to reproduce. They are found in the roots of over 3000 distinct plant species. Under favourable conditions, their soil population can easily increase due to their widespread distribution (Calderón-Urrea et al. 2016; Hajihassani et al. 2018; Subbotin and Chitambar 2018). Due to their widespread frequency and the large yield losses they cause on a range of crops, PPNs are estimated to result in around $157 billion in yearly global agricultural losses, most of which are attributable to RKN (Wesemael et al. 2011). So far, *Meloidogyne* includes 98 described species, which are obligate parasites of almost all vascular plants. Some seriously constrain agricultural production in tropical, subtropical, and temperate regions (Subbotin and Chitambar 2018). They are associated with many crops, including vegetables, soybean, rice, maize, rubber tree, ornamentals, coffee, etc. while the most frequent species include *Meloidogyne arenaria, Meloidogyne hapla, Meloidogyne incognita,* and *Meloidogyne javanica.* As they are viewed as possible causes of harm to economically significant crops, many others have been gaining interest. Berkeley first identified the RKN in 1855 (Barber 1901). Every year, RKN cause damage to about 5% of the world's total agricultural production (Karajeh 2008). RKN are well known to exhibit significant genetic variability and an extreme cytogenetic diversity. They can reproduce through either mandatory parthenogenesis or mandatory amphimixis.

Mitotic parthenogenesis is the main process of RKN reproduction and males are believed not to have any roles in reproduction. Female nematodes can lay up to 1000 eggs, each containing a juvenile (J1) in its first stage. When moisture and temperature circumstances are ideal, the primary infectious form, the second-stage juvenile

(J2), frequently hatches from the egg. The J2 employs a stylet, a body part with piercing abilities, to enter the host plant's root cell. After entering the host plant's root, these parasites move to the cortical tissue and cells to become sedentary. RKN have four stages of juvenile life; after four sequential moults, they reach adulthood. The environment significantly affects how RKN determine their gender. More males are conceived due to unfavourable circumstances and lack of nutrition. The RKN's life cycle is completed as sedentary females lay eggs on the root's surface, and males quit the host plant as they develop mobility during their third moult (Moens et al. 2009; Bhowmik et al. 2021; Ciancio 2021).

Through the use of cell wall-digesting enzymes, nematodes invade plant tissue (Quentin et al. 2013). Second-stage juveniles of the virulent form move through root cells by perforating right behind the root's tip and remaining close to the vascular cylinder (Abad et al. 2009). Then, by secreting effector proteins, the dormant RKN promote the development of a feeding site in the root. These proteins enable RKN to evade the host plant's defence mechanisms and turn them into a source of nutrients (Quentin et al. 2013). Karyokinesis occurs in the 5–7 cells surrounding the cells in which the RKN become sedentary, but no cytokinesis occurs since no cell plate is produced. As a result, a cell starts with two nuclei, and the process continues until there are roughly 100 nuclei. RKN begin to trigger the creation of giant cells, which grow rapidly, reaching their maximum size in just a few weeks (Abad et al. 2009). Due to gall formation, such damage inhibits the host plant from taking water and nutrients properly. Although a laboratory inspection was necessary for accurate species identification, these galls are the main signs of RKN infection. The symptoms of the upper part of the host plant from the ground exhibit yellowing, stunning, wilting, and premature shedding of the foliage (Wesemael et al. 2011; Ciancio 2021).

There are numerous ways to control RKN. Synthetic nematicides, primarily fumigants, have been the last century's most popular parasite control method (Brito et al. 2020). The nematicides' non-biodegradable nature evokes concerns about environmental contamination, nematode resistance, and plant toxicity. Its high cost and ban in many countries because these nematicides provoke researchers to find another alternative to RKN control (Khan et al. 2019). Additionally, synthetic chemical pesticides have been outlawed since they are carcinogenic, leave behind hazardous waste, disrupt hormone balance, are toxic to sperm, and take a long time to decompose (Barros et al. 2019). They are harmful to people, plants, animals, flora, and the fauna of agriculturally significant soils. Since eradicating RKN in the field is practically impossible, one of the main aims of nematode management is to prevent their spread to other areas (Forghani and Hajihassani 2020).

Researchers from all across the world are working to provide new, environmentally friendly methods for managing RKN. The environment-friendly management techniques include soil modification, soil treatment, application of industrial waste, biological agents etc. Controlling plant agents are better alternatives for synthetic nematicides (Khan et al. 2019). Nowadays, plants and plant-derived products are considered protective agents against various plant parasites and other pests. Since some of their metabolites can be utilized as pesticides directly or as starting points for the synthesis of better chemical structures, they are promising sources of compounds to address the issue as mentioned earlier (El-nagdi et al. 2017; Barros et al. 2019; Jardim et al. 2020a; Brito et al. 2020; Ciancio 2021). Therefore, it is essential to develop different control mechanisms using techniques that are friendly to the environment. Many studies have been conducted on the subject worldwide, producing helpful results and intriguing insights that can raise farmers' revenue. New data on these techniques' efficacy will continue to be made public as long as research on their development is ongoing. We discussed the main mechanisms of action of plant products against RKN in this chapter and the progress made in the prospecting of biocontrol of *Meloidogyne* species with a focus on the various nematicidal activities imposed by natural products and their broad-scale effectiveness in controlling RKN. Also included are molecular suppression mechanisms. Consequently, the current extended chapter offers an up-to-date, state-of-the-art report on the key elements influencing the success or failure of the use of natural products as a durable method to regulate RKN.

6.2 Overview of the Biological Control of *Meloidogyne* Species

Many researchers have discovered various strategies to manage RKN infections in different crops. For the control of nematodes, various chemical nematicides have been used; however, they have turned out to be hazardous to plants and the agricultural ecosystem (Medina-Canales et al. 2019). However, these chemical nematicides have been entirely banned and limited. There is a pressing need to find alternatives that are less expensive, environmentally responsible, and less harmful to the host plants (D'Addabbo et al. 2014; Abd-Elgawad 2016). Since biological control benefits farmers, crops, and the environment, it is the most suitable approach for inhibiting the infections of *Meloidogyne* spp. (Naz et al. 2021). There is a lot of interest in using biological control agents (BCAs) based on bacteria, fungi, actinomycetes, and other microorganisms (Table 6.1). Among these, bacteria and fungi are the most prevalent microorganisms naturally occurring in soil ecosystems and have several efficient ways to manage nematodes (Askary 2015; Blyuss et al. 2019).

Various species of fungi belonging to the genera, *Arthrobotrys, Actylellina, Aspergillus, Catenaria, Hirsutella, Monacrosporium, Dactylellina, Purpureocillium, Pochonia,* and *Trichoderma,* are excellent BCAs against PPNs, particularly for RKN control (Devi 2018; Saxena 2018; LIU Liu et al. 2019; Fan et al. 2020; Soliman et al. 2021). Endophytic fungi such as *Acremonium, Alternaria, Trichoderma, Purpureocillium,* and *Fusarium* can occupy plant roots and improve plant immunity through various mechanisms (Schouten 2016). They may drive J2 of RKN out from roots, reduce fecundity, and slow or halt RKN development (Topalović et al. 2020). *Purpureocillium* and *Trichoderma* species can destroy RKN at various stages of life in the root systems or soil. *Pochonia chlamydosporia* has also been reported to have the ability to cause systemic resistance to *M. incognita* in many crops (Abd-Elgawad and Askary 2018; Ghahremani et al. 2019).

Table 6.1 Examples of various biological control agents (BCAs) used against important root-knot nematodes (*Meloidogyne* spp.)

Biological control agents	RKN species	*Type of study*	Host plant	References
Bacteria				
Agrobactrium tumefaciens	*M. ethiopica*	In vivo	Tomato	(Lamovšek et al. 2017)
Bacillus amiloliquefaciens	*M. incognita*	In vitro, in vivo	Tomato	(Jamal et al. 2017)
B. cereus	*M. incognita*	In vitro, in vivo	Tomato	(Li et al. 2019) (Xiao et al. 2018)
B. coagulans	*M. incognita*	In vitro, in vivo	Cotton	(Xiang et al. 2018)
B. firmus	*M. incognita*	In vivo	Tomato	(d'Errico et al. 2019)
B. licheniformis	*M. incognita*	In vitro, in vivo	Tomato	(Colagiero et al. 2018)
B. megaterium	*M. incognita* *M. graminicola*	In vivo	Sugar beet	(Mostafa et al. 2018)
B. pumilus	*M. arenaria*	In vitro, in vivo	Tomato	(Lee and Kim 2016)
B. subtilis	*M. incognita*, *M. graminicola*, *M. javanica*	In vitro, in vivo In vitro, in vivo	Tomato Sugarcane	(Basyony and Abo-Zaid 2018) (de Mazzuchelli et al. 2020)
Serratia marcescens	*M. incognita*, *M. javanica*	In vitro	Tomato	(Rahul et al. 2014)
Pseudomonas aeruginosa	*M. javanica*	In vitro	Tomato	(Siddiqui et al. 2003)
P. fluorescens	*M. incognita*,	In vivo	Cowpea	(Abd-El-Khair et al. 2019)
P. stutzeri	*M. incognita*,	In vitro, in vivo	Mungbean	(Khan et al. 2016)
Pasteuria penetranse	*M. exigua*	In vivo	Coffee	(Botelho et al. 2019)
Fungi				
Trichoderma atroviride	*M. incognita*	In vivo	Pepper	(Herrera-Parra et al. 2017)
T. asperellum	*M. javanica*	In vivo	Pineapple	(Kiriga et al. 2018)
T. harzianum	*M. incognita*	In vivo	French bean	(Gogoi and Mahanta 2013)
T. longibrachiatum	*M. incognita*	In vitro, in vivo	Cucumber	(Zhang et al. 2015)
T. viride	*M. graminicola*	In vivo	Rice	(Narasimhamurthy et al. 2017)
Paecilomyces Lilacinus	*M. javanica*	In vivo	Eggplant	(Ashraf and Khan 2010)
Pochonia chlamydosporia	*M. incognita*	In vivo	Tomato	(de Silva et al. 2017)

(continued)

Table 6.1 (continued)

Biological control agents / *Bacteria*	RKN species	*Type of study*	Host plant	References
Purpureocillium lilacinum	*M. enterolobii*	In vitro, in vivo	Tomato, banana	(Silva et al. 2017)
Lecanicillium muscarium	*M. incognita*	In vitro, in vivo	Tomato	(Hussain et al. 2018)
Metarhizium guizhouense	*M. incognita*	In vitro	–	(Thongkaewyuan and Chairin 2018)
Mortierella globalpina	*M. chitwoodi*	In vitro, in vivo	Pepper	(DiLegge et al. 2019)
Xylaria grammica	*M. incognita*	In vitro, in vivo	Melon, tomato	(Kim et al. 2018)
Arbuscular mycorrhizal fungi (AMF)				
Glomus mosseae	*M. incognita*	In vitro, in vivo	Tomato	(Vos et al. 2012)
Glomeromycota fungi	*M. exigua*	In vivo	Coffee	(Alban et al. 2013)

Dactylellina and *Arthrobotrys* can trap RKN J2 in the soil via their hyphal structures, lowering the nematode's invasion capacity (Wang et al. 2014).

Trichoderma longibrachiatums were examined in vitro by Zhang et al. (2015) for their ability to suppress *M. incognita*. The results showed that J2 had a more considerable lethal impact (>88%) on the nematode when exposed for 14 days to 1×10^5 to 1×10^7 conidia/ml. The *same fungal concentrations significantly reduced the M. incognita infection in cucumbers*, and glasshouse conditions improved plant growth. *Trichoderma* species have also been demonstrated to be effective pepper *M. incognita* control agents and plant growth enhancers (Herrera-Parra et al. 2017). In pots treated with *T. virens*, *T. atroviride*, and *T. harzianum*-C2T, the galling index was decreased by 22 to 35%. In addition, *T. atroviride* reduced nematode egg and female production by 63% and 14.36%, respectively. In a commercial pineapple production setting, the effects of *Purpureocillium lilacinum* and *Trichoderma* spp. on *M. javanica* have been investigated (Kiriga et al. 2018). When treated as individual inocula, *T. atroviride* F5S21, *T. asperellum* M2RT4, *Trichoderma* sp. MK4, *Trichoderma* sp. MK4, and two strains of *P. lilacinum* (KLF2 and MR2) decreased the root galling of *M. javanica* from 61 to 82%. The most effective fungus was *T. asperellum* M2RT4, which reduced egg numbers, egg mass, and galling by more than 88, 78, and 82%, respectively. It also expanded the fresh weight of the root by 91%. *M. enterolobii*, which impacts tomato and banana crops, was tested against *P. chlamydosporia* and *P. lilacinum* by Silva et al. (2017). On tomato and banana roots, *P. chlamydosporia* caused a 34% suppression of *M. enterolobii* eggs, while *P. lilacinum* caused a 44% suppression of *M. enterolobii* eggs on tomato roots. These efficacies were noted when fewer than 500 *M. enterolobii* eggs were inoculated. It was determined that fields with low nematode pressure might use

P. lilacinum and *P. chlamydosporia* as a component of integrated pest management (IPM) strategy.

Another type of potential fungi that serve as obligate plant root symbionts are arbuscular mycorrhizal fungi (AMF) (Smith et al. 2010). The plant supplies the symbionts with photosynthetic carbon and the latter aid in boosting root nutrient uptake and enhancing root structure and growth. Additionally, they frequently compete with PPNs for nutrients and space, leading to plant systemic resistance (Singh et al. 2011; Schouteden et al. 2015). Some plant species, including coffee (*M. exigua* and *M. coffeicola*) and tomato (*M. incognita*), have shown suppressive effects of AMF against *Meloidogyne* spp. in vitro, glasshouse, and field studies (Vos et al. 2012; Alban et al. 2013).

Likewise, numerous studies have shown that rhizospheric bacteria such as *Arthrobacter, Achromobacter, Agrobacterium, Bacillus, Comamonas, Pasteuria, Burkholderia, Rhizobium, Serratia, Pseudomonas,* and *Variovorax* can control RKN and fall into nematophagous soil-borne category (Li et al. 2015b; Tiwari et al. 2017; Wolfgang et al. 2019; Huang et al. 2020). They have multiple mechanisms for controlling or combating RKN, including competition for nutrition requirements, direct parasitism, and antibiosis (Mendoza et al. 2008; Cawoy et al. 2011; Abd-Elgawad and Askary 2018). Antibiosis is among the most frequently employed action mechanisms due to synthesizing volatile organic compounds (VOCs), toxins, and some antibiotics (Saraf et al. 2014). *Bacillus cereus* BCM2 colonized the root exudates and worked as a second-stage juvenile (J2) repellant when it was employed to manage *M. incognita* in tomato crops, resulting in decreased nematode degradation (Li et al. 2019). In comparison to the control, BCM2 treatment of nematode-infected tomato plants resulted in 67.1% fewer J2. Another study found that two days before *Meloidogyne ethiopica* inoculation, treating tomato plants with *Agrobacterium tumefaciens* decreased root galling and egg counts 45 and 90 days later (Lamovšek et al. 2017). Split-root studies revealed a systemic nature of the observed *A. tumefaciens*-plant interaction. *Bacillus amyloliquefaciens* strain Y1 was examined in vitro and in vivo on tomato to suppress *M. incognita*. A substantial inhibition of RKN egg hatching and J2 mortality was brought on by bacterial culture supernatant and crude Y1 extract (Jamal et al. 2017). Supernatant concentrations between 10% and 40% decreased egg hatching by 32.5–60.6% after 5 days of in vitro exposure. J2 have a very high death rate that increases dramatically with treatment concentration and exposure period, peaking at 80% after 3 days at 40% concentration. Plants treated with Y1 had significantly greater growth parameters than untreated controls. In a different investigation, *B. cereus* Jdm1 was used to control *M. incognita* in tomato crops (Xiao et al. 2018). The culture supernatant significantly reduced J2 numbers and inhibited egg hatching under in vivo conditions. Additionally, Jdm1 treatment decreased the severity of root galling (43%), while enhancing tomato plant growth performance. Gall index 30 days post-inoculation (DPI) had a stronger control effect up to 50% in field studies. The treatment initially impacted the tomato rhizosphere bacterial community, but it quickly recovered. In a greenhouse study, *Pseudomonas fluorescens*, *Bacillus pumilus*, and *Bacillus subtilis* were efficient against

M. incognita on cowpea (Abd-El-Khair et al. 2019). *P. fluorescens* caused the most significant decrease in nematode populations (89%), followed by a mixture of *B. subtilis* and *P. fluorescens* (88.50%). The combination treatment produced the highest yield increase (70.2%), followed by *B. pumilus* at 49.3%. The obligate parasite, *Pasteuria* spp., a widely distributed endospore-forming bacterium, are incredibly safe BCAs to control RKN (Kokalis-Burelle 2015). They can act on nematodes in harsh environments with varying soil temperature, moisture, and pH. Their primary means of action include altering RKN J2 (Abd-Elgawad 2021). The J2 produces few or no eggs in host plants when infected with a small number of *Pasteauria* spores, but as the number of spores grows, the J2 becomes less mobile and loses its capacity to enter roots (Liu et al. 2017; Tapia-Vázquez et al. 2022). In addition to using BCAs to manage *Meloidogyne* species, plant products have received much attention as more environmentally friendly nematicides (Ntalli et al. 2011; Laquale et al. 2015; Grubišić et al. 2018; Atolani and Fabiyi 2020). In this regard, the current review investigates previous research on the effects of natural products used for the biological control of RKN.

6.2.1 Nematicidal Activity of Plant Extracts

Using plant extracts is one of the effective PPN control strategies (Siddiqui and Alam 1988; Ntalli et al. 2020a, b). A plant extract is a complex mixture with many chemical compounds, obtainable by chemical, physical, and/or microbiological processes from a natural source and usable in any technology field (Pino et al. 2013; Pavela 2016). The plant extracts provide an environmentally friendly option for controlling PPNs because they are safe, rapidly biodegradable, non-persistent, and less toxic (Chitwood 2002). Moreover, plant extract application encompasses several methods, including cover cropping, whole plant inclusion, concentrated essential oils, and defatted seed meal (Lazzeri et al. 2009; Laquale et al. 2015; Ntalli et al. 2018). Many plants have been investigated from which extracts of leaves, seeds, and roots are used to control PPNs associated with different crops (Table 6.2). Their nematicidal properties are directly related to the content of certain compounds such as phenols, tannins, azadirachtins, alkaloids, and glycosides. These compounds are toxic to nematodes (Eloh et al. 2020). Many of these compounds are nematoxic or have nematostatic effects on different PPN species. These compounds can be biocidal or interfere in other ways with the life cycle of nematodes (Alam et al. 1990; Sukul 1992). Among the plant species known to have nematicidal properties are *Tagetes erecta* L., *T. patula* L. (Buena et al. 2008; Faizi et al. 2011; Munhoz et al. 2017), *Verbesina encelioides* (Cav.), *Inula viscosa* Aiton (Oka et al. 2006; Oka 2012), *Taraxacum officinale* (L.) Weber, *Artemisia annua* L. (Laquale et al. 2015; D'Addabbo et al. 2017), *Ricinus communis, Lantana camara* (Wondimeneh et al. 2013), and *Jatropha curcas* (Oluwatayo et al. 2019).

Yasmin et al. (2003) reported the efficacy of various neem plant components (leaves and seeds) against *M. javanica* associated with the sweet gourd. *M. incognita* on tomato plants were found to be lowered by the aqueous extract of garlic bulbils

(Martinotti et al. 2016). Adegbite and Adesiyan (2006) evaluated the root extracts of *Ricinus communis*, *Jatropha curcas*, *Azadirachta indica,* and *Chromolaena odorata* against RKN on edible soybean, and they concluded that all extracts tested were successful in preventing *Meloidogyne*'s egg from hatching. Certain chemicals such as flavonoids, amides, alkaloids, and saponins were assumed to be responsible for the nematicidal action on the nematode egg hatching (Goswami and Vijayalakshmi 1986; Haroon et al. 2018). Vilchis-Martinez et al. (2013) examined the nematicidal activity of 22 plant species against *M. incognita*. They found that the crude aqueous extracts of *Argemone mexicana* L., *Chenopodium album* L., *Datura stramonium* L., *Nerium oleander* L., and *Raphanus raphanistrum* L. could be taken into consideration as a possible substitute for the management and control of this nematode species.

Essential oils have also been reported to work effectively against several crop PPN (Kim et al. 2018). However, their biological activities depend upon their chemical compounds, which in turn, depend on different factors, such as extraction method, plant parts used, plant age, phenological stage of the plant used for extraction, and harvesting season (Angioni et al. 2006; Isman et al. 2007). Various essential oils from botanical and medicinal plants have been used to control several PPNs associated with different crops (Pandey et al. 2000; Park et al. 2005; Ozdemir and Gozel 2017; Ntalli et al. 2020a, b) (Table 6.2). Among the promising species, the use of plants of the genus *Tagetes* (*Asteraceae*) stands out, which has been recognized for producing nematicidal compounds such as dihydrotagetone, cis-ocimene, and E-tagetone, among others (Kimpinski et al. 2000; Ploeg 2000). Some brassica species contain thiocyanates, isothiocyanates, glucosinolates, and nitriles, generating sulfurous chemicals (dos Neves et al. 2009), and have shown nematicidal properties. The highlighting species of brassica are canola (*Brassica napus*) and mustard (*Sinapis alba*), having nematicidal activity against *M. incognita* (Aballay and Insunza 2002). The essential oil of garlic (*Allium sativum* L.) has been used to treat a variety of PPNs, including *Meloidogyne* (El Saedy et al. 2014; Jardim et al. 2020b) and *Bursaphelenchus xylophilus* (Park et al. 2005). The efficacy of garlic and thyme essential oils against *M. incognita* "race 2" was reported by Cetintas and Yarba (2010). The nematicidal activity of aqueous garlic extract, a commercial product, was assessed by Abd-Elgawad et al. (2009) against *Meloidogyne* spp. They showed that aqueous garlic extract caused J2 reduction of *Meloidogyne* spp. Essential oils from *Mentha rotundifolia, M. piperita, M. citrata, M. spicata, azadirachtin, Foeniculum vulgare, Perlargonium graveolens, Ocimun basilicum, Cymbopogon grasses, C. flexuosus*, and *C. winterianus* have been shown to have high nematicidal action against *M. incognita* and *M. javanica* (Saxena et al. 1987; Oka et al. 2000; Sinha et al. 2006; Ntalli et al. 2010). Recent research conducted by Borges et al. (2018) demonstrated that using *Schinus terebinthifolius* essential oil decreased the prevalence of J2 *M. javanica* in lettuce. The highest levels of α-terpineol and terpinen-4-ol in *S. terebinthifolius* are responsible for the plant's nematicidal potential (Echeverrigaray et al. 2010). To prevent RNKs (*M. incognita*) on tomato plants, Radwan et al. (2007) developed six formulations as emulsifiable

Table 6.2 Significant examples of natural nematicides from plant extracts against plant-parasitic nematodes

Nematodes species	Plant species	Plant part used	Product type	References
Meloidogyne javanica	*Piper* *hispidinervum* *Mentha spicata* *Capsicum* *frutescens* *Melia azedarach* *Xanthium* *strumarium* *Achillea wilhelmsii* *Ficus glomerata* *Roxb* *Croton caudatus* *Geilser* *Centella asiatica* *Linn.* *Inula viscosa* *Thymus* *citriodorus* *Mentha pulegium* *Rosmarinus* *officinalis* *Azadirachta indica* *Berberis* *brevissima*	Aerial parts Aerial parts Aerial parts Aerial parts Aerial parts Aerial parts Aerial parts Aerial parts Aerial parts Aerial parts Aerial parts Aerial parts Aerial parts Aerial parts Underground parts	Essential oil Essential oil Essential oil Essential oil Extract	(Andrés et al. 2017) (Kimbaris et al. 2017) (Kepenkci and Saglam 2018) (Kepenkci and Saglam 2018) (Kepenkci and Saglam 2018) (Kepenkci and Saglam 2018) (Chanu and Mohilal 2019) (Chanu and Mohilal 2019) (Chanu and Mohilal 2019) (Oka et al. 2006) (Ntalli et al. 2020a, b) (Kimbaris et al. 2017) (Mattei et al. 2014) (Moosavi 2012) (Saqib et al. 2019)
M. incognita	*Thymus* *citriodorus* *Acacia* *niloticaAregimone* *Mexicana* *Chenopodium* *album* *Cucumis melo var.* *agrestis* *Azadirachta indica* *Eucolyptus* *microtheca* *Nicotiana tabacum* *L* *Syzygium* *aromaticum L* *Acorus calamus L* *Ocimum sanctum L* *Cymbopogon* *schoenanthus* (L.) *Cinnamomum* *zeylanicum* *Mentha canadensis* *Lavandula*	Aerial parts Aerial parts	Essential oil Essential oil Essential oil Essential oil Essential oil Essential oil Essential oil Essential oil Essential oil Extract/ Essential oil Essential oil	(Ntalli et al. 2020a, b) (Elbadri et al. 2008) (Elbadri et al. 2008) (Elbadri et al. 2008) (Elbadri et al. 2008) (Elbadri et al. 2008) (Elbadri et al. 2008) (Taniwiryono et al. 2009) (Taniwiryono et al. 2009) (Taniwiryono et al. 2009) (Eloh et al. 2020) (Eloh et al. 2020) (Eloh et al. 2020) (Ji et al. 2016) (Ozdemir and Gozelde Silva et al. 2017) (Ozdemir and Gozelde Silva et al. 2017) (Ozdemir and Gozelde Silva et al. 2017) (Ozdemir and Gozelde

(continued)

Table 6.2 (continued)

Nematodes species	Plant species	Plant part used	Product type	References
	officinalis *Artemisia absinthium* *Piper nigrum* *Citrus bergamia* *Origanum majorana* *Tagetes erecta* *Vetiveria zizanioides* (L.) *Chenopodium ambrosioides*			Silva et al. 2017) (Mervat et al. 2012) (Mervat et al. 2012) (Jindapunnapat et al. 2018) (Bai et al. 2011)
M. graminicola	*Syzygium aromaticum* *Cymbopogon flexuosus* *Cymbopogon martinii*	Aerial parts Aerial parts Aerial parts	Essential oil Essential oil Essential oil	(Ajith et al. 2020) (Ajith et al. 2020) (Ajith et al. 2020)
M. hapla	*Origanum onites* *Salvia officinalis* *Lippia citriodora* *Mentha spicata* *Mentha longifolia* *Mentha piperita* *Foeniculum vulgare* *Coriandrum sativum* *Ocimum basilicum* *Allium ursinum* L. *Artermisia absinthium* L. *Juglans regia* L. *Salvia officinalis* L. *Tagetes patula* L. *Tanacetum vulgare* L. *Artemisia annua*	Aerial parts Aerial parts Aerial parts Aerial parts Aerial parts Aerial parts Underground parts Underground parts Aerial parts Aerial parts Aerial parts Aerial parts Aerial parts Aerial parts Aerial parts Aerial parts	Essential oil Essential oil Essential oil Essential oil Essential oil Essential oil Essential oil Essential oil Extract	(Felek et al. 2019) (Felek et al. 2019) (Felek et al. 2019) (Felek et al. 2019) (Felek et al. 2019) (Felek et al. 2019) (Felek et al. 2019) (Felek et al. 2019) (Felek et al. 2019) (Samaliev et al. 2017) (Samaliev et al. 2017) (Samaliev et al. 2017) (Samaliev et al. 2017) (Samaliev et al. 2017) (Samaliev et al. 2017) D'Addabbo et al. (2017)
Bursaphelenchus xylophilus	*Cinnamomum zeylanicum* *Eclipta prostrata*	Aerial parts Aerial parts	Essential oil Essential oil	(Park et al. 2005; Kong et al. 2006) (Shin et al. 2016)
Pratylenchus coffeae	*Terminalia nigrovenulosa* *Cinnamomum camphora* *Jasminum suptriplinerve*	Aerial parts	Extract	(Nguyen and Jung 2014)

(continued)

Table 6.2 (continued)

Nematodes species	Plant species	Plant part used	Product type	References
Pratylenchus penetrans	*Lilium longiflorum* Thunb *Phaseolus vulgaris* *Medicago sativa* *Musa acuminata*	Aerial parts	Essential oil	(Westerdahl et al. 2020)
Pratylenchus brachyurus	*Rosmarinus officinalis*	Aerial parts	Essential oil	(Mattei et al. 2014)
Pratylenchus thornei	*Hyoscyamus niger* L. *Melia azedarah* L. *Xanthium strumarium* L.	Aerial parts Aerial parts Aerial parts	Extract	(Kepenekci et al. 2016) (Kepenekci et al. 2016) (Kepenekci et al. 2016)
Pratylenchus scribneri	*Phaseolus lunatus*	Underground parts	Extract	(Rich et al. 1977)
Heterodera avenae	*Avena sativa* *Kaempferia galanga* L *Mentha canadensis*	Aerial parts Underground parts Aerial parts	Essential oil Essential oil	(Soriano et al. 2004) (Li et al. 2017) (Ji et al. 2016)
Heterodera zeae	*Tagetes patula L.*	Aerial parts	Extract	(Faizi et al. 2011)
Heterodera glycines	*Glycine max* *Paeonia suffruticosa* *Paeonia rockii* *Camellia oleifera* *Artemisia absinthium* *Ambrosia artemisiifolia* *Euphorbia esula*	Underground parts Aerial parts Aerial parts Aerial parts Aerial parts Aerial parts Aerial parts	Extract	(Huang and Barker 1991; Kennedy et al. 1999; Wen et al. 2019) (Wen et al. 2019) Wen et al. (2019) (Dhital 2020) (Dhital 2020) (Dhital 2020)
Globodera rostochiensis	*Artemisia annua* *Artemisia herba-alba* *Artemisia absinthium* *Lantana camara* *Urginia maritima*	Aerial parts Aerial parts Aerial parts Aerial parts Aerial parts	Extract	(D'Addabbo et al. 2017) (Nebih and Charif 2019) (Nebih and Charif 2019) (Nebih and Charif 2019) (Nebih and Charif 2019)
Ditylenchus dipsaci	*Trifolium repens* *Medicago sativa* *Avena sativa*	Aerial parts Aerial parts Aerial parts	Extract	(Cook et al. 1995) (Edwards et al. 1995) (Soriano et al. 2004)
Ditylenchus angustus	*Oryza sativa*	Aerial parts	Extract	(Plowright et al. 1996)
Ditylenchus destructor	*Elsholtzia fruticosa*	Aerial parts	Essential oil	(Liang et al. 2020)

(continued)

Table 6.2 (continued)

Nematodes species	Plant species	Plant part used	Product type	References
Tylenchorhynchus sp.	*Quillaja saponaria*	Aerial parts	Extract	(San Martín and Magunacelaya 2005)
Criconemoides xenoplax	*Quillaja saponaria*	Aerial parts	Extract	(San Martín and Magunacelaya 2005)
Xiphinema index	*Quillaja saponaria*	Aerial parts	Extract	(San Martín and Magunacelaya 2005)
Helicotylenchus sp.	*Quillaja saponaria*	Aerial parts	Extract	(San Martín and Magunacelaya 2005)
Rotylenchulus reniformis	*Withania somnifera* *Ocimum tenuiflorum* *Mentha arvensis* *Lantana camara* *Calotropis gigantea*	Aerial parts Aerial parts Aerial parts Aerial parts Aerial parts	Extract	(Patil et al. 2017b)
Tylenchulus semipenetrans	*Calotropis procera* *Datura alba* *Azadirachta indica*	Aerial parts Aerial parts Aerial parts	Extract	(Ahmad et al. 2004)

concentrates based on various plant seed oils, including cotton, olive, soybean, canola, and sesame.

The impact of essential oils and plant extracts was further investigated for other important PPN. Numerous research on cyst nematodes has demonstrated the potential of compounds produced by plants to lower nematode populations. For instance, Soriano et al. (2004) tested leaf extracts of oat (*Avena sativa*) on cereal cyst nematodes (*Heterodera avenae*), while Ji et al. (2016) studied the efficacy of mint (*Mentha canadensis*) essential oils and proved that it could be used as BCA. Similarly, potato cyst nematodes (*Globodera rostochiensis*) were shown to be susceptible to *Artemisia* species (D'Addabbo et al. 2017; Nebih and Charif 2019). On the other side, a strong effect of plant extracts was recorded on the stem nematode (*Ditylenchus* spp.) (Cook et al. 1995; Plowright et al. 1996; Soriano et al. 2004; Liang et al. 2020), *Xiphinema* spp. (San Martín and Magunacelaya 2005), the reniform nematode (*Rotylenchulus reniformis*) (Patil et al. 2017a), and the citrus nematode (*Tylenchulus semipenetrans*) (Ahmad et al. 2004).

6.2.2 Mode of Action of Natural Products Against *Meloidogyne* Species

The most destructive nematode pest for global agricultural productivity is the root-knot nematode. The most widely used strategy is chemical control, but several very efficient nematicides are no longer used on some crops due to environmental and public health risks (Peiris et al. 2020). Duddington pioneered nematode biocontrol in 1951. Since then, the research has led to the development of numerous commercial biological control solutions that use live microorganisms or their metabolites to target specific nematode hosts (Lamovšek et al. 2013). It is known that various organisms are antagonistic to plant parasitic nematodes (Moosavi and Zare 2012). Due to the abundance of bacteria, fungi, actinomycetes, and other predators (such as mites), biological control is a widely used technique (Lamovšek et al. 2013; Peiris et al. 2020). These biological control agents work against one another in many ways (Lamovšek et al. 2013). The microorganisms used to control nematodes biologically can be split into four basic groups: (i) obligate parasites like *Hirsutella rhossiliensis* and *Pasteuria penetrans* (Minter and Brady); (ii) facultative parasites like *Paecilomyces lilacinus* and *Verticillium* spp., which trap nematodes; (iii) rhizobacteria like *Burkholderia cepacia* and *Pseudomonas fluorescens* (Trivisan) Migula, and (iv) competitors such as mycorrhizal fungi and endophytes (such as *Glomus mosseae* Nicol and Gerd.) (Whipps and Davies 2000). Some microorganisms (such as fungi) parasitize the nematodes, while others destroy the nematodes (by producing toxic chemicals) juveniles (Lamovšek et al. 2013).

Research on microbial pathogens and antagonists of RKN and other economically significant species has improved throughout a 50-year development phase. This research has included (i) the isolation and identification of organisms with the potential to act as biological control agents, (ii) ecological soil environment manipulation to increase antagonism, (iii) the clarification of parasitism and infection mechanisms, and (iv) exploration for the development of commercial products. Therefore, it is unexpected that despite years of extensive research, the impact of biological control on the field management of RKN has remained limited. The variety and density of communities and/or individual hostile microorganisms present in a particular soil determine the level of biological control. In soil, biological control action is pervasive and can range from negligible to completely nematode-suppressing (Hallman et al. 2009). Throughout the world, PPNs significantly deteriorate various vegetables and agricultural products. Nematophagous microorganisms, which are nematodes' natural enemies, present a promising strategy for nematode pest control. Some of these microbes produce traps that the nematodes can fall into and be killed by. Others carry out their functions as parasites inside the nematodes, releasing poisons and other virulence elements that cause the nematodes to die from the inside out (Li et al. 2015a).

The most varied class of nematode natural enemies, nematophagous fungi, use a variety of techniques to capture and destroy their prey ((Nordbring-hertz et al. 2006; Stirling 2014; Peiris et al. 2020). They are widespread across the fungal kingdom and belong to various taxonomic groups. Many different species have been described (Stirling 2014). Exploiting fungi to manage nematodes is gaining attention as a fascinating and fast-expanding field of study in fungal biological control (Moosavi and Zare 2012). Some nematophagous fungi are facultative or opportunistic parasites, which can survive saprophytically, while others are obligatory parasites, which require nematodes to survive, and yet others exhibit traits that fall somewhere in the middle of these two categories. The easiest way to categorize nematophagous fungi is into those with substantial hyphal growth outside of their hosts, including nematode-trapping fungi, opportunistic parasites of nematode eggs, and those mostly endoparasitic (Viaene et al. 2006). In other words, nematophagous fungi are made up of three main categories of fungi: nematode-trapping, endoparasitic, and parasitic fungi. The latter two use specialized structures to attack vermiform living nematodes, and the parasite fungi use their hyphal ends to attack eggs and cysts. These fungi are still interesting because they may be used as biocontrol agents for parasitic nematodes affecting both plants and animals. The remarkable morphological changes and the spectacular capture of nematodes by both nematode-trapping and endoparasitic fungi further contribute to the ongoing fascination with nematophagous fungi. The fact that both nematodes and fungi are easy to grow in the lab also makes it an excellent reference system for studies of interactions (Nordbring-hertz et al. 2006).

Nematophagous fungi can be categorized into four main groups based on how they kill nematodes (Swe et al. 2011). These include endoparasitic fungi, which use their spores, egg parasitic fungi, which use their hyphal tips to invade nematode eggs or females; and nematode-trapping fungi, which use mechanical or adhesive hyphal traps, as well as fungi that generate poisons to immobilize nematodes before the invasion (Swe et al. 2011). The classification of nematophagous fungi into five classes has been proposed: opportunistic or ovicidal, nematode-trapping/predators, endoparasites, toxin-producing fungi, and producers of specific attack devices (de Elias Freitas Soares et al. 2018). Nematophagous fungi can be either facultative or obligatory parasites (Bengtsson 2015). As a spore, the obligatory parasites attack the host (Hallman et al. 2009). A feeding nematode may ingest the fungal spore or stick to a migrating nematode. An infectious hypha enters the host directly from the spore by penetrating the cuticle or the digestive tract. After developing as saprotrophs in soil or the rhizosphere, facultative parasites can produce specialized spores, conidia, or hypha that attach to or trap nematodes and infect them (Barron 1977). Nematophagous fungi can be endoparasitic, wholly dependent on nematodes for nutrition, or nematode-trapping, which alternates between carnivory and saprophytism and obtains nutrients from both organic matter and nematodes. Natural populations of these nematophagous fungi may be present in the soil during agricultural conditions. However, the natural food web may be affected, leading to poor predation activity, depending on the level of land management (Peiris et al. 2020). The most studied fungal genera against RKN were *Pochonia* spp., *Trichoderma* spp., and *Paecilomyces* spp. Most fungal genera generally decreased RKN and improved plant growth and production. However, the findings showed that fungi could not independently achieve a significant level of repression. In general, fungal bio-agents may reduce RKN population and damage levels by 45% compared to untreated situations. *Arthrobotrys* spp. and *Acremonium* spp. were discovered to be more effective at lowering RKN damage than other species (Peiris et al. 2020).

Nematophagous bacteria function in various ways, including parasitizing, producing poisons, antibiotics, or enzymes, competing with other organisms for resources, causing systemic plant resistance, and promoting the health of plants (Tian et al. 2007). In other words, non-pathogenic bacteria antagonize nematodes in three ways: (i) by establishing plant resistance (induced or systemic resistance), (ii) by destroying the signalling molecules that attract the nematodes, or (iii) by simply colonizing the roots and preventing the entry of infectious juveniles (Lamovšek et al. 2013). It is important to highlight that the low field efficacy of commercial biological control products is still a problem. This is because the processes described before are all susceptible to various biotic and abiotic influences, which restricts their application in biological control (Lamovšek et al. 2013).

6.2.2.1 Predators

Nematodes are preyed upon by predatory nematodes, mites, insects, and various other invertebrates, including tardigrades. Predators are widespread in soil and can consume other living organisms. Some predatory nematodes, including *Mononchoides gaugleri*, have undergone extensive biological and feeding studies. Using their teeth, enzymes, or toxins, they can kill numerous nematodes daily. In particular, in natural habitats where they may be abundant, micro-arthropods like mites and springtails control nematode numbers. However, they are unsuitable for biological control programs targeting specific nematode pests due to their lack of specificity for plant-parasitic nematodes. Additionally, it is thought that their large production and distribution to the soil is impractical (Viaene et al. 2006).

6.2.2.2 Fungi

Nematode-trapping fungi use mechanical or adhesive hyphal traps (Swe et al. 2011). Although nematode-trapping fungi are typically considered soil inhabitants instead of root associates, they have been isolated from the rhizosphere. Genera in the order *Orbiliales* are by far the most prevalent and well-researched group of fungi that produce specialized nematode traps. These fungi, also known as nematode-trapping or predatory fungi, can occasionally have conidial traps on their own, but more often than not, they respond to the presence of nematodes by constructing mycelial traps that they use to capture and kill their prey. Adhesive networks, adhesive knobs, constricting and non-constricting rings, and adhesive branches are the five types of trapping devices (Stirling 2014).

Arthrobotrys oligospora is the most well-known and widely studied nematode-trapping fungus. In order to capture soil-dwelling nematodes, it develops a three-dimensional hyphal network (Viaene et al. 2006). Numerous studies suggested that *A. oligospora* could be used as a biocontrol agent against the RKN *M. incognita* (Bakr et al. 2014) (Soliman et al. 2021). Experiments were conducted to determine how *Arthrobotrys oligospora* impacted tomato plants infected by *M. incognita*; experiments were carried out. The in vitro trapping rate and the impact of *A. oligospora* on the capture of *M. incognita* juveniles were calculated. *A. oligospora* creates three-dimensional, adhesive networks, and its trapping organs can capture *M. Incognita* juveniles in their second stage. The nematode juveniles were subjected to *A. oligospora* culture for 24, 48, and 72 hours as part of an in vitro test. By lengthening the exposure period at the trapping organs, the juveniles' capture rate rose (72 h). In vitro test results revealed a substantial decrease in nematode criteria compared to the untreated control. Significant improvements were also made to tomato growth factors (Bakr et al. 2014). *M. incognita* was significantly suppressed and preyed upon by *Arthrobotrys oligospora*. The fungus evolved other trapping techniques besides secreting toxic substances to *M. incognita*. The fungus promotes the growth of plants (Soliman et al. 2021). *Meloidogyne hapla* on plants can be controlled in an environmentally acceptable manner by using nematode-trapping fungi as an agent for nematode biocontrol. In tomato plants, *M. hapla* may be reduced by *Arthrobotrys thaumasia* and *A. musiformis* by 93% and 97%, respectively. JPN2 treatment (tomato plants polybag containing *M. hapla* handled with *A. musiformis*) had the lowest number of *M. hapla*-caused root-knot infections in tomato, followed by JPN1 (tomato plants polybag containing *M. hapla* handled with *A. thaumasia*) treatment. It was also observed that, in comparison to JPN1 isolation treatment, JPN2 isolate treatment can produce the highest values of root length, root wet and dry weight, stem length, and stem wet and dry weight (Purba et al. 2022).

It is crucial to highlight that the vulnerability of several nematode-trapping fungi to root-knot and other nematodes (cyst nematodes) varies. For example, *Meloidogyne hapla* was found to be more vulnerable to *Arthrobotrys oligospora* than two cyst nematode species, including *Globodera pallida* and *G. rostochiensis*, in experiments on agar. When *Dactylellina ellipsospora* and *Arthrobotrys gephyropaga* were used to challenge three species of RKN, they were substantially more vulnerable to predation than *Heterodera schachtii*. It was found that *M. javanica* was more likely to be caught by *Dactylellina candidum* and *Arthrobotrys thaumasia* than by *Heterodera schachtii*. However, *Drechslerella dactyloides* was equally damaging to the two nematode species. *Dactylellina lysipaga* affected *Meloidogyne javanica* more than *Heterodera avenae* (Stirling 2014).

6.2.2.2.1 Endoparasitic Fungi

Spores from endoparasitic fungi are used to manage nematodes (Swe et al. 2011). Endoparasitic fungus uses the spores (conidia or zoospores) of vermiform PPNs to infect them. The nematode can either consume the spores and allow them to germinate in the intestines or firmly cling to the nematode cuticle when it comes in contact with the fungus (most commonly the oesophagus or mastax). A thin penetration tube injects the spore contents into the nematode under some mechanical pressure (Moosavi and Zare 2012). The internal mycelium then grows and eventually reaches the surface of the cadaver to sporulate 20 (Moosavi and Zare 2012). A few endoparasitic fungi generate zoospores, which swim towards the nematode, attach to the cuticle, usually close to the natural orifices, and then encyst. The host body's physiological openings enable the encysted zoospores to enter and start their vegetative growth. The hyphae subsequently create a sporangium with zoospores (Viaene et al. 2006). *Hirsutella rhossiliensis*, a parasite of numerous commercially significant PPNs, has been the subject of most ecological studies on endoparasitic fungi. Multiple lines of evidence demonstrate how this fungus's levels of parasitism rely on the population density of its nematode hosts (Stirling 2014).

Egg-parasitic fungi use their hyphal tips to invade nematode eggs or females (Swe et al. 2011). Their target pest is immobile, making it easier to infect. They often infect their host by simple hyphal penetration, occasionally with the formation of an appressorium. This is because they are less specialized than the fungal parasites that target soil-dwelling nematode stages (Viaene et al. 2006). Cyst and RKN females and eggs are parasitized by *Pochonia chlamydosporia*. Hyphae enter eggs after an appressorium develops on the eggshell. The eggshell-degrading enzymes, serine protease and chitinases and the nematotoxin phomalactone produced by *P. chlamydosporia*. They may contribute to pathogenicity. Chlamydospores are produced for survival and used as an inoculum to plant the fungus in the soil and rhizosphere because they can withstand harsh climatic conditions. The fungal isolates vary significantly in their capacity to form chlamydospores, colonize roots, and infect nematodes (Viaene et al. 2006). Nematophagous fungi can regulate the populations of plant-parasitic nematodes; when the nematode population within the plant host is high. The parasites feed on females or their eggs, and the parasitism eventually rises to the point where the nematode is permanently repressed (Stirling 2014). *Pochonia chlamydosporia*, which is regarded as a facultative parasite of the nematodes, has a variable ability to suppress RKN populations (Bengtsson 2015).

6.2.2.2.2 Toxin-Producing Fungi

To immobilize nematodes before the invasion, fungi known as toxin-producing fungi produce poisons (Swe et al. 2011) (Lopez-Llorca et al. 2008). More than 200 compounds with nematicidal activity have been identified from more than 280 fungal species in 150 genera of the *Ascomycota* and *Basidiomycota*. These molecules represent a wide range of chemical classes, including alkaloids, peptides, terpenoids, macrolides, oxygen heterocycle, benzo compounds, quinones, aliphatic compounds, simple aromatic compounds, and sterols (Li et al. 2015a). Experimental evidence suggests that *Verticilium leptobactrum* uses toxin- or enzyme-based procedures to kill nematodes. This is because its metabolites influence *M. incognita* eggs' integrity, capacity to hatch, and J2 viability (Regaieg et al. 2010). Seven toxins from the fungus *Coprinus comatus* can immobilize the nematodes *Meloidogyne incognita* (Luo et al. 2007).

6.2.2.2.3 Endophytic Fungi

Endophytic fungi develop within plant tissues without harming the plant. The most well-known endophytes found on plant roots are arbuscular mycorrhizal (AM) fungi, which are obligatory symbiotic parasites of plants, including *Glomus* spp. Several plant-nematode interactions have investigated their function in preventing nematode damage and lowering nematode concentrations in the soil. The majority of this research focuses on *Meloidogyne* spp. AM fungi increase plant growth by enhancing the plant's access to nutrients, notably P, especially when nutrients are few. In addition, AM fungi reduce heavy metal toxicity, improve water intake, and reduce pest and disease damage, particularly that caused by nematodes. Before nematode invasion, colonization of roots by AM fungus may have a more substantial impact on nematode multiplication rates than after nematode invasion. Nematode antagonism's precise mechanism(s) of action are not well understood, although they are likely to include both very specific processes and several mechanisms functioning together. Additionally, they might create nematotoxic substances or obstruct the synthesis of root diffusates (Viaene et al. 2006).

6.2.2.3 Bacteria

The majority of bacteria that affect nematode behaviour, feeding, or reproduction do so accidentally by producing toxins, antibiotics, or enzymes. Numerous products, including nitrogenous compounds and volatile fatty acids, are created by bacteria during the breakdown of organic materials and may have an impact on nematode populations in the soil and rhizosphere. The identification of bacterial strains with high antagonistic activities has come about as a result of screening rhizobacteria or their metabolites (extracts of their cultures) on Petri plates. It is unclear how these metabolites are produced and how important they are in the rhizosphere. By affecting nematode hatching and motility, *Burkholderia* spp., *Pseudomonas* spp., *Bacillus* spp., and *A. radiobacter* may prevent nematode penetration of roots or may promote plant resistance. A number of these bacteria also promote plant development (PGPB). *Pasteuria* spp. has the greatest potential to be used as a biological control agent. All commercially significant nematodes that parasitize plants have been observed to be attached to and parasitized by *Pasteuria* endospores (Viaene et al. 2006).

When dome-shaped endospores attach to the cuticle of nematodes as they migrate through the soil, *Pasteuria* spp. begin their life cycle. It is thought that a Velcro-like attachment mechanism occurs between the cuticle receptor and the collagen-like fibres on the surface of the endospore (Davies and Curtis 2011). When an infection peg pierces the cuticle, endospores can either germinate immediately (as in the case of *Heterodera avenae*) or later when the nematode has entered the root and established a feeding site. Small rod-shaped bacteria grow exponentially after germination to build granular masses that eventually undergo sporogenesis and produce the following generation of spores (Davies and Curtis 2011). Compared to the first generation, juvenile nematodes on the same plant in succeeding generations only travel small distances from the egg to the surrounding roots. *Pasteuria* spp. was involved in the natural reduction of RKN in tobacco fields. However, most investigations on the bacteria's effectiveness in nematode biological control were conducted in pots due to challenges in growing enough spores for large-scale experiments. In tiny plots, nematode levels have decreased, and root galling has been documented. The affected female of *Meloidogyne* spp. continues to grow and become infertile as the bacteria produces up to two million endospores, damaging the reproductive system. When infected females and roots decompose, endospores are discharged into the soil, producing fresh inoculum for the biocontrol agent. Although spores can live in air-dried soil for several years, the dispersion of the spores in the soil, which can be modified by soil type, tillage techniques, moisture, and temperature, is crucial for the successful infection of nematodes (Viaene et al. 2006).

6.3 Molecular Suppression Mechanisms

When developing biological control agents for PPNs, it is crucial to comprehend the molecular bases of the interactions between microorganisms and nematodes (Li et al. 2015a). Over the past 10 years, there has been a substantial advancement in our knowledge of the molecular processes governing the interactions between model nematodes and nematophagous microorganisms. These revelations have provided intriguing new targets and directions for the creation of potent PPN biological control methods. This field is expected to improve more in the upcoming years as molecular biology and biotechnology develop in addition to the increase in the availability of omics data from PPNs and the related microbes. Finding the functions of those crucial genes and variables in determining the mode of action of BCAs should improve the nematicidal potential of BCAs through targeted genetic modifications, enhancing the biological control efficacy of PPN management (Li et al. 2015a). The molecular mechanisms associated with particular microorganisms' suppression of nematodes are listed here.

Nematode-trapping (NT) fungi play a significant role in the biological control of PPNs. When nematodes lack nutrients, NT fungi can transform into specialized objects known as "traps" that can catch, kill, and eat the nematodes. Thus, establishing traps is a vital indicator that the NT fungus switches from a saprophytic to a predacious existence. With the advent of gene knockdown and numerous omics, such as genomics, transcriptomics, and metabolomics, the number of studies seeking to understand the regulatory mechanism of trap formation in NT fungus has expanded. Signalling pathways have been demonstrated to play a significant role in trap formation based on the phenotypes of various mutants and multi-omics research. Additionally, reports have linked the creation of traps to small molecule substances, woronin bodies, peroxisomes, autophagy, and pH-sensing receptors (Zhu et al. 2022).

The most frequent reproductive strategy used by filamentous fungi for environmental invasion, spread, and proliferation is conidiogenesis. In the future, the nematode-trapping fungus may benefit from understanding the molecular mechanisms governing conidiation and improving conidium production for commercial development (Liu et al. 2022). Liu et al. (2022) used gene knockout in *A. oligospora* to characterize three novel conidiogenesis-related genes. Conidia formation increased significantly when the genes AoCorA and AoRgsD were knocked out; however, conidiogenesis decreased when AoXlnR was absent. Additionally, they identified the *Aspergillus nidulans* homologue of the well-known conidiogenesis-related gene AbaA. Not only did the deletion of AoAbaA stop the formation of conidia, but it also impacted the creation of traps for nematodes (Liu et al. 2022).

6.4 Factors Affecting the Success/Failure of Plant Products as Nematicides

Plants are regarded as a rich source of biocidal components ideal for environmentally friendly control and can replace chemical nematicides in managing RKN (Alam 1989). Numerous findings have documented successful nematode management approaches utilizing plant products as essential oils and botanical extracts. These natural nematicides have been extensively reported to suppress nematode reproduction strongly (Khan et al. 2019). For example, a neem-based nematicide shows high efficacy for reducing the development of RKN, over and above that, this product may improve plant growth (Mohd Yaqub 2012; Yadav et al. 2018).

Furthermore, a different study showed that applying *Androctonus caucasicus* leaf meal at 0.5 and 1 g/100 cm^3 soil reduced *M. incognita* reproduction by 82.3 and 92.7% (Di Vito et al. 2010). Likewise, four plant extracts, *Azadirachta indica* (neem), *Withania somnifera* (ashwagandha), *Tagetes erecta* (marigold), and *Eucalyptus citriodora* (eucalyptus) have recorded an important potential for minimizing root-knot index and the quantity of egg masses of *M. incognita* associated with papaya (*Carica papaya*), in the in vitro studies and under field conditions as well (Khan et al. 2008). Understanding soil biological and ecological aspects will improve the efficacy and success of bionematicide management. These variables were therefore emphasized to provide more excellent guidance for their use.

However, there have been many failures for every achievement. Even though bionematicide has demonstrated promising efficacy in laboratory or field plots, success was not achieved in a number of cases (Askary and Martinelli 2015). Since it is well recognized that a variety of circumstances, such as low efficacy in field conditions and PPN-resistance development, can affect the control efficacy of plant products as nematicides. The stabilization and effectiveness of the active natural components of bionematicides can be impacted by a variety of factors, some of which are generally connected to soil microbiology, biochemistry, and environmental circumstances. It is clear that soil biological and environmental elements have a fundamental influence on the effectiveness of natural management. Significant variables such as soil texture, moisture, temperature, the population of predatory microbes, malnutrition, and the amount of organic carbon in the chemical are all related to how quickly volatile and non-volatile substances degrade. Hence the difficulty in manipulating bionematicides and controlling these factors in the field (Abd-Elgawad and Askary 2020). Timper (2014) conducted additional research that illustrated agricultural practices' crucial role in enhancing or deterring the biological management of PPNs and other soil-borne pests. This research showed that because PPNs are associated with native antagonists, biological regulation of PPNs to either protect or stimulate their suppression may be less successful across all field areas. Therefore, the application of effective bionematicide in the soil is controlled by rhizosphere biology and microflora. To increase the performance of their products in the field, scientists and manufacturers of these active chemicals must examine the kind of soil and ambient elements during testing. According to multiple earlier publications, most effective experiments are frequently conducted under strictly controlled experimental circumstances with little opportunity to influence outside variables. The effectiveness of volatile chemicals and essential oils then varies depending on the ambient circumstances, typically within the control (Mwamula et al. 2022).

Notably, *Brassica* species liberate VOCs from their macerated leaves; some molecules present in its VOC mixture have separately been found to have nematicide potential (Zasada and Ferris 2003; Ojaghian et al. 2012). For instance, it is well known that using a natural product with mustard extract and the pure volatile molecule known as allyl isothiocyanate (AITC), which is produced when the *Brassica* species hydrolyzes glucosinolates, has a strong nematicidal activity against RKN (*Meloidogyne* spp.) (Zasada and Ferris 2003; Wu et al. 2011). However, some research revealed that the soil physical properties affected the transformation and diffusion of AITC (Borek et al. 1995). According to a study carried out by Dahlin and Hallmann (2020), the kind of soil has an impact on how well allyl ITC works as nematicide to control *M. hapla*. The results of this experiment revealed that the same concentration of allyl ITC could completely suppress the nematode population in the sand, whereas in the organic potting substrate or normal soil, this active compound showed weak potency against cucumber root galling. Moreover, the degradation of the natural component high temperatures and organic soil additives may have an impact on methyl isothiocyanate after it has been incorporated into the soil (Dungan and Yates 2003).

On the other hand, factors such as composition, variabilities in quality, and efficacy persist due to some variabilities associated with extraction and product formulation methods, which may be deemed to influence the performance of biological control of PPNs (Mwamula et al. 2022). Therefore, plant species, varieties, and application rates determine biological control's effectiveness and phototoxicity (Mazzola et al. 2007). Zasada et al. (2009) reported the efficacy of seed meals as soil amendments to control the PPNs *Pratylenchulus penetrans* and *M. incognita* depend on formulations and particle size. This experiment revealed that when the seed meal was ground to a lower particle size, as opposed to when it was used as a pellet, *S. alba*'s effectiveness against *P. penetrans* increased by 47–56%. Furthermore, under field conditions, natural nematode management strategies have minimal efficiency due to the repetitive application of the few currently available commercial bionematicides and the growth of microbial biodegradation in soil (Caboni and Ntalli 2014).

A study by Barros et al. (2019) pointed out the nematicidal effect of *Phaedranassa viridiflora* essential oil, which is attributed to benzaldehyde. The later represents 98% of the total oil mixture. According to previous research, benzaldehyde has been found in multiple natural products (Barros et al. 2014; Jardim et al. 2020a) for its nematicidal potency which has been demonstrated against *M. incognita* through laboratory experiments (Jardim et al. 2018). Nevertheless, this substance exposes a lack of efficacy against *M. incognita* under field circumstances. The key problem reducing the effectiveness of field treatment is the inability of natural substances like benzaldehyde to remain in soils and provide long-term nematode control (Barros et al. 2019). The lipid layer below the chitin of the egg exerts protection against molecular entry into the developing embryo, which may be the cause of benzaldehyde's failure to reduce *M. incognita* J2 hatching (GAUGLER 2004). In addition, numerous researches have suggested using benzaldehyde in conjunction with an organic amendment to suppress *M. incognita* in tomato and soybean (Chavarría-Carvajal et al. 2001; Kokalis-Burelle et al. 2002). While Soler-Serratosa et al. (1996) demonstrated that thymol and benzaldehyde combinations had a remarkable suppressive effect on populations of *M. arenaria* and *Heterodera glycines*. To avoid all of the aforementioned issues, it is vital to look for comparable benzaldehyde molecules (analogs) that are highly stable in environmental conditions. Accordingly, it appears that active compound standardization is needed part to ensure uniformity in efficacy.

However, regularizing the formulations to use the active compounds without putting strain on non-target soil microbial communities presents challenges (Mwamula et al. 2022). In addition, some isolated active substances, when used separately, may or may not function in isolation. According to Faria et al. (2016) and Ntalli et al. (2020a, b), many essential oils become prone to conversion and degradation reactions when employed alone. As a result, some compounds may lose quality or exhibit phytotoxicity toward non-target soil microbial populations. Although botanical nematicides are highly recommended as a viable alternative for plant protection with little side effects, various studies have reported their unfavourable impacts on microbial communities that are not the intended target (Isman 2006; Miresmailli and Isman 2014).

Therefore, it is important to understand the function of each component independently in the semi-refined mixtures to neutralize the unwanted poisonous and damaging effects on plants and the soil microbial community before manipulation. Chemical insecticides are frequently preferable to using unrefined or semi-refined plant extracts since they are more environmentally friendly. However, researchers should create formulation strategies that limit plants' chemical compartmentalization and storage capacity, which are essential before commercialization to avoid sort residual life under field conditions (Miresmailli and Isman 2014). Contrarily, it is necessary to emphasize that a variety of plant materials are frequently slow-acting and that the high cost of screening and commercial production seriously questions the sustainability of scaling up production of several of the examined compounds. As a result, some of the examined plants are frequently utilized as organic manure or soil amendment (Chitwood 2002; Ntalli and Caboni 2012).

6.5 Concluding Remarks and Future Perspectives

It is apparent that new environmentally friendly strategies need to be applied to have significant control over economic losses caused by RKN. The investigations of numerous researchers that support the use of biological control as an adequate substitute for chemical nematicides have been set down in this review. For the sake of the stability of the environment, it is essential to keep developing green technologies to increase their effectiveness. For antagonistic properties of fungi/

bacteria, for instance, some issues need to be addressed: for example, the best rate, time, frequency, and mode of application for biocontrol agents, particularly in field circumstances. By altering the habitat, combining beneficial species, and combining biocontrol with other complementary alternative strategies, it is possible to make these biocontrol products more effective against RKN. To achieve efficient biocontrol, it is essential to choose effective agents in various conditions, such as soil texture, moisture, temperature extremes, and competition.

The above review has enlightened us that botanical products are essential for optimizing production qualities and enhancing plant health. Plant components may create new organic nematicidal substances as metabolites or chemicals. The plant-based treatments are more readily available and reasonably priced for small farms than chemical nematicides. Chemistry can be used to create an appropriate formulation of a plant extract with effective nematicidal properties. However, more research can be done to determine the active components of botanicals and the intricate chemical and biological processes that occur in the rhizosphere of the host plant, allowing for the proper marketing of plant extract formulations without harming our ecosystem or the farmers' economies. Further research is still needed on how all control strategies can be employed in concert, in addition to the parts mentioned above and current technological advancements. Based on the dataset, scientific research should focus on the microbiomes of RKN suppressive soils to investigate the potential for developing more comprehensive management strategies with multi-target modes of action. Therefore, it is crucial to create and perfect interdisciplinary management ways for RKN, like combining microbial strategies using bacterial and fungal agents with natural products along with cultural control techniques or host resistance.

In summary, future research should emphasize ecologically friendly approaches to build on multidisciplinary approaches and can cover the gaps left by one-sided management techniques. The synergism between RKN antagonists, environmental conditions, sustainability, investigating the effects of novel treatments on non-target organisms, and associations between particular plants and potentially useful nematode antagonists are just a few of the critical factors that should be the focus of future work.

References

Abad P, Castagnone-Sereno P, Rosso M-N, et al (2009) Invasion, feeding and development. In: Root-knot nematodes. CABI Wallingford UK, pp. 163–181

Aballay E, Insunza V (2002) Evaluación de plantas con propiedades nematicidas en el control de *Xiphinema* index en vid de mesa cv. Thompson Seedless en la zona central de Chile. Agric Técnica 62:357–365

Abd-El-Khair H, El-Nagdi WMA, Youssef MMA et al (2019) Protective effect of *Bacillus subtilis, B. pumilus, and Pseudomonas fluorescens* isolates against root knot nematode *Meloidogyne* incognita on cowpea. Bull Natl Res Cent 43. https://doi.org/10.1186/s42269-019-0108-8

Abd-Elgawad MMM (2016) Comments on the use of biocontrol agents against plantparasitic nematodes. Int J Pharm Tech Res 9:352–359. https://doi.org/10.1201/9781482277951-14

Abd-Elgawad MMM (2021) Optimizing safe approaches to manage plant-parasitic nematodes. Plan Theory 10. https://doi.org/10.3390/plants10091911

Abd-Elgawad MMM, Askary TH (2018) Fungal and bacterial nematicides in integrated nematode management strategies. Egypt J Biol Pest Control 28. https://doi.org/10.1186/s41938-018-0080-x

Abd-Elgawad MMM, Askary TH (2020) Factors affecting success of biological agents used in controlling the plant-parasitic nematodes. Egypt J Biol Pest Control 30. https://doi.org/10.1186/s41938-020-00215-2

Abd-Elgawad MMM, Kabeil SSA, Abd El-Wahab AE (2009) Changes in protein content and enzymatic activity of tomato plants in response to nematode infection. In: journal of nematology. Soc nematologists po box 311, marceline, mo 64658 USA, p 300

Adegbite AA, Adesiyan SO (2006) Root extracts of plants to control root-knot nematode on edible soybean. J Veg Sci 12:5–12

Ahmad MS, Mukhtar T, Ahmad R (2004) Some studies on the control of citrus nematode (*Tylench ulus semipenetrans*) by leaf. Asian J Plant Sci 3:544–548

Ajith M, Pankaj SNA et al (2020) Chemical composition and nematicidal activity of essential oils and their major compounds against *Meloidogyne graminicola* (Rice root-knot nematode). J Essent Oil Res 32:526–535

Alam MM (1989) Control of plant–parasitic nematodes JBoctor of science. Dep Bot ALIGARH MUSLI M Univ ALIGARH

Alam MM, Siddiqui MA, Ahmad A (1990) Antagonistic plants. Nematode Bio-Control:41–50

Alban R, Guerrero R, Toro M (2013) Interactions between a root knot nematode (*Meloidogyne exigua*) and arbuscular mycorrhizae in coffee plant development (*Coffea arabica*)

Andrés MF, Rossa GE, Cassel E et al (2017) Biocidal effects of *piper hispidinervum* (*Piperaceae*) essential oil and synergism among its main components. Food Chem Toxicol 109:1086–1092

Angioni A, Barra A, Coroneo V et al (2006) Chemical composition, seasonal variability, and antifungal activity of *Lavandula stoechas* L. ssp. stoechas essential oils from stem/leaves and flowers. J Agric Food Chem 54:4364–4370

Ashraf M, Khan T (2010) Integrated approach for the management of *Meloidogyne javanica* on eggplant using oil cakes and biocontrol agents. Arch Phytopathol Plant Prot 43:609–614

Askary TH, Martinelli PRP (2015) Biocontrol agent of phytonemathode CAB Int, Wallingford

Askary TH (2015) Nematophagous fungi as biocontrol agents of phytonematodes. Biocontrol agents phytonematodes, Wallingford CAB Int 81–125

Atolani O, Fabiyi OA (2020) Plant parasitic nematodes management through natural products: current progress and challenges. Manag Phytonematodes Recent Adv Futur Challenges:297–315. https://doi.org/10.1007/978-981-15-4087-5_13

Bai CQ, Liu ZL, Liu QZ (2011) Nematicidal constituents from the essential oil of *Chenopodium ambrosioides* aerial parts. E-Journal Chem 8:S143

Bakr RA, Mahdy ME, Mousa EM (2014) Biological control of root-knot nematode *Meloidogyne incognita* by *arthrobotrys oligospora*. Egypt J Crop Prot 9:1–11

Barber CA (1901) A tea-eelworm disease in South India. Superintendent, Government Press

Bardgett RD, van der Putten WH (2014) Belowground biodiversity and ecosystem functioning. Nature 515:505–511. https://doi.org/10.1038/nature13855

Barron GL (1977) The nematode-destroying fungi. In: Topics in mycobiology. pp. 1–140

Barros AF, Campos VP, da Silva JCP et al (2014) Nematicidal activity of volatile organic compounds emitted by *Brassica juncea, Azadirachta indica, Canavalia ensiformis, Mucuna pruriens* and *Cajanus cajan* against *Meloidogyne incognita*. Appl Soil Ecol 80:34–43. https://doi.org/10.1016/j.apsoil.2014.02.011

Barros AF, Campos VP, De Oliveira DF et al (2019) Activities of essential oils from three Brazilian plants and benzaldehyde analogues against *Meloidogyne incognita*. Nematology 21:1081–1089. https://doi.org/10.1163/15685411-00003276

Basyony AG, Abo-Zaid GA (2018) Biocontrol of the root-knot nematode, *Meloidogyne incognita*, using an eco-friendly formulation from *Bacillus subtilis*, lab. And greenhouse studies. Egypt J Biol Pest Control 28:1–13

Bengtsson T (2015) Biological control of root-knot nematodes (*Meloidogyne* spp.) by the fungus *Pochonia chlamydosporia*

Bhowmik B, Ghosh S, Dutta S, Dey B (2021) Nematicidal efficiency of some ethnomedicinal plants against *Meloidogyne incognita*: a review nematicidal efficiency of some ethnomedicinal plants against *Meloidogyne* incognita: a review. Uttar Pradesh J Zool 42:1021–1029

Blyuss KB, Fatehi F, Tsygankova VA et al (2019) RNAi-based biocontrol of wheat nematodes using natural poly-component biostimulants. Front Plant Sci 10. https://doi.org/10.3389/fpls.2019.00483

Borek V, Morra MJ, Brown PD, McCaffrey JP (1995) Transformation of the Glucosinolate-derived Allelochemicals allyl Isothiocyanate and Allylnitrile in soil. J Agric Food Chem 43:1935–1940. https://doi.org/10.1021/jf00055a033

Borges DF, Lopes EA, Côrtes FR et al (2018) Nematicidal potential of essential oils of *Ageratum fastigiatum, Callistemon viminalis* and Schinus terebinthifolius. Biosci J 34

Botelho AO, Campos VP, da Silva JCP et al (2019) Physicochemical and biological properties of the coffee (*Coffea arabica*) rhizosphere suppress the root-knot nematode *Meloidogyne exigua*. Biocontrol Sci Tech 29:1181–1196. https://doi.org/10.1080/09583157.2019.1670781

Buena AP, Díez-Rojo MÁ, López-Pérez JA et al (2008) Screening of *Tagetes patula* L. on different populations of *Meloidogyne*. Crop Prot 27:96–100

Caboni P, Ntalli NG (2014) Botanical nematicides, recent findings. ACS Symp Ser 1172:145–157. https://doi.org/10.1021/bk-2014-1172.ch011

Calderón-Urrea A, Vanholme B, Vangestel S et al (2016) Early development of the root-knot nematode *Meloidogyne incognita*. BMC Dev Biol 16:1–14

Cawoy H, Wagner B, Patrick F, Marc O (2011) Bacillus-based biological control of plant diseases. Pestic Mod World-Pesticides Use Manag 1849:273–302

Cetintas R, Yarba MM (2010) Root-Knot Nematode, *Meloidogne incognita* Race 2. J Anim Vet Adv 9:222–225

Chanu LB, Mohilal N (2019) Management of *Meloidogyne javanica* (treub, 1885) Chitwood, 1949 using ethanol extracts of some medicinal plants from Manipur, India. J Exp Zool India 22:891–895

Chavarría-Carvajal JA, Rodríguez-Kabana R, Kloepper JW, Morgan-Jones G (2001) Changes in populations of microorganisms associated with organic amendments and benzaldehyde to control plant-parasitic nematodes. Nematropica 31:165–180

Chitwood DJ (2002) Phytochemical based strategies for nematode control. Annu Rev Phytopathol 40:221–249. https://doi.org/10.1146/annurev.phyto.40.032602.130045

Ciancio A (2021) Observations on a novel bacterial pathogen of root-knot nematodes (*Meloidogyne* spp.). Pathogens 10. https://doi.org/10.3390/pathogens10101226

Colagiero M, Rosso LC, Ciancio A (2018) Diversity and biocontrol potential of bacterial consortia associated to root-knot nematodes. Biol Control 120:11–16. https://doi.org/10.1016/j.biocontrol.2017.07.010

Cook R, Tiller SA, Mizen KA, Edwards R (1995) Isoflavonoid metabolism in resistant and susceptible cultivars of white clover infected with the stem nematode *Ditylenchus dipsaci*. J Plant Physiol 146:348–354

D'Addabbo T, Argentieri MP, Radicci V et al (2017) *Artemisia annua* compounds have potential to manage root-knot and potato cyst nematodes. Ind Crop Prod 108:195–200

D'Addabbo T, Laquale S, Lovelli S et al (2014) Biocide plants as a sustainable tool for the control of pests and pathogens in vegetable cropping systems. Ital J Agron 9:137–145. https://doi.org/10.4081/ija.2014.616

d'Errico G, Marra R, Crescenzi A et al (2019) Integrated management strategies of *Meloidogyne incognita* and *Pseudopyrenochaeta lycopersici* on tomato using a *Bacillus* firmus-based product

and two synthetic nematicides in two consecutive crop cycles in greenhouse. Crop Prot 122: 159–164. https://doi.org/10.1016/j.cropro.2019.05.004
Dahlin P, Hallmann J (2020) New insights on the role of allyl isothiocyanate in controlling the root knot nematode *Meloidogyne hapla*. Plan Theory 9:1–13. https://doi.org/10.3390/plants9050603
Davies KG, Curtis RHC (2011) Cuticle surface coat of plant-parasitic nematodes. Annu Rev Phytopathol 49:135–156. https://doi.org/10.1146/annurev-phyto-121310-111406
Devi G (2018) Utilization of nematode destroying fungi for Management of Plant-Parasitic Nematodes-a Review. Biosci Biotechnol Res Asia 15:377–396. https://doi.org/10.13005/bbra/2642
Dhital BP (2020) Determining the effects of plant extracts and Saltro Nematicide on hatching, Mortality and Reproduction of the Soybean Cyst Nematode (Heterodera glycines)
Di Vito M, Catalano F, Pecchia P et al (2010) Effects of meal and saponins of aster caucasicus and of *a. sedifolius* on the control of nematodes. Acta Hortic 883:361–368. https://doi.org/10.17660/actahortic.2010.883.45
DiLegge MJ, Manter DK, Vivanco JM (2019) A novel approach to determine generalist nematophagous microbes reveals *Mortierella globalpina* as a new biocontrol agent against *Meloidogyne* spp. nematodes. Sci Rep 9:1–9. https://doi.org/10.1038/s41598-019-44010-y
Dungan RS, Yates SR (2003) Degradation of fumigant pesticides: 1,3-Dichloropropene, methyl Isothiocyanate, chloropicrin, and methyl bromide. Vadose Zo J 2:279–286. https://doi.org/10.2136/vzj2003.2790
Echeverrigaray S, Zacaria J, Beltrão R (2010) Nematicidal activity of monoterpenoids against the root-knot nematode *Meloidogyne incognita*. Phytopathology 100:199–203
Edwards R, Cook R, Mizen T (1995) Isoflavonoid conjugate accumulation in the roots of lucerne (*Medicago sativa*) seedlings following infection by the stem nematode (*Ditylenchus dipsaci*). Nematologica 41:51–66
El-nagdi WMA, Youssef MMA, El-ghonaimy AM (2017) Efficacy of some bioproducts and plant defense elicitors against root- knot nematode, *Meloidogyne incognita* infesting dry common bean, *Phaseolus vulgaris*. Sci Agric 18. https://doi.org/10.15192/pscp.sa.2017.18.2.3844
El-Saedy MAM, Mokbel AA, Hammad SE (2014) Efficacy of plant oils and garlic cultivation on controlling *Meloidogyne incognita* infected tomato plants. Pakistan J Nematol 32
Elbadri GA, Lee DW, Park JC et al (2008) Evaluation of various plant extracts for their nematicidal efficacies against juveniles of *Meloidogyne incognita*. J Asia Pac Entomol 11:99–102
de Elias Freitas Soares F, Sufiate BL, de Queiroz JH (2018) Nematophagous fungi : far beyond the endoparasite, predator and ovicidal groups. Agric Nat Resour 52:1–8. https://doi.org/10.1016/j.anres.2018.05.010
Eloh K, Kpegba K, Sasanelli N et al (2020) Nematicidal activity of some essential plant oils from tropical West Africa. Int J Pest Manag 66:131–141
Faizi S, Fayyaz S, Bano S et al (2011) Isolation of nematicidal compounds from Tagetes patula L. yellow flowers: structure–activity relationship studies against cyst nematode Heterodera zeae infective stage larvae. J Agric Food Chem 59:9080–9093
Fan H, Yao M, Wang H et al (2020) Isolation and effect of *Trichoderma citrinoviride* Snef1910 for the biological control of root-knot nematode, *Meloidogyne incognita*. BMC Microbiol 20:1–11. https://doi.org/10.1186/s12866-020-01984-4
Faria JMS, Sena I, Ribeiro B et al (2016) First report on *Meloidogyne chitwoodi* hatching inhibition activity of essential oils and essential oils fractions. J Pest Sci 89:207–217. https://doi.org/10.1007/s10340-015-0664-0
Felek AF, Ozcan MM, Akyazi F (2019) Effects of essential oils distilled from some medicinal and aromatic plants against root knot nematode (*Meloidogyne hapla*). J Appl Sci Environ Manag 23: 1425–1430
Forghani F, Hajihassani A (2020) Recent advances in the development of environmentally benign treatments to control root-knot nematodes. Front Plant Sci 11:1–13. https://doi.org/10.3389/fpls.2020.01125

Gaugler ALBAR (2004) Feeding behavior department of entomology, Rutgers university, New Brunswick, NJ 08901–8524, USA. Encycl Insects 357–362. https://doi.org/10.1016/B978-0-12-374144-8.00107-7

Geng C, Nie X, Tang Z et al (2016) A novel serine protease, Sep1, from *Bacillus firmus* DS-1 has nematicidal activity and degrades multiple intestinal-associated nematode proteins. Sci Rep 6:1–12

Ghahremani Z, Escudero N, Saus E et al (2019) *Pochonia chlamydosporia* induces plant-dependent systemic resistance to *Meloidogyne incognita*. Front Plant Sci 10. https://doi.org/10.3389/fpls.2019.00945

Gogoi D, Mahanta B (2013) Comparative efficacy of *Glomus fasciculatum, Trichoderma harzianum*, carbofuran and carbendazim in management of *Meloidogyne incognita* and *Rhizoctonia solani* disease complex on French bean. Ann Plant Prot Sci 21:172–175

Goswami BK, Vijayalakshmi K (1986) Nematicidal properties of some indigenous plant materials against root-knot nematodes, *Meloidogyne incognita* on tomato. Indian J Nematol 16:65–68

Grubišić D, Uroić G, Ivošević A, Grdiša M (2018) Nematode control by the use of antagonistic plants. Agric Conspec Sci 83:269–275

Hajihassani A, Lawrence KS, Jagdale GB (2018) Plant parasitic nematodes in Georgia and Alabama. In: Plant parasitic nematodes in sustainable agriculture of North America. Springer, pp 357–391

Hallman J, Davies KG, Sikora R (2009) Biological control using microbial pathogens, endophytes and antagonists. In: Perry RN, Moens M, Starr JL (eds) Root-knot nematodes, pp 380–411

Haroon SA, Hassan BAA, Hamad FMI, Rady MM (2018) The efficiency of some natural alternatives in root-knot nematode control. Adv Plants Agric Res 8:355–362

Herrera-Parra E, Cristóbal-Alejo J, Ramos-Zapata JA (2017) *Trichoderma* strains as growth promoters in *Capsicum annuum* and as biocontrol agents in *Meloidogyne incognita*. Chil J Agric Res 77:318–324. https://doi.org/10.4067/S0718-58392017000400318

Huang D, Yu C, Shao Z et al (2020) Identification and characterization of nematicidal volatile organic compounds from deep-sea *Virgibacillus dokdonensis* MCCC 1A00493. Molecules 25: 744

Huang J-S, Barker KR (1991) Glyceollin I in soybean-cyst nematode interactions: spatial and temporal distribution in roots of resistant and susceptible soybeans. Plant Physiol 96:1302–1307

Hussain M, Zouhar M, Ryšánek P (2018) Suppression of *Meloidogyne incognita* by the entomopathogenic fungus *Lecanicillium muscarium*. Plant Dis 102:977–982. https://doi.org/10.1094/PDIS-09-17-1392-RE

Isman MB (2006) Pesticides based on plant essential oils: phytochemical and practical considerations. ACS Symp Ser 1218:13–26. https://doi.org/10.1021/bk-2016-1218.ch002

Isman MB, Machial CM, Miresmailli S, Bainard LD (2007) Essential oil-based pesticides: new insights from old chemistry. Pestic Chem Crop Prot public Heal Environ Saf 201–209

Jamal Q, Cho JY, Moon JH et al (2017) Identification for the first time of cyclo(D-pro-L-Leu) produced by *bacillus amyloliquefaciens* y1 as a nematocide for control of *Meloidogyne incognita*. Molecules 22:1–16. https://doi.org/10.3390/molecules22111839

Jardim I, Oliveira DF, Silva GH et al (2018) (E)-cinnamaldehyde from the essential oil of *Cinnamomum cassia* controls *Meloidogyne incognita* in soybean plants. J Pest Sci 91:479–487. https://doi.org/10.1007/s10340-017-0850-3

Jardim IN, Oliveira DF, Campos VP et al (2020a) Garlic essential oil reduces the population of *Meloidogyne incognita* in tomato plants. Eur J Plant Pathol 157:197–209. https://doi.org/10.1007/s10658-020-02000-1

Jardim IN, Oliveira DF, Campos VP et al (2020b) Garlic essential oil reduces the population of *Meloidogyne incognita* in tomato plants. Eur J Plant Pathol 1:13

Ji H, Li YC, Wen Z et al (2016) Gc-ms analysis of nematicidal essential oil of *Mentha canadensis* aerial parts against heterodera avenae and *Meloidogyne incognita*. J Essent Oil Bear Plants 19: 2056–2064

Jindapunnapat K, Reetz ND, MacDonald MH et al (2018) Activity of vetiver extracts and essential oil against *Meloidogyne incognita*. J Nematol 50:147

Karajeh MR (2008) Interaction of root-knot nematode (*Meloidogyne javanica*) and tomato as affected by hydrogen peroxide. J Plant Prot Res 48:181

Kennedy MJ, Niblack TL, Krishnan HB (1999) Infection by Heterodera glycines elevates isoflavonoid production and influences soybean nodulation. J Nematol 31:341

Kepenekci İ, Toktay H, Saglam HD, et al (2016) Effects of some indigenous plant extracts on mortality of the root lesion nematode, *Pratylenchus thornei* Sher & Allen

Kepenkci I, Saglam HD (2018) Effects of some indigenous plant extracts on *Meloidogyne javanica* infesting eggplant and pepper under greenhouse condition

Khan A, Sayed M, Shaukat SS, Handoo ZA (2008) Efficacy of four plant extracts on nematodes associated. Nematol Medit 36:93–98

Khan F, Asif M, Khan A et al (2019) Evaluation of the nematicidal potential of some botanicals against root-knot nematode, *Meloidogyne incognita* infected carrot: *in vitro* and greenhouse study. Curr Plant Biol 20:100115. https://doi.org/10.1016/j.cpb.2019.100115

Khan MR, Mohidin FA, Khan U, Ahamad F (2016) Native *pseudomonas* spp. suppressed the root-knot nematode in *in vitro* and *in vivo*, and promoted the nodulation and grain yield in the field grown mungbean. Biol Control 101:159–168

Kim TY, Jang JY, Yu NH et al (2018) Nematicidal activity of grammicin produced by *Xylaria grammica* KCTC 13121BP against *Meloidogyne incognita*. Pest Manag Sci 74:384–391

Kimbaris AC, González-Coloma A, Andrés MF et al (2017) Biocidal compounds from *Mentha* sp. essential oils and their structure–activity relationships. Chem Biodivers 14:e1600270

Kimpinski J, Arsenault WJ, Gallant CE, Sanderson JB (2000) The effect of marigolds (Tagetes spp.) and other cover crops on *Pratylenchus penetrans* and on following potato crops. J Nematol 32:531

Kiriga AW, Haukeland S, Kariuki GM et al (2018) Effect of *Trichoderma* spp. and *Purpureocillium lilacinum* on *Meloidogyne javanica* in commercial pineapple production in Kenya. Biol Control 119:27–32. https://doi.org/10.1016/j.biocontrol.2018.01.005

Kokalis-Burelle N (2015) Pasteuria penetrans for control of *Meloidogyne incognita* on tomato and cucumber, and *M. arenaria* on snapdragon. J Nematol 47:207–213

Kokalis-Burelle N, Martinez-Ochoa N, Rodríguez-Kábana R, Kloepper JW (2002) Development of multi-component transplant mixes for suppression of *Meloidogyne incognita* on tomato (*Lycopersicon esculentum*). J Nematol 34:362–369

Kong J-O, Lee S-M, Moon Y-S et al (2006) Nematicidal activity of plant essential oils against *Bursaphelenchus xylophilus* (Nematoda: *Aphelenchoididae*). J Asia Pac Entomol 9:173–178

Lamovšek J, Stare BG, Pleško IM et al (2017) Agrobacteria enhance plant defense against root-knot nematodes on tomato. Phytopathology 107:681–691. https://doi.org/10.1094/PHYTO-07-16-0269-R

Lamovšek J, Urek U, Trdan S (2013) Biological control of root-knot nematodes (*Meloidogyne* spp.): microbes against the pests. Acta Agric Slov 101:263–275. https://doi.org/10.2478/acas-2013-0022

Laquale S, Candido V, Avato P et al (2015) Essential oils as soil biofumigants for the control of the root-knot nematode *Meloidogyne incognita* on tomato. Ann Appl Biol 167:217–224. https://doi.org/10.1111/aab.12221

Lazzeri L, Curto G, Dallavalle E et al (2009) Nematicidal efficacy of biofumigation by defatted Brassicaceae meal for control of *Meloidogyne incognita* (Kofoid et white) Chitw. On a full field zucchini crop. J Sustain Agric 33:349–358

Lee YS, Kim KY (2016) Antagonistic potential of *Bacillus pumilus* L1 against root-knot nematode, *Meloidogyne arenaria*. J Phytopathol 164:29–39. https://doi.org/10.1111/jph.12421

Li J, Zou C, Xu J et al (2015a) Molecular mechanisms of nematode-Nematophagous microbe molecular mechanisms of nematode-Nematophagous microbe interactions : basis for biological control of plant-parasitic nematodes. Annu Rev Phytopathol 53:67–95. https://doi.org/10.1146/annurev-phyto-080614-120336

Li J, Zou C, Xu J et al (2015b) Molecular mechanisms of nematode-nematophagous microbe interactions: basis for biological control of plant-parasitic nematodes. Annu Rev Phytopathol 53:67–95

Li X, Hu HJ, Li JY et al (2019) Effects of the endophytic bacteria *Bacillus cereus* BCM2 on tomato root exudates and *Meloidogyne incognita* infection. Plant Dis 103:1551–1558. https://doi.org/10.1094/PDIS-11-18-2016-RE

Li YC, Ji H, Li XH et al (2017) Isolation of nematicidal constituents from essential oil of *Kaempferia galanga* L rhizome and their activity against *Heterodera avenae* Wollenweber. Trop J Pharm Res 16:59–65

Liang J, Ning A, Lu P et al (2020) Chemical composition and biological activity of essential oil extracted from the aerial part of *Elsholtzia fruticosa* against Ditylenchus destructor. J Essent Oil Bear Plants 23:575–582

Liu C, Timper P, Ji P (2017) Influence of root exudates and soil on attachment of *Pasteuria penetrans* to root-knot nematode *Meloidogyne arenaria*. In: phytopathology. Amer Phytopathological SOC 3340 Pilot Knob Road, ST Paul, MN 55121 USA, pp 146–147

Liu X, Miao Q, Zhou Z et al (2022) Identification of three novel Conidiogenesis-related genes in the nematode-trapping fungus *Arthrobotrys oligospora*. Pathogens 11:1–13. https://doi.org/10.3390/pathogens11070717

Liu Y, Ding Z, Peng D-l et al (2019) Evaluation of the biocontrol potential of *aspergillus welwitschiae* against the root-knot nematode *Meloidogyne graminicola* in rice (*Oryza sativa* L.). J Integr Agric 18:2561–2570. https://doi.org/10.1016/S2095-3119(19)62610-9

Lopez-Llorca LV, Macia-Vicente JG, Jansson HB (2008) Mode of action and interactions of nematophagous fungi. In: Ciancio A, Mukerji KG (eds) Integrated management and biocontrol of vegetable and grain crops nematodes. Springer, pp 51–76

Luo H, Liu Y, Fang L et al (2007) *Coprinus comatus* damages nematode cuticles mechanically with spiny balls and produces potent toxins to immobilize nematodes Coprinus comatus damages nematode cuticles mechanically with spiny balls and produces potent toxins to immobilize nematodes. Appl Environ Microbiol 73:3916–3923. https://doi.org/10.1128/AEM.02770-06

Martinotti MD, Castellanos SJ, Gonzalez R et al (2016) Nematicidal effects of extracts of garlic, grape pomace and olive mill waste, on *Meloidogyne incognita*, on grapevine cv chardonnay. Rev la Fac Ciencias Agrar Univ Nac Cuyo 48:211–224

Mattei D, Dias-Arieira CR, Biela F et al (2014) Essential oil of *Rosmarinus officinalis* in the control of *Meloidogyne javanica* and *Pratylenchus brachyurus* in soybean. Biosci J 30

Mazzola M, Brown J, Izzo AD, Cohen MF (2007) Mechanism of action and efficacy of seed meal-induced pathogen suppression differ in a brassicaceae species and time-dependent manner. Phytopathology 97:454–460. https://doi.org/10.1094/PHYTO-97-4-0454

de Mazzuchelli RCL, Mazzuchelli EHL, de Araujo FF (2020) Efficiency of *Bacillus subtilis* for root-knot and lesion nematodes management in sugarcane. Biol Control 143:104185. https://doi.org/10.1016/j.biocontrol.2020.104185

McCarter JP (2008) Molecular approaches toward resistance to plant-parasitic nematodes. Springer

Medina-Canales MG, Terroba-Escalante P, Manzanilla-López RH, Tovar-Soto A (2019) Assessment of three strategies for the management of *Meloidogyne* arenaria on carrot in Mexico using *Pochonia chlamydosporia* var. mexicana under greenhouse conditions. Biocontrol Sci Tech 29:671–685. https://doi.org/10.1080/09583157.2019.1582267

Mendoza AR, Kiewnick S, Sikora RA (2008) In vitro activity of *Bacillus firmus* against the burrowing nematode *Radopholus similis*, the root-knot nematode *Meloidogyne incognita* and the stem nematode *Ditylenchus dipsaci*. Biocontrol Sci Tech 18:377–389. https://doi.org/10.1080/09583150801952143

Mervat AA, Shawky SM, Shaker GS (2012) Comparative efficacy of some bioagents, plant oil and plant aqueous extracts in controlling *Meloidogyne incognita* on growth and yield of grapevines. Ann Agric Sci 57:7–18

Miresmailli S, Isman MB (2014) Botanical insecticides inspired by plant-herbivore chemical interactions. Trends Plant Sci 19:29–35. https://doi.org/10.1016/j.tplants.2013.10.002

Moens M, Perry RN, Starr JL (2009) *Meloidogyne* species–a diverse group of novel and important plant parasites. Root-knot nematodes 1:483

Mohd Yaqub WAHB (2012) Juglans regia L. and its constituents. Ann Plant Prot Sci. Last page: (204)

Moosavi MR (2012) Nematicidal effect of some herbal powders and their aqueous extracts against *Meloidogyne* javanica. Nematropica:48–56

Moosavi MR, Zare R (2012) Fungi as biological control agents of plant-parasitic nematodes. In: Mérillon JM, Ramawat KG (eds) Plant defence: biological control. Springer Science+Business Media B. V, pp. 67–107

Mostafa FAM, Khalil AE, Nour El-Deen AH, Ibrahim DS (2018) The role of bacillus megaterium and other bio-agents in controlling root-knot nematodes infecting sugar beet under field conditions. Egypt J Biol Pest Control 28:1–6

Munhoz VM, Baida FC, Lopes GC et al (2017) Extracts and semi-purified fractions of Tagetes patula flowers in the control of root-knot nematodes. Semin Ciências Agrárias 38:3529–3538

Mwamula AO, Kabir MF, Lee D (2022) A review of the potency of plant extracts and compounds from key families as an alternative to synthetic Nematicides: history, efficacy, and current developments. Plant Pathol J 38:53–77. https://doi.org/10.5423/PPJ.RW.12.2021.0179

Narasimhamurthy HB, Ravindra H, Sehgal M et al (2017) Bio-management of rice root-knot nematode (*Meloidogyne* graminicola). J Entomol Zool Stud 5:1433–1439

Naz I, Khan RAA, Masood T et al (2021) Biological control of root knot nematode, *Meloidogyne* incognita, in vitro, greenhouse and field in cucumber. Biol Control 152:104429. https://doi.org/10.1016/j.biocontrol.2020.104429

Nebih HSD, Charif F (2019) Toxicity of plant extracts on larvae (L2) of potato cyst nematode Globodera spp. (Nematoda, Heteroderidae). AgroBiologia 9:1234–1241

dos Neves WS, de Freitas LG, Coutinho MM et al (2009) Ação nematicida de extratos de alho, mostarda, pimenta malagueta, de óleo de mostarda e de dois produtos à base de capsainóides e alil isotiocianato sobre juvenis de *Meloidogyne* javanica,(treub) Chitwood, 1949, em casa de vegetação. Summa Phytopathol 35:255–261

Nguyen DMC, Jung WJ (2014) Nematicidal properties of crude extracts obtained from medicinal plants against root-lesion nematode Pratylenchus coffeae. J Vietnamese Environ 6:264–269

Nordbring-hertz B, Jansson HB, Tunlid A (2006) Nematophagous Fungi. Encycl Life Sci 1–12. https://doi.org/10.1038/npg.els.0004293

Ntalli N et al (2020a) Thymus Citriodorus (Schreb) botanical products as. Plants MDPI

Ntalli N, Bratidou Parlapani A, Tzani K et al (2020b) Thymus citriodorus (Schreb) botanical products as ecofriendly nematicides with bio-fertilizing properties. Plan Theory 9:202

Ntalli N, Monokrousos N, Rumbos C et al (2018) Greenhouse biofumigation with Melia azedarach controls *Meloidogyne* spp. and enhances soil biological activity. J Pest Sci 91:29–40

Ntalli NG, Caboni P (2012) Botanical nematicides: A review. J Agric Food Chem 60:9929–9940. https://doi.org/10.1021/jf303107j

Ntalli NG, Ferrari F, Giannakou I, Menkissoglu-Spiroudi U (2011) Synergistic and antagonistic interactions of terpenes against *Meloidogyne* incognita and the nematicidal activity of essential oils from seven plants indigenous to Greece. Pest Manag Sci 67:341–351. https://doi.org/10.1002/ps.2070

Ntalli NG, Ferrari F, Giannakou I, Menkissoglu-Spiroudi U (2010) Phytochemistry and nematicidal activity of the essential oils from 8 Greek Lamiaceae aromatic plants and 13 terpene components. J Agric Food Chem 58:7856–7863

Ojaghian MR, Cui ZQ, Xie GL et al (2012) Brassica green manure rotation crops reduce potato stem rot caused by Sclerotinia sclerotium. Australas Plant Pathol 41:347–349. https://doi.org/10.1007/s13313-012-0142-6

Oka Y (2012) Nematicidal activity of Verbesina encelioides against the root-knot nematode *Meloidogyne* javanica and effects on plant growth. Plant Soil 355:311–322

Oka Y, Ben-Daniel B-H, Cohen Y (2006) Control of *Meloidogyne* javanica by formulations of Inula viscosa leaf extracts. J Nematol 38:46

Oka Y, Nacar S, Putievsky E et al (2000) Nematicidal activity of essential oils and their components against the root-knot nematode. Phytopathology 90:710–715

Brito ODC, Ferreira JCA, Hernandes I, da Silva EJ, Dias-Arieira CR et al (2020) Management of *Meloidogyne* javanica on tomato using agro-industrial wastes. Nematology 22:1141–1154. https://doi.org/10.1163/15685411-bja10018

Oluwatayo JI, Jidere CI, Nwankiti A (2019) Nematicidal effect of some botanical extracts for the management of *Meloidogyne* incognita and on growth of tomato. Asian J Agric Hortic Res:1–8

Ozdemir E, Gozel U (2017) Efficiency of some plant essential oils on root-knot nematode *Meloidogyne* incognita. J Agric Sci Technol A 7:178–183

Pandey R, Kalra A, Tandon S et al (2000) Essential oils as potent source of nematicidal compounds. J Phytopathol 148:501–502

Park I-K, Park J-Y, Kim K-H et al (2005) Nematicidal activity of plant essential oils and components from garlic (Allium sativum) and cinnamon (Cinnamomum verum) oils against the pine wood nematode (Bursaphelenchus xylophilus). Nematology 7:767–774

Patil J, Rangasamy V, Lakshmi L (2017a) Efficacy of entomopathogenic Heterorhabditis and Steinernema nematodes against the white grub, Leucopholis lepidophora Blanchard (Coleoptera: Scarabaeidae). Crop Prot 101:84–89. https://doi.org/10.1016/j.cropro.2017.07.021

Patil J, Yadav S, Sharma MK (2017b) Effect of plant extracts on egg hatching and juvenile mortality of Reniform nematode, Rotylenchulus reniformis. Int J Pure App Biosci 5:158–164

Pavela R (2016) History, presence and perspective of using plant extracts as commercial botanical insecticides and farm products for protection against insects–a review. Plant Prot Sci 52:229–241

Peiris PUS, Li Y, Brown P, Xu C (2020) Fungal biocontrol against *Meloidogyne* spp. in agricultural crops: a systematic review and meta-analysis. Biol Control 144:104235. https://doi.org/10.1016/j.biocontrol.2020.104235

Pino O, Sánchez Y, Rojas MM (2013) Metabolitos secundarios de origen botánico como una alternativa en el manejo de plagas. I: Antecedentes, enfoques de investigación y tendencias. Rev Protección Veg 28:81–94

Ploeg A (2000) Effects of amending soil with Tagetes patula cv. Single gold on *Meloidogyne* incognita infestation of tomato. Nematology 2:489–493

Plowright RA, Grayer RJ, Gill JR et al (1996) The induction of phenolic compounds in rice after infection by the stem nematode Ditylenchus angustus. Nematologica 42:564–578

Purba RTT, Fauzi F, Sari RW et al (2022) Arthrobotrys thaumasia and A. musiformis as biocontrol agents against *Meloidogyne* hapla on tomato plant. Biodiversitas 23:3659–3666. https://doi.org/10.13057/biodiv/d230743

Quentin M, Abad P, Favery B (2013) Plant parasitic nematode effectors target host defense and nuclear functions to establish feeding cells. Front Plant Sci 4:53

Radwan MA, Ibrahim Kassem SM, Abu-Elamayem MM, El-Maadawy EK (2007) Use of some emulsified plant seed oils as a safe alternative for the management of *Meloidogyne* incognita infecting tomato. Arch Phytopathol Plant Prot 40:345–352

Rahul S, Chandrashekhar P, Hemant B et al (2014) Nematicidal activity of microbial pigment from Serratia marcescens. Nat Prod Res 28:1399–1404

Regaieg H, Ciancio A, Raouania NH et al (2010) Effects of culture filtrates from the nematophagous fungus Verticillium leptobactrum on viability of the root-knot nematode *Meloidogyne* incognita. World J Microbiol Biotechnol 26:2285–2289. https://doi.org/10.1007/s11274-010-0397-4

Rich JR, Keen NT, Thomason IJ (1977) Association of coumestanss with tee hypersensitivity of Lima bean roots to Pratylenchus scribneri. Physiol Plant Pathol 10:105–116

Samaliev H, Markova D, Nikolova M, Baicheva O (2017) Management of root-knot nematode *Meloidogyne* hapla on strawberry plant with some plant extracts. Российский паразитологический журнал

San Martín R, Magunacelaya JC (2005) Control of plant-parasitic nematodes with extracts of Quillaja saponaria. Nematology 7:577–585

Saqib A, İshrat N, Alamzeb M, Urrashidd M (2019) Activity guided isolation of nematicidal constituents from the roots of Berberis brevissima Jafri and Berberis parkeriana Schneid. J Agric Sci 25:108–114

Saraf M, Pandya U, Thakkar A (2014) Role of allelochemicals in plant growth promoting rhizobacteria for biocontrol of phytopathogens. Microbiol Res 169:18–29. https://doi.org/10.1016/j.micres.2013.08.009

Saxena DB, Goswami BK, Tomar SS (1987) Nematicidal activity of some essential oils against *Meloidogyne* incognita. Indian Perfum 3:150

Saxena G (2018) Biological control of root-knot and cyst nematodes using Nematophagous fungi. In: Root biology, pp 221–237

Schouteden N, De Waele D, Panis B, Vos CM (2015) Arbuscular mycorrhizal fungi for the biocontrol of plant-parasitic nematodes: a review of the mechanisms involved. Front Microbiol 6:1280

Schouten A (2016) Mechanisms involved in nematode control by endophytic fungi. Annu Rev Phytopathol 54:121–142. https://doi.org/10.1146/annurev-phyto-080615-100114

Shin JH, Kwon O, Lee CM et al (2016) Nematicidal activity of Eclipta prostrata extract and terthiophene against pine wood nematode, Bursaphelenchus xylophilus. Korean J Pestic Sci 20:56–65

Siddiqui IA, Shaukat SS, Khan GH, Ali NI (2003) Suppression of *Meloidogyne* javanica by Pseudomonas aeruginosa IE-6S+ in tomato: the influence of NaCl, oxygen and iron levels. Soil Biol Biochem 35:1625–1634

Siddiqui MA, Alam MM (1988) Effect of seed dressing with plant latex on Tylenchorhynchus brassicae and plant growth of cabbage and cauliflower. Pakistan J Nematol 6:65–71

de Silva JO, Santana MV, Freire LL et al (2017) Agentes de biocontrole no manejo de *Meloidogyne* incognita em tomateiro. Ciência Rural 47. https://doi.org/10.1590/0103-8478cr20161053

Silva SD, Carneiro RMDG, Faria M et al (2017) Evaluation of Pochonia chlamydosporia and Purpureocillium lilacinum for suppression of *Meloidogyne* enterolobii on tomato and banana. J Nematol 49:77–85. https://doi.org/10.21307/jofnem-2017-047

Singh LP, Gill SS, Tuteja N (2011) Unraveling the role of fungal symbionts in plant abiotic stress tolerance. Plant Signal Behav 6:175–191

Sinha A, Maheshwari RC, Dureja P, Mojumder V (2006) Nematicidal activity of essential oils and their components against the root-knot nematode *Meloidogyne* incognita. Indian J Nematol 36: 109–114

Smith SE, Facelli E, Pope S, Andrew Smith F (2010) Plant performance in stressful environments: interpreting new and established knowledge of the roles of arbuscular mycorrhizas. Plant Soil 326:3–20

Soler-Serratosa A, Kokalis-Burelle N, Rodríguez-Kábana R et al (1996) Allelochemicals for control of plant-parasitic nematodes. 1. In vivo nematicidal efficacy of thymol and thymol/benzaldehyde combinations. Nematropica 26:57–71

Soliman MS, El-Deriny MM, Ibrahim DSS et al (2021) Suppression of root-knot nematode *Meloidogyne* incognita on tomato plants using the nematode trapping fungus Arthrobotrys oligospora Fresenius. J Appl Microbiol 131:2402–2415. https://doi.org/10.1111/jam.15101

Soriano IR, Asenstorfer RE, Schmidt O, Riley IT (2004) Inducible flavone in oats (Avena sativa) is a novel defense against plant-parasitic nematodes. Phytopathology 94:1207–1214

Stirling GR (2014) Natural ennemies of nematodes. In: Stirling GR (ed) Nematophagous fungi and oomycetes, 2nd edn, pp 101–156

Subbotin SA, Chitambar JJ (2018) Plant parasitic nematodes in sustainable agriculture of North America. Springer

Sukul NC (1992) Plants antagonistic to plant-parasitic nematodes. Ind Rev Life Sci 12:23–52

Swe A, Li J, Zhang KQ et al (2011) Nematode-Trapping Fungi. Curr Res Environ Appl Mycol 1:1–26

Taniwiryono D, Berg H, Riksen JAG et al (2009) Nematicidal activity of plant extracts against the root-knot nematode, *Meloidogyne* incognita. Open Nat Prod J 2:77

Tapia-Vázquez I, Montoya-Martínez AC, los Santos-Villalobos D et al (2022) Root-knot nematodes (*Meloidogyne* spp.) a threat to agriculture in Mexico: biology, current control strategies, and perspectives. World J Microbiol Biotechnol 38:1–18

Thongkaewyuan A, Chairin T (2018) Biocontrol of *Meloidogyne* incognita by Metarhizium guizhouense and its protease. Biol Control 126:142–146. https://doi.org/10.1016/j.biocontrol.2018.08.005

Tian B, Yang J, Zhang KQ (2007) Bacteria used in the biological control of plant-parasitic nematodes : populations, mechanisms of action, and future prospects. FEMS Microbiol Ecol 61:197–213. https://doi.org/10.1111/j.1574-6941.2007.00349.x

Timper P (2014) Conserving and enhancing biological control of nematodes. J Nematol 46:75–89

Tiwari S, Pandey S, Chauhan PS, Pandey R (2017) Biocontrol agents in co-inoculation manages root knot nematode [*Meloidogyne* incognita (Kofoid & White) Chitwood] and enhances essential oil content in Ocimum basilicum L. Ind Crop Prod 97:292–301

Topalović O, Hussain M, Heuer H (2020) Plants and associated soil microbiota cooperatively suppress plant-parasitic nematodes. Front Microbiol 11:1–15. https://doi.org/10.3389/fmicb.2020.00313

van den Hoogen J, Geisen S, Routh D et al (2019) Soil nematode abundance and functional group composition at a global scale. Nature 572:194–198. https://doi.org/10.1038/s41586-019-1418-6

Viaene N, Coyne DL, Kerry BR (2006) Biological and cultural management. In: Perry RN, Moens M (eds) Plant nematology, pp 346–369

Vilchis-Martinez K, Manzanilla-Lopez RH, Powers SJ, Montes-Belmont R (2013) Effect of the addition of crude plant extracts on the parasitism of Pochonia chlamydosporia var. chlamydosporia on *Meloidogyne* incognita. Nematropica 43:254–260

Vos C, Claerhout S, Mkandawire R et al (2012) Arbuscular mycorrhizal fungi reduce root-knot nematode penetration through altered root exudation of their host. Plant Soil 354:335–345

Wang X, Li GH, Zou CG et al (2014) Bacteria can mobilize nematode-trapping fungi to kill nematodes. Nat Commun 5:5776. https://doi.org/10.1038/ncomms6776

Wen Y, Meyer SLF, MacDonald MH et al (2019) Nematotoxicity of Paeonia spp. extracts and Camellia oleifera tea seed cake and extracts to Heterodera glycines and *Meloidogyne* incognita. Plant Dis 103:2191–2198

Wesemael W, Viaene N, Moens M (2011) Root-knot nematodes (*Meloidogyne* spp.) in Europe. Nematology 13:3–16

Westerdahl BB, Giraud D, Riddle LJ, Anderson CA (2020) Essential oils for managing Pratylenchus penetrans on Easter lilies. J Nematol 52:1

Whipps JM, Davies KG (2000) Success in biological control of plant pathogens and nematodes by microorganisms. In: Gurr G, Wratten S (eds) Biological control: measures o/success. Kluwer Academic Publishers, pp. 231–269

Wolfgang A, Taffner J, Guimarães RA et al (2019) Novel strategies for soil-borne diseases: exploiting the microbiome and volatile-based mechanisms toward controlling *Meloidogyne*-based disease complexes. Front Microbiol 10:1296

Wondimeneh T, Sakhuja PK, Tadele T (2013) Root-knot nematode (*Meloidogyne* incognita) management using botanicals in tomato (Lycopersicon esculentum). Acad J Agric Res 1:9–16

Wu H, Wang CJ, Bian XW et al (2011) Nematicidal efficacy of isothiocyanates against root-knot nematode *Meloidogyne* javanica in cucumber. Crop Prot 30:33–37. https://doi.org/10.1016/j.cropro.2010.09.004

Xiang N, Lawrence KS, Donald PA (2018) Biological control potential of plant growth-promoting rhizobacteria suppression of *Meloidogyne* incognita on cotton and Heterodera glycines on soybean: a review. J Phytopathol 166:449–458. https://doi.org/10.1111/jph.12712

Xiao L, Wan JW, Yao JH et al (2018) Effects of Bacillus cereus strain Jdm1 on *Meloidogyne* incognita and the bacterial community in tomato rhizosphere soil. 3 Biotech 8(0):319. https://doi.org/10.1007/s13205-018-1348-2

Yadav S, Patil J, Saroj Yadav C, Kumar A (2018) Bio-nematicidal effect of Azadirachta indica, against *Meloidogyne* incognita in tomato. ~ 2757 ~. Int J Chem Stud 6:2757–2761

Yasmin L, Rashid MH, Uddin MN et al (2003) Use of neem extract in controlling root-knot nematode (*Meloidogyne* javanica) of sweet-gourd. Pakistan. J Plant Pathol

Zasada IA, Ferris H (2003) Sensitivity of *Meloidogyne* javanica and Tylenchulus semipenetrans to isothiocyanates in laboratory assays. Phytopathology 93:747–750. https://doi.org/10.1094/PHYTO.2003.93.6.747

Zasada IA, Meyer SLF, Morra MJ (2009) Brassicaceous seed meals as soil amendments to suppress the plant-parasitic nematodes Pratylenchus penetrans and *Meloidogyne* incognita. J Nematol 41:221–227

Zhang S, Gan Y, Xu B (2015) Biocontrol potential of a native species of Trichoderma longibrachiatum against *Meloidogyne* incognita. Appl Soil Ecol 94:21–29. https://doi.org/10.1016/j.apsoil.2015.04.010

Zhu MC, Li XM, Zhao N et al (2022) Regulatory mechanism of trap formation in the nematode-trapping fungi. J Fungi 8:1–14. https://doi.org/10.3390/jof8040406

Epigenetic Mechanisms and Their Role in Root Gall Formation

7

Arshad Khan, Amir Khan, Faryad Khan, Mohammad Shariq, Saba Fatima, Saeeda Zaima Zeb, and Mansoor Ahmad Siddiqui

Abstract

Root-knot nematodes (RKNs), *Meloidogyne* spp., are obligatory plant parasites that affect crop productivity by infecting many plant species. They induce the redifferentiation of vascular tissues of the root into a pseudo-organ termed a gall, where some cells are changed into incredibly metabolically active giant cells (GCs), which serve as their feeding sites. Epigenetic mechanisms play a significant role in the development of gall formation; however, their key role in the interactions between RKNs and plants is not well known. Epigenetic components such as small RNAs, DNA methylation, and histone modification play important roles in host plants' gall development triggered by RKNs. Furthermore, the epigenetic machinery is thought to play a vital role in forming nematode-feeding sites or galls. The developmental reprogramming of host root cells by RKNs causes these feeding sites to have hypertrophied GCs. Effectors are secreted by RKNs, which are involved in the formation of specialized feeding sites or GCs and are responsible for the numerous morphological and physiological changes that occur during the development of galls. The epigenetic mechanism underlying the development of GCs will be the main focus of this study because it is crucial to gall formation. We also described the role of small RNAs, including miRNAs and siRNAs can be involved in epigenetic mechanisms during galls development.

Keywords

Root-knot nematodes · Epigenetic · Giant cells · Effectors · Galls · small RNAs

A. Khan (✉) · A. Khan · F. Khan · M. Shariq · S. Fatima · S. Z. Zeb · M. A. Siddiqui
Department of Botany, Aligarh Muslim University, Aligarh, India

F. Ahmad, G. Nombela (eds.), *Root-Galling Disease of Vegetable Plants*,
https://doi.org/10.1007/978-981-99-3892-6_7

7.1 Introduction

In recent decades, exploring the epigenetic regulation of gene function has acquired a key role in the biological sciences. The word "epigenetics" was introduced by Conrad Waddington in 1942 (Waddington 1942). Epigenetic mechanisms in diverse developmental and environmental scenarios regulate various biological activities. During the life cycle of all organisms, including plants and animals, epigenetic mechanisms play a crucial role (Duan et al. 2018). Numerous studies have shown that nematode infection activates various epigenetic regulatory mechanisms (Atighi Quchan Atigh 2020). The mechanisms of epigenetic change have been widely investigated in both healthy and pathological processes (Poças-Fonseca et al. 2020). Plant epigenetic configuration is altered by biotic factors, which ultimately affect biotic interactions by influencing plant responses. Perfus-Barbeoch et al. (2014) reported that the pathogenicity of *Meloidogyne* is governed by epigenetic regulation. In *Meloidogyne* spp., DNA folding into chromatin significantly impacts cellular functions that use DNA as a template, such as replication, repair, recombination, and transcription (Pratx et al. 2018). The nucleosome is the building block of chromatin, consisting of 147 base pairs of DNA encased around an octamer of histones (Luger et al. 1997). Epigenetic codes, including DNA methylation, histone modifications, histone variations, and noncoding RNAs (lncRNA), regulate chromatin's structure and biological function (Duan et al. 2018).

Nematodes are the most abundant animals on earth (Van den Hoogen et al. 2019) and a major biotic component of soil (Bardgett and Van Der Putten 2014). More than 4100 plant-parasitic nematode species are thought to exist (Decraemer and Hunt 2006). PPNs are responsible for approximately 12.3% of annual global agricultural production losses to an estimated $157 billion annually (Singh et al. 2015). RKNs are polyphagous sedentary endoparasites that seriously threaten agricultural production (Machado 2015; Peiris et al. 2020). The RKNs are in the genus *Meloidogyne*, which has about 100 described species, including four of the most important species, *M. incognita*, *M. javanica*, *M. hapla*, and *M. arenaria*, which are responsible for substantial losses in agriculture around the world (Coyne et al. 2018; Sikandar et al. 2020).

DNA methylation is an epigenetic process that regulates gene expression by modifying DNA chemically. It involves incorporating methyl groups (CH_3) into cytosine bases at the C_5 position to produce 5-methylcytosines. DNA methylation is a ubiquitous, remarkably persistent, and heritable epigenetic marker. In plants, DNA methylation occurs in the nucleotide contexts CG, CHG, and CHH, which are generated by distinct enzymes. The first universal DNA methylation assessments validated and expanded by whole genome bisulfite sequencing indicated significant DNA hypomethylation of cytosine in the context of CHH sequences of DNA in the cells of root galls (Kyndt et al. 2019). DNA methylation modulates the expression of target genes via modifying the binding affinity of DNA-binding proteins (transcriptional apparatus) to DNA or by procuring proteins implicated in gene suppression. Bennett and Meredith (2021) developed 15 transgenic *Arabidopsis* GUS reporter lines to study genes associated with DNA methylation and demethylation pathways.

Researchers looked at how these genes were spatially and temporally expressed in different plant organs during development in response to exogenous phytohormones and diseases caused by PPNs. The findings indicate distinct and consistent expression profiles in roots, shoots, and reproductive organs, highlighting the significance of a proportion between DNA methylation and demethylation.

Moreover, promoter activity shows that hormone-associated methylome control systems enhance tissue differentiation. At distinct stages of infection, CG and non-CG methyltransferases had comparable and unique expression profiles in syncytia and galls produced by *Heterodera schachtii* and *M. incognita*, respectively. As compared to *H. schachtii*, DNA demethylases were more active in response to *M. incognita*. In addition, hypermethylated mutants deficient in active DNA demethylation displayed contrasting reactions to infection that can be substantially understood by the contradictory regulation of pathogenesis-related genes by *H. schachtii* and *M. incognita*. These findings demonstrate that methylation-dependent mechanisms control how plants respond to infection by two different types of nematodes in similar and different ways (Bennett and Meredith 2021).

The DNA-histone complex is formed in eukaryotic cells when the DNA molecule coils surrounding histone proteins. Post-translational histone alterations are another epigenetic method of gene expression regulation. The histone tails that extend from the nucleosome core can be altered by incorporating other groups, most commonly methyl and acetyl groups, which regulate transcription factors such as DNA binding proteins on the surface of DNA (Lawrence et al. 2016). The methylation and acetylation of lysine (K) residues are two of the most common biochemical modifications that affect histone proteins. Considered markers are the genome-wide patterns of three histone proteins, H3K27me3, H3K9ac, and H3K9me2. H3K9ac is usually thought of as a gene activation marker, while H3K9me2 and H3K27me3 are thought of as gene repression markers (Armstrong and Spencer 2021). Although histone-modifying enzymes are downregulated in *M. graminicola* initiated galls in rice, neither their impact on plant defense nor their genome-wide influence has been adequately examined (Atighi et al. 2021). ChIP-seq demonstrates that nematode-induced galls had a lot of highly methylated histones. This is consistent with the observation that histone lysine methyltransferases were tightly activated during transcription. Experiments covering several generations demonstrated that the progeny of nematode-infected rice plants are substantially tolerant. These findings suggest that epigenetic alterations are an important regulator of rice nematode defenses and that these modifications might be heritable (Kyndt et al. 2019). Nonetheless, nematodes may utilize epigenetic processes for various regulatory units to combat plant defenses through molecular pathways.

MicroRNAs (miRNAs) and siRNAs constitute a large class of small regulatory RNAs found in all plants and animals. Plants use both miRNAs and siRNAs to respond to pathogen infections. Pathogen infection altered the expression of numerous miRNAs in plant species (Gualtieri et al. 2020). Several miRNAs (miR159, miR172, and miR390) participating in *Arabidopsis* plant developmental stages are crucial to GC/gall formation (Diaz-Manzano et al. 2018; Jaubert-Possamai et al. 2019; Hewezi 2020). Medina et al. (2017) identified that 24 miRNAs differentially

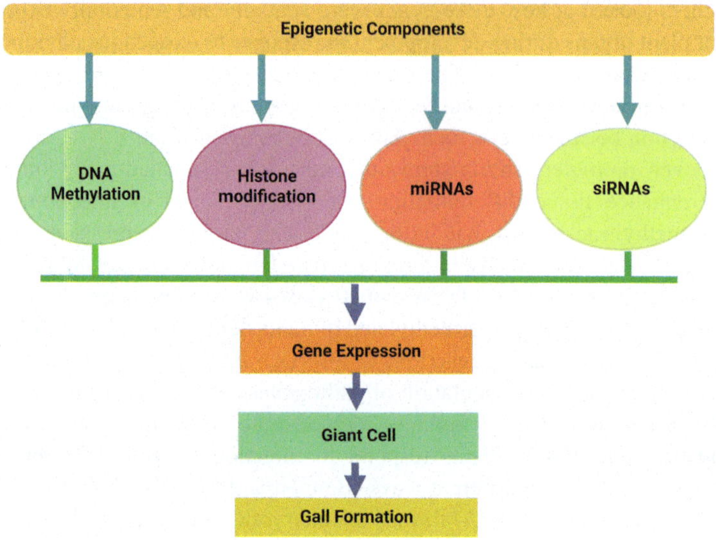

Fig. 7.1 Various epigenetic components and their role in giant cell/gall formation

expressed in gall as plausible regulators of gall development through sequencing small RNAs (sRNAs) in non-infected root of *Arabidopsis* and from galls with *M. incognita*. Sixty-two miRNAs were found to have different levels of expression between roots that were non-infected and early galls (Medina et al. 2017). In furthermore, large-scale sequencing of sRNAs has demonstrated the accumulation of siRNAs in *Arabidopsis* during early and post-infection (Cabrera et al. 2016) and moderate/late (Medina et al. 2017) infection periods. The epigenetic regulatory systems of DNA methylation, sRNAs, and histone alterations very effectively illustrate epigenetic profiling inside plants. Since past few years, several investigations have shown several intricacies regarding the dynamic nature of epigenetic modulations in gall formation (Fig. 7.1). In this chapter, we have discussed the functional aspects of sRNAs in causing root gall development during parasitism, as well as the putative role of miRNAs and siRNAs, genomic information, and nematode feeding site formation of nematode.

7.2 From Gene to Genome

Genomics information, along with downstream functional genomics and proteomics, can provide knowledge of the key role of parasitism in establishing nematode-feeding sites (NFSs) or galls caused by RKNs. Due to the availability of reasonably well-annotated genome reference sequences for both tomato and RKN, the tomato-RKN system has become an ideal crop model for researching host-pathogen interactions (Shukla et al. 2018). The first genomic strategy of EST sequence

Table 7.1 Genomic information of *Meloidogyne* species

RKN species	Strain designation	Number of predicted genes	Assembly size (Mb)	Protein-coding region (Mb)	References
M. hapla	VW9	14,220	53.01	–	Opperman et al. (2008)
M. incognita	W1	24,714	121.96	43.7	Szitenberg et al. (2017)
M. javanica	VW4	26,917	150.35	75.2	Szitenberg et al. (2017)
M. incognita	V3	45,351	183.53	–	Blanc-Mathieu et al. (2017)
M. arenaria	HarA	30,308	163.75	82.2	Szitenberg et al. (2017)

This table is adopted from Szitenberg et al. (2017) and Blanc-Mathieu et al. (2017)

analysis of pre-parasitic *M. incognita* J2s indicated multiple cell wall hydrolytic enzymes (McCarter et al. 2003). The first draft of the genome of *M. incognita* was published in 2008, and it identified several putative effectors (Abad et al. 2008). The 86 Mbp genome of *M. incognita*, on the other hand, encodes nearly 19,200 genes. This species reproduces via obligate mitotic parthenogenesis and has a complicated aneuploidy pattern (Bird et al. 2009). Numerous different nematode genomes, such as *Caenorhabditis*, free-living nematodes, and nematode parasites of humans and animals, have been sequenced to varying degrees of coverage (http://www.ncbi.nlm.nih.gov/projects/genome/guide/nematode/index.html). Recently, a first draft of the *M. graminicola* genome with a 35 Mb genome assembly size was published (Somvanshi et al. 2018). Despite this, the assembly was highly fragmented, including over 4300 contigs with an N50 length of 20 kb. Blanc-Mathieu et al. (2017) sequenced the genomes of three asexually reproducing RKN species, with the assemblies for *M. incognita*, *M. javanica*, and *M. arenaria* reaching 184, 236, and 258 Mb, respectively (Table 7.1).

7.3 Nematode Feeding Sites (NFSs) or Galls

RKNs are obligatory sedentary endoparasites of plants with pronounced sexual dimorphism, i.e., females are pyriform or saccate and males are vermiform. The second-stage juveniles (J2s) become sedentary, feed on special nurse cells, and undergo further morphological modifications. They have a hollow, protruding stylet at the anterior end of the body, which is used to inject secretions into infected root cells and extract nutrients from those cells. They have developed incredibly complex ways to interact with their host during evolution (Abad et al. 2003). At the beginning of parasitism, infected J2s enter the root tip and move between cells to target the vascular tissues of the host root. Each J2 then triggers the redifferentiation of 5–7 cells of root into highly metabolically active GCs, which are hypertrophied and

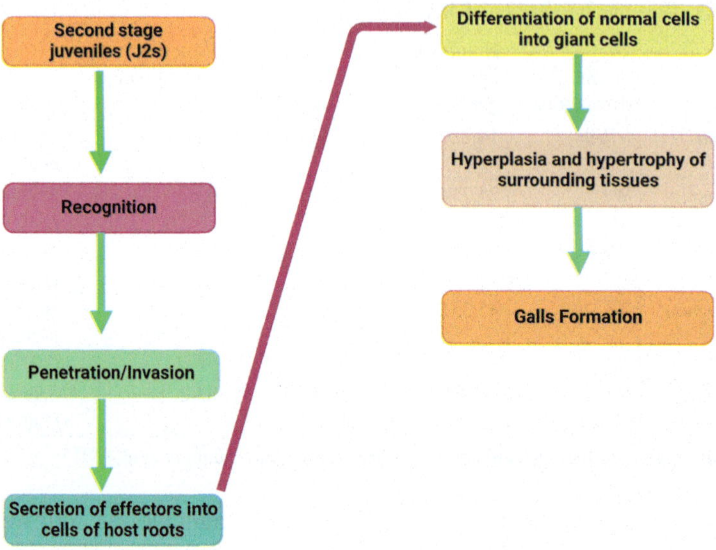

Fig. 7.2 Life cycle of RKN and formation of nematode feeding sites or galls

multinucleated. Synchronous repetitive karyokinesis without cell division produced GCs. It changed into an irregular outgrowth termed as galls (Crespi and Frugier 2008) (Fig. 7.2). The differentiation of GCs involves many rounds of mitosis without cytokinesis, followed by many cycles of endoreduplication that make the nuclei and cells bigger (Vieira et al. 2013). GCs also have rearranged cytoskeletons, a ruptured vacuolar system, and a lot of organelles in their cytoplasm, including mitochondria, Golgi apparatus, endoplasmic reticulum, ribosomes, and plastids (Banora et al. 2011; Rodiuc et al. 2014). During the development of GCs, several physiological and morphological alterations occur and transform into NFSs, providing nutrients for the nematode's growth and reproduction (Palomares-Rius et al. 2017). RKNs are quite advanced parasites that hijack host machinery by secreting effector chemicals to activate and sustain feeding cells inside the host roots (Abad and Williamson 2010). After becoming sedentary and commencing the development of NFS, the sub-ventral glands (SvGs) gradually lose activity, and the dorsal gland (DG) becomes active. DG effectors provide two important functions: modulation of the plant cell cycle and mitigating cell death (Jagdale et al. 2021).

7.4 Epigenetic Changes and Galls Formation

Transcriptomic studies of galls have shown the key role of epigenetic regulation and significant reprogramming during gall formation. Studies conducted over the previous decade have highlighted the mechanisms responsible for the formation of GCs.

Research conducted on two hosts, *Arabidopsis* and tomato, has shown that the development of galls is accompanied by a significant suppression in gene expression (da Silva 2020). There are two primary categories of short RNAs, the miRNAs that are 21 nucleotides long and the epigenetically active sRNAs that are 21 to 24 nucleotides long (Simon and Meyers 2011).

7.4.1 Small RNAs and Galls Formation

In 1999, David Baulcombe's team was the first to discover sRNAs in plants (Hamilton and Baulcombe 1999). sRNAs in plants are usually made up of 21–24 nucleotides (nt). They are produced from double-stranded RNAs (dsRNAs) by DICER-LIKE proteins (DCLs) (Xie et al. 2004; Kasschau et al. 2007). Several categories of sRNAs have been described, and new classes are continually being extracted from other species, hence enhancing the diversity and complexity of the populations of sRNAs. Recent research has demonstrated that host sRNAs and RNA silencing mechanisms regulate the plant immune system against potential pathogens, particularly PPNs (Hewezi and Baum 2013; Katiyar-Agarwal and Jin 2010). Although the mechanisms underlying this transcriptome analysis are little understood, many investigations have shown that this gene suppression may be controlled by epigenetic processes, such as short sRNAs (Da Silva 2020). Recent findings showed an enormous and variable accumulation of sRNAs in the early development of galls, which may be involved in epigenetic processes like RNA-directed DNA methylation (RdDM). The repeat-associated small interfering RNAs (rasiRNAs) enriched in early galls, which targeted retrotransposons, some of whose important members were suppressed, indicating an epigenetic process, such as RdDM, is involved during gall development. Therefore, we have chosen to explore the DNA methylation modifications that take place in the galls developed after contact between *M. javanica* and *Arabidopsis thaliana* as the host plant (da Silva 2020). RdDM pathway mutations and azacitidine treatment demonstrated that loss of DNA methylation reduces disease susceptibility (Kyndt et al. 2019). RdDM is governed by lncRNAs and sRNAs. Kyndt et al. (2019) identified over 1000 noncoding rice transcripts that are differentially expressed in response to nematode infection using whole RNA-sequencing. These transcripts contain both poly-adenylated and non-adenylated lncRNAs. sRNAs are divided into two separate classes based on their biogenesis and precursor structure: miRNAs and siRNAs. Throughout gall and GC developmental processes, gene expression undergoes a profound reprogramming, as revealed by microarray-based transcriptome analyses (Portillo et al. 2009; Barcala et al. 2010) and extensive sequencing analysis (Ji et al. 2013). It permitted the usurpation of the complexity of the gall transcriptome, which contained all the various tissues present in this pseudo-organ, and the establishment of distinctions between the overall gall transcriptome and the GC-specific transcriptome. Over 20 genes are known to play key roles in the small and miRNA biosynthesis pathways of *M. incognita* (Iqbal et al. 2016).

7.4.1.1 miRNA and Galls Formation

MiRNAs are a vast group of short regulatory RNAs that broadly occur throughout all animals and plants (Ha and Kim 2014; Zhang et al. 2016b). MiRNAs are related to gene silencing by base pairing with complementary or substantially identical target gene sequences. Its binding to complementary base pairs induces mRNA degradation or post-translational silencing (Bartel 2004). MiRNAs are synthesized by RNA polymerase II from transcripts of miRNA genes. The primary miRNA (pri-miRNA) transcript folds back and is transformed into the stem-loop precursor termed as precursor miRNA (pre-miRNA). Dicer-like 1 (DCL1), in association with hyponastic leaves 1 (HYL1) and serrate (SE), cleaves the pri-miRNA in the nucleus to form a precursor miRNA (pre-miRNA). Subsequently, DCL1 and its cofactors break the pre-miRNA, producing a duplex of the mature miRNA and its complementary strand. After then, to prevent the miRNA duplex from being degraded, the HUA ENHANCER 1 protein (HEN1) adds a methyl group to the OH ends of both strands and then moves to the cytoplasm from the nucleus (Jaubert-Possamai et al. 2019). In the cytoplasm, post-translational modifications play an important role in epigenetic regulatory systems (Fig. 7.3). However, its length varies from 18 to 25 nucleotides, and most miRNAs are between 20–22 nucleotides long (Zhang et al. 2006). The miRNAs implicated in RKN-induced gall development have been studied in *Arabidopsis* excised galls and non-infected roots (Cabrera et al. 2016: Medina et al. 2017). According to the findings of the investigations, mRNAs serve an important role in controlling gene expression, translational repression, and mRNA degeneration (Borges and Martienssen 2015). Numerous studies on several

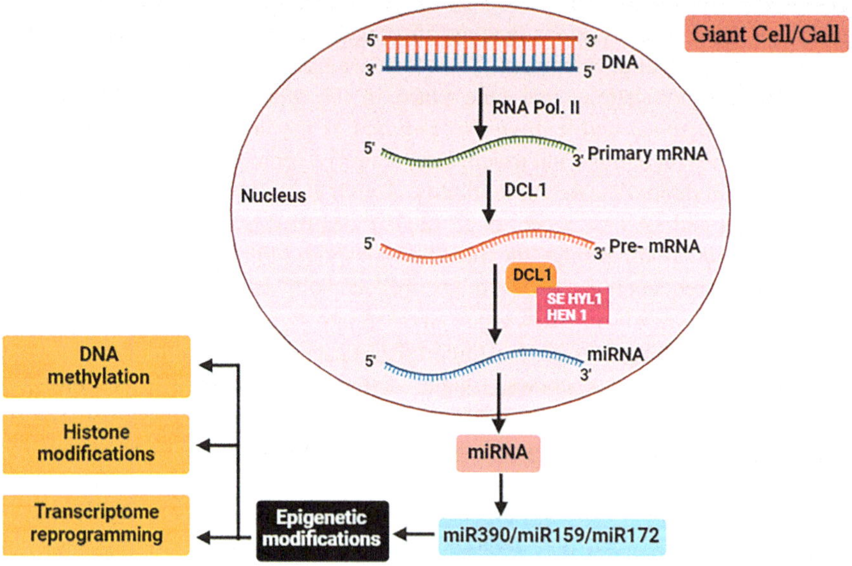

Fig. 7.3 miRNA biosynthetic pathway and their role in epigenetic modifications within the giant cell

species of plants, including *Arabidopsis*, tomato, soybean, cotton, and brinjal, indicate that nematode invasion modifies the expression of miRNA genes (Koter et al. 2018; Pan et al. 2019). Furthermore, several conserved miRNAs play an active role in the development of feeding sites in many plant species. These miRNAs may serve as fundamental regulators of the translational reprogramming that occurs during nematode NFSs. After piercing the root, the J2s choose one or more target cells and then the nematode injects a variety of effector secretions into root cells, transforming them into hypertrophied multinucleate GCs that serve as a feeding site to provide the nutrients necessary for nematode proliferation. RKN J2s pick five to seven parenchymatous cells and cause their proliferation and differentiation into GCs by sequential mitosis without cytokinesis (Caillaud et al. 2008). The GCs are situated within a root protrusion termed a gall which is resulting from the hyperplasia and hypertrophy of the surrounding tissues. In addition, the investigations have turned their focus to the processes underpinning miRNA-mediated transcriptome regulation during the formation of syncytium and GCs. Gene silencing is related to the activation of miRNAs during nematode parasitism of susceptible plants via modification of phytohormone pathways (Hewezi and Baum 2015; Gheysen and Mitchum 2019). RKNs prompt the GCs inside the vascular tissues to develop into galls. miRNAs and/or rasiRNAs driven epigenetic processes may play an important role in the particular gene regulation in early-developing GCs. Consequently, the sRNA abundance and the involvement of the miR390/TAS3/ARFs component throughout early gall/GC development were investigated. The sRNA population differs markedly between galls and controls, with a great validation rate and consistency with their target gene expression: miRNAs were significantly suppressed, but rasiRNAs were predominantly elevated in galls. The promoters of MIR390a and TAS3, which are prominent in galls, as well as the pARF3:ARF3-GUS line, suggest that TAS3-derived tsiRNAs have a role in galls. Early-developing GCs and galls exhibit generalized gene suppression, which is a marker of early-developing GCs (Barcala et al. 2010; Portillo et al. 2013) which comprises genes associated with plant defense (Hewezi and Baum 2015).

In addition, several miRNA genes were identified in *M. incognita*-infested tomato plants at various phases of growth (Kaur et al. 2017). Given all of these perspectives and findings, it is possible to conclude that miRNAs are important regulators of genetic circuits in gall development. It is noteworthy that these miRNAs regulate genomic patterns following the occurrence of infection and respond to different nematode species. These findings may therefore presumptively conclude that this parasitism emphasizes the differences in gene regulation pathways between syncytium and GCs. Recent investigations of numerous miRNAs indicate that epigenetic control of gene expression has a role in gall/GCs organogenesis (Table 7.2).

7.4.1.2 siRNA and Galls Formation

SiRNAs can be used as an emerging technique for the genetics of parasitism genes in nematodes if they could target genes expressed in the internal organs of the nematode and during parasitism. siRNAs might be an effective technique for reverse genetics of nematode parasitism genes if they were to (i) target genes expressed in

Table 7.2 List of several miRNAs involved in giant cells/galls formation

Nematode	Host Plant	Expression Site	miRNA designation	Target	References
Meloidogyne incognita	*Arabidopsis*	Root galls	miRNA390, miRNA775, miRNA839	Production of TAS3 tasiRNAs	Cabrera et al. (2016)
M. javanica	*Arabidopsis*, tomato, pea	Galls and giant-cells	miR172	TOE1	Diaz-Manzano et al. (2018)
Heterodera schachtii	*Arabidopsis*	Syncytium	miR396, miR858	GRF1 or GRF3	Hewezi (2020)
H. schachtii	*A. thaliana*	Syncytium	miR396-GRF1/GRF3	GRF1 and GRF3	Hewezi et al. (2012)
M. incognita	Soybean	Syncytium	miR159, miR396, miR858	GRF MYB33	Jaubert-Possamai et al. (2019)
M. incognita	*Solanum lycopersicum*	Roots	miR156, miR159, miR164 and miR396	GRF1, GRF2 and GRF3	Kaur et al. (2017)
M. javanica	*Arabidopsis*	Root galls	miR390	TAS3	Marin et al. (2010)
M. incognita	*Arabidopsis*	Root galls	miR159, miR319, miR398, miR408	MYB33	Medina et al. (2017)
M. incognita	*Gossypium hirsutum*	Roots	miR159-MYB, miR319-TCP4 and miR167-ARF8	TCP4, ARF8	Pan et al. (2019)
H. schachtii	*Arabidopsis*	Syncytium	miR858	MYB83	Piya et al. (2017)
M. graminicola	*Oryza sativa*	Root galls	miR5149, miR156l-5p, miR164e		Verstraeten et al. (2021)
M. incognita	*Cucumis metuliferus*	Roots	miR156-SBP, ath-miR159a-MYB104 and aly-miR827- 3p-PTI, miR390-ARF3	SBP, ARF3, MYB104	Ye et al. (2020)
M. incognita	*G. hirsutum, Lycopersicon esculentum*	Roots, galls, feeding sites	miR100, miR124, miR228, miR71, miR92, miR34, miR50, miR279	–	Zhang et al. (2016a, b)
M. incognita	*S. lycopersicum*	Roots	miR319/TCP4 and miR396/GRF	TCP4	Zhao et al. (2015)

the inner organ tissues of infective nematodes and (ii) target genes induced by parasite infection. The spontaneous generation of secondary siRNAs by RNA-dependent RNA polymerases (RDRP) amplifies RNA interference in nematodes (Sijen et al. 2001). The RNA interference (RNAi) apparatus has been extensively explored in the free-living nematode, *Caenorhabditis elegans* (Mello and Conte 2004), and the RNAi pathway effector proteins are highly conserved in the RKN *M. incognita* (Abad et al. 2008; Dalzell et al. 2010).

The regulating mechanism of sRNAs in plant-nematode interaction was initially discovered in *Arabidopsis* mutants wherein sRNA generation was repressed during cyst-nematode virulence to host plants. Dicer and RNA-dependent RNA-polymerases (RDRP) mutants exhibited lower susceptibility to *M. incognita* and *H. schachtii,* respectively (Ruiz-Ferrer et al. 2018). Furthermore, Argonaute mutants, ago1–25, ago1–27, etc., that showed significantly reduced susceptibility to *M. incognita* were evaluated (Medina et al. 2017). In addition, elevated sequence alignment experiments revealed that sRNA isolated from root galls of *Arabidopsis* contributed to the identification of siRNA clusters from galls post-infection (Medina et al. 2018). Unlike gene locations in the body, gene promoters had a lot of different types of heterochromatic siRNA. As there were more siRNA arrays in galls than in normal roots, this suggests that nematodes play a role in the biogenesis and suppression of siRNA (Medina et al. 2018). This method, in conjunction with gene expression analysis, concluded that siRNA groups play a fundamental role in the control of galls through the RdDM mechanism. Similarly, siRNAs were detected in *M. javanica*-infested *Arabidopsis* root galls, indicating a unique relationship between sRNA generation biogenesis and accumulation inside galls (Cabrera et al. 2016). Recent research revealed the effectiveness of discrete 21 bp siRNAs as gene-silencing agonists in RKN J2s when targeting neuropeptide genes essential for neuromuscular function and effectors of the miRNA pathway (Dalzell et al. 2011). Arguel et al. (2012) demonstrated that siRNAs may reach and concentrate in the esophagus, amphidial sacks, and associated neurons of the nematode during soaking of infectious juveniles. The fundamental concept is to incorporate into host plants an expression cassette that generates double-stranded RNAs (dsRNAs) that target one or more nematode genes that are crucial for parasitic infections. The collected sequencing data must now be analyzed using a specialized algorithm to detect siRNAs formed in galls and examine siRNA-mediated regulatory networks and their involvement in gall development (Medina et al. 2017). *Arabidopsis* and cyst nematode interactions (Hewezi 2020) and root galls of rice caused by *M. graminicola* have both been linked to DNA methylation changes (Atighi Quchan Atigh 2020). Moreover, DNA methylation and its interaction with the dynamic regulation of sRNAs have not been characterized as exerting a function in the regulation of gene expression in the galls of dicotyledonous plants. Early galls were found to have hypomethylation, whereas GCs were the major cause of hypermethylation, which is associated with an incredibly high level of gene suppression. In contrast, intermediate or late galls demonstrated a large-scale redistribution of differentially methylated regions (DMRs), but no massive increase in DNA methylation in comparison to non-infected roots. Following these observations,

DNA methylation and demethylation mutants exhibited poor nematode reproduction and gall/GCs formation (Silva et al. 2022).

7.5 RKN Effectors and Alteration in Cell Functions

Numerous multifaceted strategies have also been used to get a broad view of the large changes at the transcription, protein, or metabolite levels that happen when nematodes infect different species of host plants (Liu et al. 2016; Kumari et al. 2016; Jain et al. 2016; Ye et al. 2017). It focuses primarily on highlighting the integrated network of genetic toolboxes responsible for initiating the development of feeding mechanisms inside host plants. Recent research has increasingly emphasized acquiring reasonably valuable information regarding plant-RKN interactions, emphasizing the molecular foundation. Changes in gene products and alterations in the cell cycle (De Almeida et al. 2015), cell wall (Gheysen and Mitchum 2011; Wieczorek 2015), and cell metabolism (Miyara et al. 2015; Siddique and Grundler 2015) or growth pathophysiological pathways (Cabrera et al. 2015) have laid the foundation for enhanced knowledge of the molecular mechanisms developing GCs and galls (Fig. 7.4).

In addition, nematodes expel many secretary molecules that they use to manipulate the metabolic machinery of their host cells. Most of the genes encoding cellulose-pectinolytic enzymes are implicated in weakening plant cell walls after nematode penetration, resulting in cell alteration. Earlier research showed that many PPNs have the same effector proteins that disintegrate the plant cell wall (e.g.,

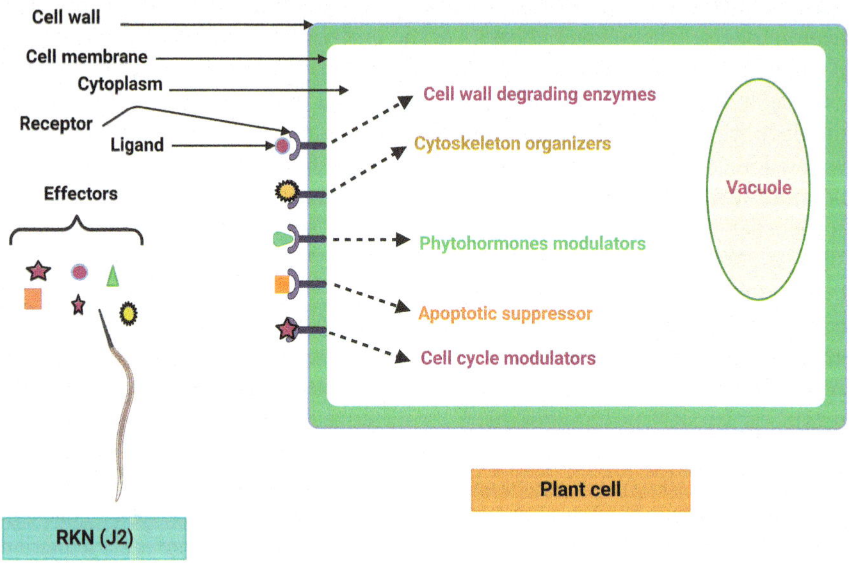

Fig. 7.4 Effector and their role in the alteration of cell functions

cellulases, pectinases). Gene expression in infected root cells is altered, indicating the complex morphological and physiological modifications during GC establishment (Gheysen and Fenoll 2002). These proteins seem to have been transmitted from one species to another through horizontal gene transfer (HGT) from fungi and bacteria (Danchin et al. 2010; Haegeman et al. 2011).

Nonetheless, most recognized and feasible effector genes are called "orphan" proteins because they do not have any known homology in species other than the PPNs (Mitchum et al. 2013). Most of these potential effectors lack sequence homology to proteins in the public domain; hence, their roles in nematode parasitism are largely a mystery. Nevertheless, only a small proportion of these potential effectors exhibited considerable sequence homology with major epigenetic modification factors (Noon et al. 2015; Eves-van den Akker et al. 2016; Gardner et al. 2018).

7.6 Conclusions

RKNs are prominent biotrophic parasites that infect plants and induce remarkable morphological and physiological alterations. RKNs rely on specialized host cells that develop from their vascular cells in the early root to complete their life cycle. Undoubtedly, NFSs exhibit tremendous gene expression changes, most of which are ultimately downregulation activities. GCs have a high rate of metabolism, many organelles, and enlarged nuclei and nucleoli due to their thick cytoplasm. Epigenetically, the fate of GCs is controlled by the differential production of miRNAs and siRNAs, the methylation of DNA, and the alteration of histone proteins. Concurrently, galls develop around GCs due to accelerated vascular cell proliferation and hypertrophy of the endodermis and the cortex. In addition, it is becoming increasingly evident that sRNA molecules play crucial roles in regulating these alterations. Various short RNAs, including miRNAs and siRNAs, are formed at NFSs, where they may play an important role in root gall formation. Despite great advances in the comprehension of the regulatory roles of multiple epigenetic components in gall development, the coordinated roles of these components have yet to be investigated. It is becoming obvious that several epigenetic alterations are closely interrelated. The molecular mechanism by which RKNs trigger epigenetic modifications in host plants is still partially understood. It is quite probable that nematode effectors play crucial roles in inducing epigenetic responses to gall development.

References

Abad P, Favery B, Rosso MN, Castagnone-Sereno P (2003) Root-knot nematode parasitism and host response: molecular basis of a sophisticated interaction. Mol Plant Pathol 4:217–224. https://doi.org/10.1046/j.1364-3703.2003.00170.x

Abad P, Williamson VM (2010) Plant nematode interaction: a sophisticated dialogue. Adv Bot Res 53:147–192. https://doi.org/10.1016/S0065-2296(10)53005-2

Abad P, Gouzy J, Aury JM, Castagnone-Sereno P, Danchin EG, Deleury E, Perfus-Barbeoch L, Anthouard V, Artiguenave F, Blok VC, Wincker P (2008) Genome sequence of the metazoan plant-parasitic nematode *Meloidogyne incognita*. Nat Biotechnol 26:909–915. https://doi.org/10.1038/nbt.1482

Armstrong C, Spencer SL (2021) Replication-dependent histone biosynthesis is coupled to cell-cycle commitment. Proceed Nation Acade Sci 118:e2100178118. https://doi.org/10.1073/pnas.2100178118

Arguel MJ, Jaouannet M, Magliano M, Abad P, Rosso MN (2012) siRNAs trigger efficient silencing of a parasitism gene in plant parasitic root-knot nematodes. Genes 3(3):391–408. https://doi.org/10.3390/genes3030391

Atighi Quchan Atigh M (2020) The role of epigenetics in the interaction between rice and root-knot nematodes. Doctoral dissertation,. Ghent University

Atighi MR, Verstraeten B, De Meyer T, Kyndt T (2021) Genome-wide shifts in histone modifications at early stage of rice infection with *Meloidogyne graminicola*. Mol Plant Pathol 22:440–455. https://doi.org/10.1111/mpp.13037

Banora MY, Rodiuc N, Baldacci-Cresp F, Smertenko A, Bleve-Zacheo T, Mellilo MT, Karimi M, Hilson P, Evrard JL, Favery B, Engler G, Abad P, de Almeida EJ (2011) Feeding cells induced by phytoparasitic nematodes requireg-tubulin ring complex for microtubule reorganisation. PLoS Pathog 7:e1002343. https://doi.org/10.1371/journal.ppat.1002343

Barcala M, García A, Cabrera J, Casson S, Lindsey K, Favery B, Escobar C (2010) Early transcriptomic events in microdissected *Arabidopsis* nematode-induced giant cells. Plant J 61: 698–712. https://doi.org/10.1111/j.1365-313X.2009.04098.x

Bardgett RD, Van Der Putten WH (2014) Belowground biodiversity and ecosystem functioning. Nature 515:505–511. https://doi.org/10.1038/nature13855

Bartel DP (2004) MicroRNAs: genomics, biogenesis, mechanism, and function. Cell 116:281–297. https://doi.org/10.1016/S0092-8674(04)00045-5

Bennett MM, Meredith M (2021) Epigenetic mechanisms governing plant growth, development, and responses to nematode parasitism. Doctoral dissertation,. University of Tennessee–Knoxville. https://doi.org/10.1146/annurev-phyto-010820-012805

Bird DM, Williamson VM, Abad P, McCarter J, Danchin EG, Castagnone-Sereno P, Opperman CH (2009) The genomes of root-knot nematodes. Annu Rev Phytopathol 47:333–351. https://doi.org/10.1146/annurev-phyto-080508-081839

Blanc-Mathieu R, Perfus-Barbeoch L, Aury JM, Da Rocha M, Gouzy J, Sallet E, Danchin EG (2017) Hybridisation and polyploidy enable genomic plasticity without sex in the most devastating plant-parasitic nematodes. PLoS Gen 13:e1006777. https://doi.org/10.1371/journal.pgen.1006777

Borges F, Martienssen RA (2015) The expanding world of small RNAs in plants. Nat Rev Mol Cell Biol 16:727–741. https://doi.org/10.1038/nrm4085

Cabrera J, Barcala M, García A, Rio-Machín A, Medina C, Jaubert-Possamai S, Escobar C (2016) Differentially expressed small RNA s in Arabidopsis galls formed by *Meloidogyne javanica*: a functional role for miR390 and its TAS 3-derived tasi RNA s. New Phytol 209:1625–1640. https://doi.org/10.1111/nph.13735

Cabrera J, Diaz-Manzano FE, Barcala M, Arganda-Carreras I, de Almeida-Engler J, Engler G, Escobar C (2015) Phenotyping nematode feeding sites: three-dimensional reconstruction and volumetric measurements of giant cells induced by root-knot nematodes in Arabidopsis. New Phytol 206:868–880. https://doi.org/10.1111/nph.13249

Caillaud MC, Abad P, Favery B (2008) Cytoskeleton reorganisation: a key process in root-knot nematode-induced giant cell ontogenesis. Plant Signal Behav 3:816–818. https://doi.org/10.4161/psb.3.10.5889

Coyne DL, Cortada L, Dalzell JJ, Claudius-Cole AO, Haukeland S, Luambano N, Talwana H (2018) Plant-parasitic nematodes and food security in sub-Saharan Africa. Annu Rev Phytopathol 56:381. https://doi.org/10.1146/annurev-phyto-080417-045833

Crespi M, Frugier F (2008) De novo organ formation from differentiated cells: root nodule organogenesis. Sci Sign 1:re11-re11. https://doi.org/10.1126/scisignal.149re11
Da Silva ACP (2020) Epigenetic changes in galls induced by the nematode *Meloidogyne javanica* in arabidopsis. Doctoral dissertation,. Universidad de Castilla-La Mancha
Dalzell JJ, McVeigh P, Warnock ND, Mitreva M, Bird DM, Abad P, Maule AG (2011) RNAi effector diversity in nematodes. PLoS Negl Trop Dis 5:e1176. https://doi.org/10.1371/journal.pntd.0001176
Dalzell JJ, Warnock ND, Stevenson MA, Mousley A, Fleming CC, Maule AG (2010) Short interfering RNA-mediated knockdown of drosha and pasha in undifferentiated *Meloidogyne incognita* eggs leads to irregular growth and embryonic lethality. Int J Parasitol 40:1303–1310. https://doi.org/10.1016/j.ijpara.2010.03.010
Danchin EG, Rosso MN, Vieira P, de Almeida-Engler J, Coutinho PM, Henrissat B, Abad P (2010) Multiple lateral gene transfers and duplications have promoted plant parasitism ability in nematodes. Proceed Nat Acad Sci 107:17651–17656. https://doi.org/10.1073/pnas.1008486107
De Almeida EJ, Vieira P, Rodiuc N, de Sa MFG, Engler G (2015) The plant cell cycle machinery: usurped and modulated by plant-parasitic nematodes. Adv Bot Res 73:91–118. https://doi.org/10.3390/plants7040085
Decraemer W, Hunt DJ (2006) Structure and classification. Plant Nematol:3–32
Diaz-Manzano FE, Cabrera J, Ripoll JJ, Del Olmo I, Andrés MF, Silva AC, Escobar C (2018) A role for the gene regulatory module microRNA172/target of early activation tagged 1/flowering locus T (mi RNA 172/TOE 1/FT) in the feeding sites induced by *Meloidogyne javanica* in *Arabidopsis thaliana*. New Phytol 217:813–827. https://doi.org/10.1111/nph.14839
Duan CG, Zhu JK, Cao X (2018) Retrospective and perspective of plant epigenetics in China. J Genet Genomics 45:621–638. https://doi.org/10.1016/j.jgg.2018.09.004
Eves-van den Akker S, Laetsch DR, Thorpe P, Lilley CJ, Danchin EG, Da Rocha M, Jones JT (2016) The genome of the yellow potato cyst nematode, *Globodera rostochiensis*, reveals insights into the basis of parasitism and virulence. Genom Biol 17:1–23. https://doi.org/10.1186/s13059-016-0985-1
Gardner M, Dhroso A, Johnson N, Davis EL, Baum TJ, Korkin D, Mitchum MG (2018) Novel global effector mining from the transcriptome of early life stages of the soybean cyst nematode *Heterodera glycines*. Sci Rep 8:1–15. https://doi.org/10.1038/s41598-018-20536-5
Gheysen G, Fenoll C (2002) Gene expression in nematode feeding sites. Annu Rev Phytopathol 40:191. https://doi.org/10.1146/annurev.phyto.40.121201.093719
Gheysen G, Mitchum MG (2011) How nematodes manipulate plant development pathways for infection. Curr Opin Plant Biol 14:415–421. https://doi.org/10.1016/j.pbi.2011.03.012
Gheysen G, Mitchum MG (2019) Phytoparasitic nematode control of plant hormone pathways. Plant Physiol 179:1212–1226. https://doi.org/10.1104/pp.18.01067
Gualtieri C, Leonetti P, Macovei A (2020) Plant miRNA cross-kingdom transfer targeting parasitic and mutualistic organisms as a tool to advance modern agriculture. Front Plant Sci 11:930. https://doi.org/10.3389/fpls.2020.00930
Ha M, Kim VN (2014) Regulation of microRNA biogenesis. Nat Rev Mol Cell Biol 15:509–524. https://doi.org/10.1038/s41580-018-0059-1
Haegeman A, Jones JT, Danchin EG (2011) Horizontal gene transfer in nematodes: a catalyst for plant parasitism? Mol Plant-Micr Inter 24:879–887. https://doi.org/10.1094/MPMI-03-11-0055
Hamilton AJ, Baulcombe DC (1999) A species of small antisense RNA in posttranscriptional gene silencing in plants. Sci 286:950–952. https://doi.org/10.1126/science.286.5441.950
Hewezi T (2020) Epigenetic mechanisms in nematode–plant interactions. Annu Rev Phytopathol 58:119–138. https://doi.org/10.1146/annurev-phyto-010820-012805
Hewezi T, Baum TJ (2013) Manipulation of plant cells by cyst and root-knot nematode effectors. Mol Plant-Micr Inter 26:9–16. https://doi.org/10.1094/MPMI-05-12-0106-FI
Hewezi T, Baum TJ (2015) Gene silencing in nematode feeding sites. Adv Bot Res 73:221–239

Hewezi T, Maier TR, Nettleton D, Baum TJ (2012) The *Arabidopsis* microRNA396-GRF1/GRF3 regulatory module acts as a developmental regulator in the reprogramming of root cells during cyst nematode infection. Plant Physiol 159:321–335. https://doi.org/10.1104/pp.112.193649

Iqbal S, Fosu-Nyarko J, Jones MG (2016) Genomes of parasitic nematodes (*Meloidogyne hapla, Meloidogyne incognita, Ascarissuum* and *Brugiamalayi*) have a reduced complement of small RNA interference pathway genes: knockdown can reduce host infectivity of *M. incognita*. Funct Integ Gen 16:441–457. https://doi.org/10.1007/s10142-016-0495-y

Jagdale S, Rao U, Giri AP (2021) Effectors of root-knot nematodes: an arsenal for successful parasitism. Front Plant Sci 12:800030. https://doi.org/10.3389/fpls.2021.800030

Jain S, Chittem K, Brueggeman R, Osorno JM, Richards J, Nelson BD Jr (2016) Comparative transcriptome analysis of resistant and susceptible common bean genotypes in response to soybean cyst nematode infection. PLoS One 11:e0159338. https://doi.org/10.1371/journal.pone.0159338

Jaubert-Possamai S, Noureddine Y, Favery B (2019) MicroRNAs, new players in the plant–nematode interaction. Front Plant Sci 10:1180. https://doi.org/10.3389/fpls.2019.01180

Ji H, Gheysen G, Denil S, Lindsey K, Topping JF, Nahar K, Kyndt T (2013) Transcriptional analysis through RNA sequencing of giant cells induced by *Meloidogyne graminicola* in rice roots. J Exp Bot 64:3885–3898. https://doi.org/10.1093/jxb/ert219

Kasschau KD, Fahlgren N, Chapman EJ, Sullivan CM, Cumbie JS, Givan SA, Carrington JC (2007) Genome-wide profiling and analysis of *Arabidopsis* siRNAs. PLoS Biol 5:e57. https://doi.org/10.1371/journal.pbio.0050057

Katiyar-Agarwal S, Jin H (2010) Role of small RNAs in host-microbe interactions. Annu Rev Phytopathol 48:225. https://doi.org/10.1146/annurev-phyto-073009-114457

Kaur P, Shukla N, Joshi G, VijayaKumar C, Jagannath A, Agarwal M, Kumar A (2017) Genome-wide identification and characterisation of miRNAome from tomato (*Solanum lycopersicum*) roots and root-knot nematode (*Meloidogyne incognita*) during susceptible interaction. PLoS One 12:e0175178. https://doi.org/10.1371/journal.pone.0175178

Koter MD, Swięcicka M, Matuszkiewicz M, Pacak A, Derebecka N, Filipecki M (2018) ThemiRNAome dynamics during developmental and metabolic reprogramming of tomato root infected with potato cyst nematode. Plant Sci 268:18–29. https://doi.org/10.1371/journal.pone.0175178

Kumari C, Dutta TK, Banakar P, Rao U (2016) Comparing the defence-related gene expression changes upon root-knot nematode attack in susceptible versus resistant cultivars of rice. Sci Rep 6:1–13. https://doi.org/10.1038/srep22846

Kyndt T, AtighiQuchanAtigh M, Verstraeten B, Meijer A, De Meyer T (2019) The role of epigenetics in the interaction between rice and parasitic nematodes. In IS-MPMI XVIII, Bienn Cong Mol Plant-Micr Int 32:142. https://doi.org/10.1111/nph.15408

Lawrence M, Daujat S, Schneider R (2016) Lateral thinking: how histone modifications regulate gene expression. Trends Genet 32:42–56. https://doi.org/10.1016/j.tig.2015.10.007

Liu T, Chen X, Hu F, Ran W, Shen Q, Li H, Whalen JK (2016) Carbon-rich organic fertilisers to increase soil biodiversity: evidence from a meta-analysis of nematode communities. Agri Eco Env 232:199–207. https://doi.org/10.1016/j.agee.2016.07.015

Luger K, Mäder AW, Richmond RK, Sargent DF, Richmond TJ (1997) Crystal structure of the nucleosome core particle at 2.8 a resolution. Nature 389:251–260. https://doi.org/10.1038/38444

Machado ACZ, Silva SA, Dorigo OF, Riede CR, Garbuglio DD (2015) Phenotypic variability and response of Brazilian oat genotypes to different species of root-knot and root-lesion nematodes. Europ J Plant Pathol 141:111–117. https://doi.org/10.21307/jofnem-2017-086

Marin E, Jouannet V, Herz A, Lokerse AS, Weijers D, Vaucheret H, Nussaume L, Crespi MD, Maizel A (2010) miR390, *Arabidopsis*TAS3tasiRNAs, and their auxin response factor targets define an autoregulatory network quantitatively regulating lateral root growth. Plant Cell 22: 1104–1117. https://doi.org/10.1105/tpc.109.072553

McCarter JP, Dautova Mitreva M, Martin J, Dante M, Wylie T, Rao U, Waterston RH (2003) Analysis and functional classification of transcripts from the nematode *Meloidogyne incognita*. Genom Biol 4:1–19. https://doi.org/10.1186/gb-2003-4-4-r26

Medina C, Da Rocha M, Magliano M, Raptopoulo A, Marteu N, Lebrigand K, Jaubert-Possamai S (2018) Characterisation of siRNAs clusters in *Arabidopsis thaliana* galls induced by the root-knot nematode *Meloidogyne incognita*. BMC Genomics 19:1–16. https://doi.org/10.1186/s12864-018-5296-3

Medina C, Da Rocha M, Magliano M, Ratpopoulo A, Revel B, Marteu N, Jaubert-Possamai S (2017) Characterisation of microRNAs from Arabidopsis galls highlights a role for miR159 in the plant response to the root-knot nematode *Meloidogyne incognita*. New Phytol 216:882–896. https://doi.org/10.1111/nph.14717

Mello CC, Conte D (2004) Revealing the world of RNA interference. Nature 431:338–342. https://doi.org/10.1038/nature02872

Mitchum MG, Hussey RS, Baum TJ, Wang X, Elling AA, Wubben M, Davis EL (2013) Nematode effector proteins: an emerging paradigm of parasitism. New Phytol 199:879–894. https://doi.org/10.1111/nph.12323

Miyara SB, Ionit I, Buki P, Kolomiets M (2015) The role of lipid signalling in regulating plant–nematode interactions. Adv Bot Res 73:139–166. https://doi.org/10.1371/journal.pone.0145660

Noon JB, Hewezi T, Maier TR, Simmons C, Wei JZ, Wu G, Baum TJ (2015) Eighteen new candidate effectors of the phytonematode *Heterodera glycines* produced specifically in the secretory esophageal gland cells during parasitism. Phyto Pathol 105:1362–1372. https://doi.org/10.1094/PHYTO-02-15-0049-R

Opperman CH, Bird DM, Williamson VM, Rokhsar DS, Burke M, Cohn J, Windham E (2008) Sequence and genetic map of *Meloidogyne hapla*: a compact nematode genome for plant parasitism. Proceed Nat Acad Sci 105:14802–14807. https://doi.org/10.1073/pnas.0805946105

Palomares-Rius JE, Escobar C, Cabrera J, Vovlas A, Castillo P (2017) Anatomical alterations in plant tissues induced by plant-parasitic nematodes. Front Plant Sci 8:1987. https://doi.org/10.3389/fpls.2017.01987

Pan X, Nichols RL, Li C, Zhang B (2019) MicroRNA-target gene responses to root knot nematode (*Meloidogyne incognita*) infection in cotton (*Gossypium hirsutum* L.). Genom 111:383–390. https://doi.org/10.1016/j.ygeno.2018.02.013

Peiris PUS, Li Y, Brown P, Xu C (2020) Fungal biocontrol against *Meloidogyne* spp. in agricultural crops: a systematic review and meta-analysis. Biol Control 144:104235. https://doi.org/10.3390/microorganisms11010137

Perfus-Barbeoch L, Castagnone-Sereno P, Reichelt M, Fneich S, Roquis D, Pratx L, Abad P (2014) Elucidating the molecular bases of epigenetic inheritance in non-model invertebrates: the case of the root-knot nematode *Meloidogyne incognita*. Front Physiol 5:211. https://doi.org/10.3389/fphys.2014.00211

Piya S, Kihm C, Rice JH, Baum TJ, Hewezi T (2017) Cooperative regulatory functions of miR858 and MYB83 during cyst nematode parasitism. Plant Physiol 174:1897–1912. https://doi.org/10.1104/pp.17.00273

Poças-Fonseca MJ, Cabral CG, Manfrao-Netto JHC (2020) Epigenetic manipulation of filamentous fungi for biotechnological applications: a systematic review. Biotechnol Lett 42:885–904. https://doi.org/10.1007/s10529-020-02871-8

Portillo M, Cabrera J, Lindsey K, Topping J, Andrés MF, Emiliozzi M, Escobar C (2013) Distinct and conserved transcriptomic changes during nematode-induced giant cell development in tomato compared with *Arabidopsis*: a functional role for gene repression. New Phytol 197:1276–1290. https://doi.org/10.1111/nph.12121

Portillo M, Lindsey K, Casson S, Gloria G-C, Solano R, Fenoll C, Escobar C (2009) Isolation of RNA from laser-capture-microdissected giant cells at early differentiation stages suitable for differential transcriptome analysis. Mol Plant Pathol 10:523–535. https://doi.org/10.1111/j.1364-3703.2009.00552.x

Pratx L, Rancurel C, Da Rocha M, Danchin EG, Castagnone-Sereno P, Abad P, Perfus-Barbeoch L (2018) Genome-wide expert annotation of the epigenetic machinery of the plant-parasitic nematodes *Meloidogyne* spp., with a focus on the asexually reproducing species. BMC Genomics 19:1–21. https://doi.org/10.1186/s12864-018-4686-x

Rodiuc N, Vieira P, Banora MY, de Almeida EJ (2014) On the track of transfer cell formation by specialised plant-parasitic nematodes. Front Plant Scie 5:160. https://doi.org/10.3389/fpls.2014.00160

Ruiz-Ferrer V, Cabrera J, Martinez-Argudo I, Artaza H, Fenoll C, Escobar C (2018) Silenced retrotransposons are major rasiRNAs targets in Arabidopsis galls induced by *Meloidogyne javanica*. Mol Plant Pathol 19:2431–2445. https://doi.org/10.1111/mpp.12720

Shukla N, Yadav R, Kaur P, Rasmussen S, Goel S, Agarwal M, Kumar A (2018) Transcriptome analysis of root-knot nematode (*Meloidogyne incognita*)-infected tomato (*Solanum lycopersicum*) roots reveals complex gene expression profiles and metabolic networks of both host and nematode during susceptible and resistance responses. Mol Plant Pathol 19:615–633. https://doi.org/10.1111/mpp.12547

Siddique S, Grundler FM (2015) Metabolism in nematode feeding sites. Adv Bot Res 73:119–138. https://doi.org/10.1016/j.mib.2018.09.004

Sijen T, Fleenor J, Simmer F, Thijssen KL, Parrish S, Timmons L, Fire A (2001) On the role of RNA amplification in dsRNA-triggered gene silencing. Cell 107:465–476. https://doi.org/10.1016/s0092-8674(01)00576-1

Sikandar A, Zhang M, Wang Y, Zhu X, Liu X, Fan H, Duan Y (2020) In vitro evaluation of *Penicillium chrysogenum* Snef1216 against *Meloidogyne incognita* (root-knot nematode). Sci Rep 10:1–9. https://doi.org/10.1038/s41598-020-65262-z

Silva AC, Ruiz-Ferrer V, Müller SY, Pellegrin C, Abril-Urías P, Martinez-Gomez A, Escobar C (2022) The DNA methylation landscape of the root-knot nematode-induced pseudo-organ, the gall, in *Arabidopsis*, is dynamic, contrasting over time, and critically important for successful parasitism. New Phytol 236:1888. https://doi.org/10.1111/nph.18395

Simon SA, Meyers BC (2011) Small RNA-mediated epigenetic modifications in plants. Curr Opin Plant Biol 14:148–155. https://doi.org/10.1016/j.pbi.2010.11.007

Singh S, Singh B, Singh AP (2015) Nematodes: a threat to sustainability of agriculture. Procedia Environ Sci 29:215–216. https://doi.org/10.1016/j.proenv.2015.07.270

Somvanshi VS, Tathode M, Shukla RN, Rao U (2018) Nematode genome announcement: a draft genome for Rice root-knot nematode. J Nematol 50:111–116. https://doi.org/10.21307/jofnem-2018-018

Szitenberg A, Salazar-Jaramillo L, Blok VC, Laetsch DR, Joseph S, Williamson VM, Lunt DH (2017) Comparative genomics of apomictic root-knot nematodes: hybridisation, ploidy, and dynamic genome change. Genome Biol Evol 9:2844–2861. https://doi.org/10.1093/gbe/evx201

Van Den Hoogen J, Geisen S, Routh D, Ferris H, Traunspurger W, Wardle DA, Crowther TW (2019) Soil nematode abundance and functional group composition at a global scale. Nature 572:194–198. https://doi.org/10.1038/s41586-019-1418-6

Verstraeten B, Atighi MR, Ruiz-Ferrer V, Escobar C, De Meyer T, Kyndt T (2021) Noncoding RNAs in the interaction between rice and *Meloidogyne graminicola*. BMC Genomics 22:1–19. https://doi.org/10.1186/s12864-021-07735-7

Vieira P, Escudero C, Rodiuc N, Boruc J, Russinova E, Glab N, de Almeida EJ (2013) Ectopic expression of K ip-related proteins restrains root-knot nematode-feeding site expansion. New Phytol 199:505–519. https://doi.org/10.1111/nph.12255

Waddington CH (1942) The epigenotype. Endeavour 1:18–20. https://doi.org/10.1093/ije/dyr184

Wieczorek K (2015) Cell wall alterations in nematode-infected roots. Adv Bot Res 73:61–90

Xie Z, Johansen LK, Gustafson AM, Kasschau KD, Lellis AD, Zilberman D, Carrington JC (2004) Genetic and functional diversification of small RNA pathways in plants. PLoS Biol 2:e104. https://doi.org/10.1371/journal.pbio.0020104

Ye DY, Qi YH, Cao SF, Wei BQ, Zhang HS (2017) Histopathology combined with transcriptome analyses reveals the mechanism of resistance to *Meloidogyne incognita* in *Cucumis metuliferus*. J Plant Physiol 212:115–124. https://doi.org/10.1016/j.jplph.2017.02.002

Ye D, Jiang Y, Wang C, Roberts PA (2020) Expression analysis of microRNAs and their target genes in *Cucumis metuliferus* infected by the root-knot nematode *Meloidogyne incognita*. Physiol Mol Plant Pathol 111:101491. https://doi.org/10.5511/plantbiotechnology.20.0716a

Zhang BH, Pan XP, Cox SB, Cobb GP, Anderson TA (2006) Evidence that miRNAs are different from other RNAs. CMLS 63:246–254. https://doi.org/10.1007/s00018-005-5467-7

Zhang H, Li C, Davis EL, Wang J, Griffin JD, Kofsky J, Song BH (2016a) Genome-wide association study of resistance to soybean cyst nematode (*Heterodera glycines*) HG type 2.5. 7 in wild soybean (*Glycine soja*). Front Plant Sci 7:1214. https://doi.org/10.3389/fpls.2016. 01214

Zhang H, Li Y, Liu Y, Liu H, Wang H, Jin W, Xu D (2016b) Role of plant MicroRNA in cross-species regulatory networks of humans. BMC Syst Biol 10:1–10. https://doi.org/10.1186/ s12918-016-0292-1

Zhao W, Li Z, Fan J, Hu C, Yang R, Qi X, Wang S (2015) Identification of jasmonic acid-associated microRNAs and characterisation of the regulatory roles of the miR319/TCP4 module under root-knot nematode stress in tomato. J Exp Bot 66:4653–4667. https://doi.org/10.1093/jxb/ erv238

Mass Spectrometry Imaging (MSI) and Root Gall Elucidation

8

Saeeda Zaima Zeb, Hera Nadeem, and Faheem Ahmad

Abstract

Nematodes are the most destructive pest that is responsible for significant agricultural losses all over the world. Every plant species has at least one species of nematode that parasitizes them in their lifetime. It is essential to understand the metabolic modifications generated during the interaction between nematodes and plants to produce resistant plants or elucidate more effective molecules in the fight against this pathogen. The use of mass spectrometry (MS) to classify nematodes has a history that spans over two decades and is replete with a wide variety of applications that have met with varying degrees of commercialization. The matrix-assisted laser desorption/ionization mass spectrometry imaging technique, abbreviated as MALDI-MSI, has been applied for in situ identification and mapping of endogenous polypeptides and secondary metabolites originating from nematode-induced gall tissue. In addition, during the past few years, molecular networking has developed as an important tool for monitoring and interpreting the chemical domain available in MS data that is not targeted. As a result, the MSI-based galls explication is the primary emphasis of this chapter. Moreover, a description of a considerably more advanced analysis carried out by employing molecular networking is included.

Keywords

Metabolomics · GNPS · Hypertrophy · Secondary Metabolites · *Meloidogyne* spp.

S. Z. Zeb · H. Nadeem · F. Ahmad (✉)
Department of Botany, Aligarh Muslim University, Aligarh, India
e-mail: faheem.bt@amu.ac.in

F. Ahmad, G. Nombela (eds.), *Root-Galling Disease of Vegetable Plants*,
https://doi.org/10.1007/978-981-99-3892-6_8

8.1 Introduction

Mass spectrometry imaging (MSI) is a key technique that has arisen over the past 20 years for the label-free, untargeted spatiochemical characterization of biological systems (Spengler 2015; Buchberger et al. 2018). The most widely used MSI method for molecular imaging of both mammalian (Schwamborn and Caprioli 2010) and plant tissues (Kaspar et al. 2011) is matrix-assisted laser desorption/ionization imaging (MALDI). Mass spectrometry is the most helpful method for identifying components and has been widely used in plant research (Dong et al. 2016). Because it can determine both molecular compositions and spatial distributions, mass spectrometry imaging has recently made significant strides in plant analysis (Ehrhardt and Frommer 2012). Mass spectrometry imaging makes comprehending specific plant component's functions and regulatory mechanisms possible. Widespread plant applications result from technological advancements such as sample preparation, ionization technology development, innovative matrix design, and single-cell MSI (Hansen and Lee 2018). MSI encompasses a wide range of platform types (given in Fig. 8.1), the most well-known of which are matrix-assisted laser desorption/ionization (MALDI), desorption electrospray ionization (DESI), and secondary ion mass spectrometry (SIMS). The Table 8.1 shows comparative description of ionization methods with the help of applications, performance, advantages and disadvantages of the techniques; MALDI, SIMS and DESI. Saliently, the spatiochemical information provided by MSI is significantly quite accurate compared to various microscopic imaging techniques. It is substantially more instinctive when compared to colorimetric imaging, as MSI works in a manner that is quite identical to colorimetric imaging.

Regarding individual numbers, nematodes make up the largest group of multicellular animals on the planet. It is estimated that over 4100 species of plant-parasitic nematodes have been identified (ITIS, accessed on 27 Aug 2023) New species are constantly being discovered, and some that were once considered harmless or non-harmful are now becoming parasites as cropping patterns change (Nicol et al. 2011). Every year, crop productivity is significantly decreased by plant diseases brought on by plant pathogens, resulting in enormous economic losses around the globe. Specially nematodes are the most destructive pest infecting most cultivated plant species and contributing significantly to global agricultural losses. It has been calculated that plant nematodes inflict $US80 billion in damage annually (Nicol et al. 2011). Even with modern technology, nematodes still cause developed countries to lose 5 to 10% of their crop output.

Nematodes that parasitize plants engage in a broad range of interactions with their hosts. Each has a hollow, protruding stylet or mouth spear that can pierce cells to enable feeding. The root-knot nematodes (RKN) are obligatory endoparasites and spread around the globe. Invading endoparasitic forms lead to root-gall disease (Fig. 8.1). These galls are referred to as "root-knot-like" because they resemble the appearance of knots or lumps on the roots, which is how these nematodes got their common name (Fig. 8.2). The majority of these nematodes can be found in soils that

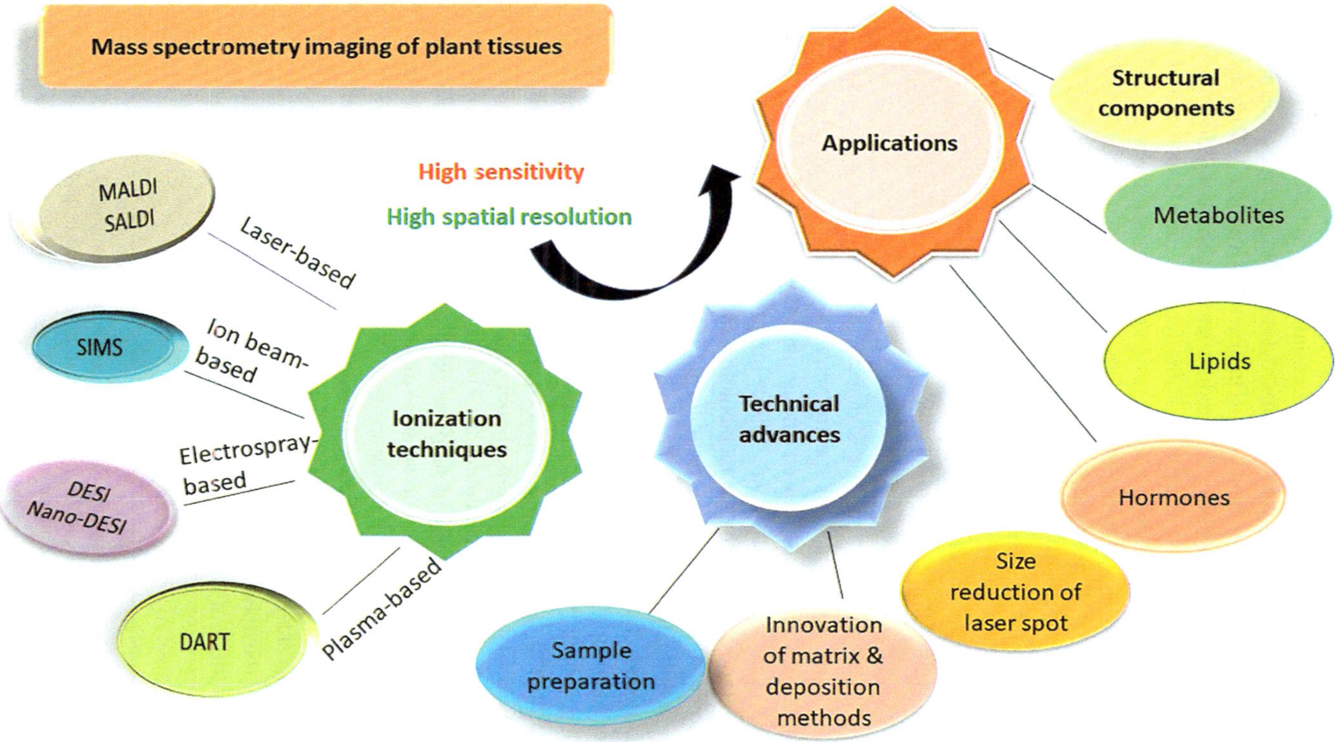

Fig. 8.1 This figure explains via a flow chart diagram about different ionization techniques, their technical advancement, and various applications of mass spectrometry imaging. The ionization techniques are classified based on beams used, e.g., laser, ion, electrospray, and plasma, hence named accordingly. MSI has become an established tool, and its advancement includes improvements in reproducible sample preparation to facilitate reliable data interpretation and

are only a few feet deep. Notable tropical species include *Meloidogyne arenaria*, *M. incognita*, and *M. javanica*, while *M. hapla* is a temperate species (Moens et al. 2009). These four most prevalent species account for up to 95% of all RKN (Dong et al. 2012). RKN enters the plant vascular cylinder and becomes sedentary after moving between cortical tissue and cells (Hussey and Grundler 1998). Nematodes inject discharges regularly to keep the feeding site in good condition. Nematodes also consume the substances already inside the feeding cells (Jones and Northcote 1972) to grow and produce eggs. Swellings or galls form on the roots of infected plants and are caused by hypertrophy and hyperplasia of root cells induced by nematode feeding. The galls range from slight thickenings to lumps of 5 to 10 cm in diameter. Compared to typical nodules, which form after infection by helpful, symbiotic bacteria that fix ambient nitrogen for the plant and receive photosynthates in return, nematode-induced galls are globular, irregular deformations and are not surface-attached. In giant cells, developing galls, and surrounding tissues, there has been a significant drop in defense-related hormones, primarily ethylene and salicylate.

During the past 10 years, the fast advancement of matrix-assisted laser-desorption/ionization time-of-flight mass spectrometry (MALDI-TOF MS) techniques for microbe characterization has allowed for significantly better microorganism detection and identification (Morris et al. 1996). In conjunction with newly developed computational tools for mining the metabolome, the interdisciplinary field of omics science known as metabolomics presents unrivalled opportunities to provide a comprehensive qualitative and quantitative description of all metabolites in a biological system (Dhanasekaran et al. 2015; Booth et al. 2013). Secondary metabolites like isoprenoids, phenylpropanoids, alkaloids, and fatty acids, are capable of acting in both the constitutive and inducible defensive mechanisms of plants against the natural pests that affect plants, and among these, several have been studied in MALSI MSI elucidation of root galls (Cheng et al. 2007; Ziegler and Facchini 2008; Vogt 2010; Fujimoto et al. 2015). According to Wang et al. (2009), it is hypothesized that secondary metabolites have a role in both discouraging root-knot nematodes and luring them to an area (in a species-specific manner). Finding a method to study metabolomic pathways without interfering with them is a significant technical challenge when researching biological systems (Prell and Poole 2006). Direct tissue analysis using MALDI-MSI makes it possible to identify analytes in individual organs (Kutz et al. 2004; Stemmler et al. 2007) and even single cells with high sensitivity (Neupert and Predel 2005; Rubakhin et al. 2006). Molecules derived from a wide variety of biological inputs, including peptides from frog skin efflux

Fig. 8.1 (continued) instrumentation, allowing for high acquisition speeds and enhanced spatial resolution improving throughput and depth. The application of MSI helps in situ investigation of various endogenous molecules accumulated in different organs of plants and helps to visualize the spatial distribution of molecules, as metabolites, peptides, or proteins, by their molecular masses

Table 8.1 MS imaging (MSI) analysis of plant samples: A comparison of ionization methods (*DESI* desorption ESI; *SIMS* secondary ion MS) (Hu et al. 2021)

	Probe beam	Spatial resolution	Pressure	Advantage	Disadvantage	Application	Analyte	Reference
MALDI	Laser	~ 50 μm	Vacuum or ambient	Cover a wide mass range	The complexity of the spectrum is caused by matrix background signals, while the matrix crystals restrict the spatial resolution of the spectrum	Seed, stem, leaf, fruit, petal, root and wood tissue	Cellulose, lignin, hemicellulose, oligosaccharides and metabolites	Chaurand et al. (1999)
SIMS	Ion	50 nm -5 μm	Vacuum	High spatial resolution	High risk of fragmentation for analytes	Wood tissue	Cellulose, polysaccharide, lignin, and metabolite	Hu et al. (2021)
DESI	Solvent	50–500 μm	Ambient	Simple sample pre-treatment	Poor spatial resolution; disadvantageous for non-polar chemicals	Leaf, seed and petal	Metabolites	Janfelt (2015)

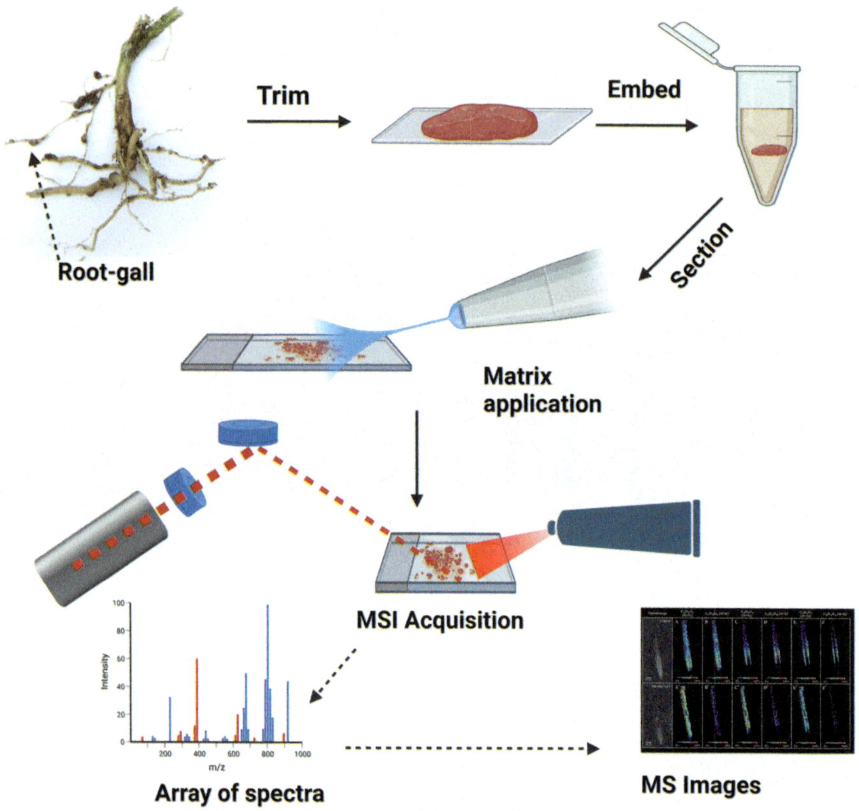

Fig. 8.2 Fig. 2 presents a high-level overview of the MSI operation. Galls from the root are isolated and encased in gelatin, frozen in a cryostat, and finally mounted on an ITO-coated glass slide. The type of analytes identified will depend on the matrix choice and application technique. A combination of matrices could provide complementary outcomes. The MALDI-TOF/TOF mass spectrometer is used for MSI acquisition, and the MSI software is used to assemble MS spectra into pictures

(Brand et al. 2006; Magalhães et al. 2008; Barbosa et al. 2018), pituitary cells (Sosnowski et al. 2015), as well as the product of plants, including hesperidin along with rutin (Kaspar et al. 2011; Soares et al. 2015) are identified. Therefore, to understand the molecular mechanisms underlying the infection and maintenance of the feeding sites during nematode parasitism, secondary metabolites, peptides, and proteins in complex plant tissues like galls are being identified using MALDI-MSI technology, which is a promising instrument (Barbosa et al. 2018).

8.2 Nematological Mass Spectrometry Imaging

The enormous variety in the taxonomic properties of plant parasitic nematodes (PPN) makes microscopic examination a frequently unreliable and time-consuming technique because these nematodes are specific. It is also challenging to quantify a particular species of concern among the populations comprising various kinds of PPN in soil samples collected or plant parts, including roots. In recent years, there has been a rise in the MALDI-MSI application to the tissues of plants, and so this method is rapidly developing into a helpful instrument for identifying molecules originating through any tissue. Although MSI analysis of proteins, as well as peptides in plants, has broadly regarded as being particularly difficult (Dong et al. 2016), only a limited number of research explain such a use (Grassl et al. 2011; Kaspar et al. 2011; Peukert et al. 2014; Gemperline et al. 2016).

MALDI is a mass spectrometry method frequently utilized in proteomics research. This technology has also been effectively implemented in directly examining peptides and proteins in bacteria and nematodes, straight down to the level of individual organelles in these organisms (Rubakhin et al. 2000; Ahmad and Wu 2011; Kuehl et al. 2011; Ahmad et al. 2012). The MSI analytical technique makes it possible to do label-free, high-resolution spatial mapping of a wide variety of biomolecules in a single experiment and provide qualitative and quantitative chemical information (Petras et al. 2017). The term "multimodal imaging" refers to an integrated method for acquiring pictures that combines structural and chemical information from more than or equal to two imaging modalities to create a single image (Neumann et al. 2020; Tuck et al. 2020). The spatial resolutions now achievable with commercial MALDI imaging systems range from 5 to 20 mm.

Root nodules formed due to symbiotic interactions between *Medicago truncatula* and *Sinorhizobium meliloti* were analyzed by MALDI-MSI, which allowed for identifying and mapping amino acids, flavonoids, carbohydrates, lipids, organic acid, as well as the conjugates of these acids. The work stated above demonstrates the usefulness of MALDI experiments in investigating the relationships between plants and microbes (Ye et al. 2013). It was discovered that glycerophospholipids were only found in the feeding site of the nematode, whereas other compositions were only found in the roots that were not affected by the nematode. This finding lends credence to the hypothesis that glycerophospholipids play a role in nematode infection and the continued growth of roots (Barbosa et al. 2018).

8.2.1 Steps Required to Generate Mass Spectrometric Imaging for Root Galls Elucidation

8.2.1.1 Galls Sample Preparation

In MALDI-MSI investigations, galls from the nematode-infected roots are taken for subsequent sectioning. In the initial step of the process, a vibratome is utilized to produce thick slices to conduct the most accurate morphological observation feasible. Galls were collected at various time points for sectioning. From the previously reported information, these slices had a thickness that ranged from 50 μm to 300 μm, while the thickness of 120 μm exhibited the most favorable morphological results (Barbosa et al. 2018). To test the quality of the tissue and the sections, bright-field microscopy was used on slices that included giant cells that had been mildly fixed. These slices were placed on glass slides, allowed to float in distilled water, and then cover-slipped and evaluated. Before undergoing cryosectioning at a temperature of −15 degrees Celsius, galls have been buried in egg yolk, agarose, or gelatin. This procedure was performed to get around the concern regarding diffusion and reduction of cellular constituents seen in moderately frozen then vibro-sliced tissue of galls. Even though these treatments are more delicate, the integrity of the tissue might be preserved with far less risk of component diffusion. When cooled to a low temperature, blocks that had galls implanted in gelatin or agarose became immediately brittle and unusable for cryosectioning.

8.2.1.2 Matrix Application

The tissue preparation is followed by matrix application. Imaging mass spectrometry relies on two key criteria to determine its spatial resolution: the laser beam diameter and the matrix crystal size. The newly developed method of ambient ionization, also known as laser ablation electrospray ionization (LAESI), is the examination of biological materials done directly in an environment free from the matrix, natural atmospheric condition with limited to no preparation of the sample and in substantially shorter time span than what is required by conventional methods of analysis (Kulkarni et al. 2018). Using a stainless steel sieve to apply solid matrices is a method that is efficient, low-cost, and commonly utilized for uniform dry-coating. It also gives excellent spatial resolution (about 100 μm or larger), making it a popular choice (Puolitaival et al. 2011; Yang et al. 2009; Yang et al. 2012). It has been discovered that applying matrix with either a 20-or 53-m stainless steel test sieve (Puolitaival et al. 2011; Yang et al. 2009) is an efficient, inexpensive, and reliable method to evenly dry coat as well as saturate samples without discernible variations in the quality of IMS data at a spatial resolution of 100 μm by 100 μm or higher. For both positive and negative modes of galls elucidation mass spectrometric imaging (MSI), a 1:1 mixture of 2,5-dihydroxybenzoic acid (DHB) and-cyano-4-hydroxycinnamic acid (CHCA) matrices is utilized, which is marketed by Sigma-Aldrich under the name Universal MALDI Matrix. There are several methods of MALDI Matrix application.

8.2.1.2.1 Airbrush Application

Every step of the airbrush process should be carried out inside a fume hood. DHB matrix solution (150 mg/ml in 50% methanol/0.1% TFA v/v) is to be used for intensely cleaning the airbrush solution container and nozzle after using methanol and then holding the airbrush a reasonable distance away. On the surface of the slide, 10–15 coats of matrix need to be applied, with a duration of 10 seconds of spray and 30 seconds of drying time in between each coat; the end result ought to be a transparent matrix layer. It is required that the airbrush be thoroughly cleaned with methanol once the application of the matrix solution has been completed in order to prevent blockage from occurring.

8.2.1.2.2 Sublimation Application of MALDI Matrix

The magnitude of the matrix transferred to the glass slide will be directly proportional to the size of the sublimation chamber. The larger sublimation compartments (the size of a flask holding 400 ml) use approximately 300 mg of DHB, but the smaller compartments (the size of a flask containing 150 mL) use approximately 100 mg of DHB and require the glass slide to be cut down for it to fit in the compartments.

8.2.1.2.3 Automatic Sprayer

This sprayer system has a heating element built into the nozzle, allowing the solvent to evaporate more quickly. The concentration of the matrix soon becomes higher as the solvent evaporates. The matrix applied to the sample using the airbrush and the matrix sprayed using the automatic sprayer contain the same concentrations of the substance being applied.

The matrix application step is followed. The dehydration of the sample at 37 degrees Celsius is a crucial step in the MSI process. After this point, the sample can be put into a vacuum desiccator to prevent it from drying out. This step ensures that the components of the MALDI mass spectrometer (the source, mass analyzer, and detector) achieve the required standard of vacuum pressure for operating condition and that the width of the dehydrated sample is compatible. In addition, this step ensures that the MALDI mass spectrometer is calibrated correctly. However, contradicting and thereby eliminating the step of dehydration, the technique known as nano DESI MSI (desorption electrospray ionization mass spectrometry imaging) makes it possible to analyze the sample through the use of mass spectrometry without the need for dehydration or the application of a matrix (Watrous et al. 2012; Watrous et al. 2012).

8.2.1.3 Image Acquisition

Following the creation of "teach points" on the sample using a WiteOut correction fluid pen to draw a plus sign on each corner of the sample, insert a glass slide into the adapter plate of the MALDI slide, and the sample optical image is to be recorded using a scanner. The teaching points will be used after the sample is marked with a WiteOut correction fluid pen. The software provided by the firm that made the instrument needs to be used to create a file for picture acquisition. This file needs to be set up while considering the step size of the raster and the diameter of the laser that is either comparable to or shorter than the size of the raster step. The optical picture must be loaded into the software, and the plate must be aligned to correspond correctly with the optical image. Before commencing the acquisition, it is essential to calibrate the instrument by employing either internal standards, a calibration mixture, or cluster ions of the standard matrix. Mark the areas of tissue going for the analyses with MS Imaging, along with the dot of the absolute matrix that will be placed on the slide and utilized as a "blank" in the analysis. The imaging file will be opened in the commercially available program from the manufacturer, followed by extracting ion photographs from the file. This allows the generation of the image. Open-source software is also available for processing MSI data (Robichaud et al. 2013).

8.2.2 MSI-Based Molecular Networking

Molecular networking, also known as MN, is a computational method that can potentially assist in displaying and interpreting the complex data produced by MS analysis. Molecular networking has also been integrated with the two- and three-dimensional viewing of metabolites through imaging mass spectrometry and real-time mass spectrometry (Fang and Dorrestein 2014). Because of its usefulness in visualizing and annotating data from non-targeted mass spectrometry (MS) (Quinn et al. 2017; Traxler and Kolter 2012), molecular networking has emerged as an important tool in the field of bioinformatics since its debut in 2012 (Watrous et al. 2012). When it comes to comparing metabolite profiles and intricate, high-resolution mass spectrometry data, the molecular networking approach is one of the most effective, sensitive, and efficient methods available. The method of molecular networking is unique in that it goes beyond the common practice of matching the spectra against the spectra of reference. Instead, it compares the observational spectrum to each other and correlates related molecules based on the similarities in their spectral signatures. With the first release of GNPS in 2013, which is a web-validated MS knowledge collecting and analysis platform, molecular networking became accessible to the general public for the very first time. (Wang et al. 2016). Since then, it has seen significant use in mass spectrometry-based metabolomics as an aid in annotating molecule families based on the fragmentation spectra of the molecules in those families (MS2). Within living organisms, metabolic networks are used to explain interconnected paths of transport mechanisms and biochemical processes of chemical species with a relatively small molecular weight (secondary metabolites, hormones, metabolic intermediates, and signaling molecules) (Wagner and Fell 2001; Jeong et al. 2000; Ma and Zeng 2003). Not only may the analytes of interest be detected, identified, and seen concurrently, but hundreds of additional chemical species can also be detected, identified, and visualized. The goal of this is to make an attempt to establish a connection between the structures of molecules and the activities and origins of biological systems (Petras et al. 2017). Molecular identification and the elucidation of chemical structures are primarily restricted to substances (such as chemicals that are commercially accessible) for which data of mass spectrometric reference are recorded in spectral library resources. This is the case since these substances are easier to analyze (Vinaixa et al. 2016; Kind et al. 2018; Montenegro-Burke et al. 2020). MALDI-MSI-based molecular networking study has been applied to nematode-induced gall tissue in order to detect and map endogenous polypeptides and secondary metabolites in situ. This technique is utilized to understand better how nematodes cause galls (Barbosa et al. 2018). The principal application is the detection as well as tracking of small molecules, the majority of which are metabolites (Lee et al. 2012; Dong et al. 2016), including alkaloids (Lu et al. 2010), carbohydrates (Veličković et al. 2014), phenolics (Franceschi et al. 2012; Becker et al. 2014), and lipids (Zaima et al. 2010; Horn et al. 2012; Horn et al. 2013). Though MSI analysis of plant proteins, as well as peptides, has been regarded as being problematic (Dong et al. 2016), comparatively limited reports explain one such application (Grassl et al. 2011; Kaspar et al. 2011; Peukert et al. 2014; Gemperline et al. 2016).

Based on the relative abundances of m/z and spatial distribution of secondary metabolites, the powerful and valuable technology known as mass spectrometry imaging (MSI) can describe the functional roles that secondary metabolites play in a biological environment. MSI accomplishes this by analyzing the geographical distribution of secondary metabolites and the m/z values of those compounds (Lei et al. 2011). MSI makes it possible to do analyses and identify secondary metabolites involved at various phases of an illness caused by a phytopathogen. Compounds with low molecular weight, particularly secondary metabolites, can play key roles in the defense mechanisms that plants use against natural pests by repulsion and attraction of root-knot nematodes in a species-specific manner (Wang et al. 2009). The MALDI-MSI technology was applied to analyze significant components, such as proteins, secondary metabolites, and peptides, which contribute to the infection development and stability of galls generated by these helminths in tomato (Barbosa et al. 2018). Inspecting plants by invaders promptly activates ion flow mechanisms, controls the generation of reactive oxygen species (ROS), creates secondary metabolites and the primary metabolite modification, and favors the expression of defender genes (Dat et al. 2000; Yang et al. 2018). Hormones such as salicylic acid, jasmonic acid, and ethylene are capable of controlling a plant's defenses against pathogens and generating secondary metabolites with antimicrobial properties (Hasegawa et al. 2010; Meng et al. 2013; Couto and Zipfel 2016; Mhlongo et al. 2018). During the nitrogen fixation process, MALDI-imaging has indeed been utilized to trace metabolites that are prevalent in the roots and nodules of *Medicago truncatula* that helped identify molecules that have been deposited during the formation of DHB coating (Ye et al. 2013). In addition, studies utilizing mass spectrometry, nuclear magnetic resonance (NMR), or gas chromatography have also been conducted to identify compounds extracted from the stem and leaves of tomato plants (Eloh et al. 2016). These studies also included the roots of *M. truncatula* (Baldacci-Cresp et al. 2015).

GNPS, which stands for Global Natural Products Social Molecular Networking, is an ecosystem for mass spectrometry that is based on the Internet and has the goal of becoming an open-access knowledge base for the community-wide organization and sharing of raw, processed, or annotated fragmentation mass spectrometry data (MS/MS). In order to process and evaluate the spectrum data collected, MN techniques, which are part of the GNPS ecosystem (Peng et al. 2017), are used. The workflows employed in this sort of molecular networking are predicated on the basics that the metabolites found inside complicated mixes are diverse forms that originate from the same building blocks (Beniddir et al. 2021).

8.3 Conclusion

MALDI MSI is a powerful method for exhaustively analyzing metabolic networks in the context of the spatial organization of cells and tissues, which has garnered an increasing amount of attention in the field of plant research and, here in particular, for root gall elucidation. Root galls can be studied relatively quickly, and in order to discriminate or identify nematodes, their spectra can be used through comparisons with reference spectra or with spectra acquired concurrently with the benchmark species. When considered together, molecular networking and MSI offer a comprehensive view of the myriad of natural connections and the infectious process caused by pathogens. It has been demonstrated beyond a reasonable doubt that the MSI methodology efficiently comprehends plant metabolites' distribution, function, and migration pathway.

References

Ahmad F, Gopal J, Wu HF (2012) Rapid and highly sensitive detection of single nematode via direct MALDI mass spectrometry. Talanta 93:182–185. https://doi.org/10.1016/j.talanta.2012.02.009

Ahmad F, Wu HF (2011) Characterization of pathogenic bacteria using ionic liquid via single drop microextraction combined with MALDI-TOF MS. Analyst 136:4020–4027. https://doi.org/10.1039/c1an15350a

Baldacci-Cresp F, Maucourt M, Deborde C, Pierre O, Moing A, Brouquisse R, Favery B, Frendo P (2015) Maturation of nematode-induced galls in *Medicago truncatula* is related to water status and primary metabolism modifications. Plant Sci 232:77–85. https://doi.org/10.1016/j.plantsci.2014.12.019

Barbosa EA, Bonfim MF Jr, Bloch C Jr, Engler G, Rocha T, de Almeida EJ (2018) Imaging mass spectrometry of endogenous polypeptides and secondary metabolites from galls induced by root-knot nematodes in tomato roots. Mol Plant-Microbe Interact 31:1048–1059. https://doi.org/10.1094/MPMI-02-18-0049-R

Becker L, Carré V, Poutaraud A, Merdinoglu D, Chaimbault P (2014) MALDI mass spectrometry imaging for the simultaneous location of resveratrol, pterostilbene and viniferins on grapevine leaves. Molecules 19:10587–10600. https://doi.org/10.3390/molecules190710587

Beniddir MA, Kang KB, Genta-Jouve G, Huber F, Rogers S, Van Der Hooft JJ (2021) Advances in decomposing complex metabolite mixtures using substructure-and network-based computational metabolomics approaches. Nat Prod Rep 38:1967–1993. https://doi.org/10.1039/D1NP00023C

Booth SC, Weljie AM, Turner RJ (2013) Computational tools for the secondary analysis of metabolomics experiments. Comput Struct Biotechnol J 4:e201301003. https://doi.org/10.5936/csbj.201301003

Brand GD, Krause FC, Silva LP, Leite JR, Melo JA, Prates MV, Pesquero JB, Santos EL, Nakaie CR, Costa-Neto CM, Bloch C Jr (2006) Bradykinin-related peptides from Phyllomedusa hypochondrialis. Peptides 27:2137–2146. https://doi.org/10.1016/j.peptides.2006.04.020
Buchberger AR, DeLaney K, Johnson J, Li L (2018) Mass spectrometry imaging: a review of emerging advancements and future insights. Anal Chem 90:240. https://doi.org/10.1021/2Facs.analchem.7b04733
Chaurand P, Stoeckli M, Caprioli RM (1999) Direct profiling of proteins in biological tissue sections by MALDI mass spectrometry. Anal Chem 71:5263–5270. https://doi.org/10.1021/ac990781q
Cheng AX, Lou YG, Mao YB, Lu S, Wang LJ, Chen XY (2007) Plant terpenoids: biosynthesis and ecological functions. J Integr Plant Biol 49:179–186. https://doi.org/10.1111/j.1744-7909.2007.00395.x
Couto D, Zipfel C (2016) Regulation of pattern recognition receptor signalling in plants. Nat Rev Immunol 16:537–552. https://doi.org/10.1038/nri.2016.77
Dat J, Vandenabeele S, Vranova EV, Van Montagu M, Inzé D, Van Breusegem F (2000) Dual action of the active oxygen species during plant stress responses. Cellular and Molecular Life Sciences CMLS 57:779–795. https://doi.org/10.1007/s000180050041
Dhanasekaran AR, Pearson JL, Ganesan B, Weimer BC (2015) Metabolome searcher: a high throughput tool for metabolite identification and metabolic pathway mapping directly from mass spectrometry and using genome restriction. Bmc Bioinformatics 16:1–3. https://doi.org/10.1186/s12859-015-0462-y
Dong L, Huang C, Huang L, Li X, Zuo Y (2012) Screening plants resistant against Meloidogyne incognita and integrated management of plant resources for nematode control. Crop Protection 33:34–39. https://doi.org/10.1016/j.cropro.2011.11.012
Dong Y, Li B, Malitsky S, Rogachev I, Aharoni A, Kaftan F, Svatoš A, Franceschi P (2016) Sample preparation for mass spectrometry imaging of plant tissues: a review. Front Plant Sci 7:60. https://doi.org/10.3389/fpls.2016.00060
Ehrhardt DW, Frommer WB (2012) New technologies for 21st century plant science. Plant Cell 24: 374–394. https://doi.org/10.1105/tpc.111.093302
Eloh K, Sasanelli N, Maxia A, Caboni P (2016) Untargeted metabolomics of tomato plants after root-knot nematode infestation. J Agric Food Chem 64:5963–5968. https://doi.org/10.1021/acs.jafc.6b02181
Fang J, Dorrestein PC (2014) Emerging mass spectrometry techniques for the direct analysis of microbial colonies. Curr Opin Microbiol 19:120–129. https://doi.org/10.1016/j.mib.2014.06.014
Franceschi P, Dong Y, Strupat K, Vrhovsek U, Mattivi F (2012) Combining intensity correlation analysis and MALDI imaging to study the distribution of flavonols and dihydrochalcones in Golden delicious apples. J Exp Bot 63:1123–1133. https://doi.org/10.1093/jxb/err327
Fujimoto T, Mizukubo T, Abe H, Seo S (2015) Sclareol induces plant resistance to root-knot 24 nematode partially through ethylene-dependent enhancement of lignin accumulation. https://doi.org/10.1094/MPMI-10-14-0320-R
Gemperline E, Keller C, Jayaraman D, Maeda J, Sussman MR, Ané JM, Li L (2016) Examination of endogenous peptides in Medicago truncatula using mass spectrometry imaging. J Proteome Res 15:4403–4411. https://doi.org/10.1021/acs.jproteome.6b00471
Grassl J, Taylor NL, Millar A (2011) Matrix-assisted laser desorption/ionization mass spectrometry imaging and its development for plant protein imaging. Plant Methods 7:1–1
Hansen RL, Lee YJ (2018) High-spatial resolution mass spectrometry imaging: toward single cell metabolomics in plant tissues. Chem Rec 18:65–77. https://doi.org/10.1002/tcr.201700027
Hasegawa M, Mitsuhara I, Seo S, Imai T, Koga J, Okada K, Yamane H, Ohashi Y (2010) Phytoalexin accumulation in the interaction between rice and the blast fungus. Mol Plant-Microbe Interact 23:1000–1011. https://doi.org/10.1094/MPMI-23-8-1000

Horn PJ, James CN, Gidda SK, Kilaru A, Dyer JM, Mullen RT, Ohlrogge JB, Chapman KD (2013) Identification of a new class of lipid droplet-associated proteins in plants. Plant Physiol 162: 1926–1936. https://doi.org/10.1104/pp.113.222455

Horn PJ, Korte AR, Neogi PB, Love E, Fuchs J, Strupat K, Borisjuk L, Shulaev V, Lee YJ, Chapman KD (2012) Spatial mapping of lipids at cellular resolution in embryos of cotton. Plant Cell 24:622–636. https://doi.org/10.1105/tpc.111.094581

Hu W, Han Y, Sheng Y, Wang Y, Pan Q, Nie H (2021) Mass spectrometry imaging for direct visualization of components in plants tissues. J Separation Sci 44:3462–3476. https://doi.org/10.1002/jssc.202100138

Hussey RS, Grundler FM (1998) Nematode parasitism of plants. The physiology and biochemistry of free-living and plant-parasitic nematodes. 213–243

ITIS, Integrated Taxonomic Information System (2023) Ditylenchus Report. Data Base. Available online: https://www.itis.gov/servlet/SingleRpt/SingleRpt?search_topic=TSN&search_value=196461#null. Accessed 27 Aug 2023

Janfelt C (2015) Imaging of plant materials using indirect desorption electrospray ionization mass spectrometry. Methods Mol Biol 2:91–97. https://doi.org/10.1007/978-1-4939-1357-2_9

Jeong H, Tombor B, Albert R, Oltvai ZN, Barabási AL (2000) The large-scale organization of metabolic networks. Nature 7:651–654. https://doi.org/10.1038/35036627

Jones MG, Northcote DH (1972) Multinucleate transfer cells induced in coleus roots by the root-knot nematode, Meloidogyne arenaria. Protoplasma 75:381–395. https://doi.org/10.1007/BF01282117

Kaspar S, Peukert M, Svatos A, Matros A, Mock HP (2011) MALDI-imaging mass spectrometry–an emerging technique in plant biology. Proteomics 11:1840–1850. https://doi.org/10.1002/pmic.201000756

Kind T, Tsugawa H, Cajka T, Ma Y, Lai Z, Mehta SS, Wohlgemuth G, Barupal DK, Showalter MR, Arita M, Fiehn O (2018) Identification of small molecules using accurate mass MS/MS search. Mass Spectrom Rev 37:513–532. https://doi.org/10.1002/mas.21535

Kuehl B, Marten SM, Bischoff Y, Brenner-Weiß G, Obst U (2011) MALDI-ToF mass spectrometry–multivariate data analysis as a tool for classification of reactivation and non-culturable states of bacteria. Anal Bioanal Chem 401:1593–1600. https://doi.org/10.1007/s00216-011-5227-5

Kulkarni P, Wilschut RA, Verhoeven KJ, van der Putten WH, Garbeva P (2018) LAESI mass spectrometry imaging as a tool to differentiate the root metabolome of native and range-expanding plant species. Planta 248:1515–1523. https://doi.org/10.1007/s00425-018-2989-4

Kutz KK, Schmidt JJ, Li L (2004) In situ tissue analysis of neuropeptides by MALDI FTMS in-cell accumulation. Anal Chem 76:5630–5640. https://doi.org/10.1021/ac049255b

Lee YJ, Perdian DC, Song Z, Yeung ES, Nikolau BJ (2012) The use of mass spectrometry for imaging metabolites in plants. Plant J 70:81–95. https://doi.org/10.1111/j.1365-313X.2012.04899.x

Lei Z, Huhman DV, Sumner LW (2011) Mass spectrometry strategies in metabolomics. J Biol Chem 286:25435–25442. https://doi.org/10.1074/jbc.R111.238691

Lu L, Yue H, Song FR, Tsao R, Liu ZQ, Liu SY (2010) Rapid profiling of alkaloids in several medicinal herbs by matrix-assisted laser desorption/ionization mass spectrometry. Chem Res Chin Uni 26:11–16

Ma H, Zeng AP (2003) Reconstruction of metabolic networks from genome data and analysis of their global structure for various organisms. Bioinformatics 19:270–277. https://doi.org/10.1093/bioinformatics/19.2.270

Magalhães BS, Melo JA, Leite JR, Silva LP, Prates MV, Vinecky F, Barbosa EA, Verly RM, Mehta A, Nicoli JR, Bemquerer MP (2008) Post-secretory events alter the peptide content of the skin secretion of Hypsiboas raniceps. Biochem Biophys Res Commun 377:1057–1061. https://doi.org/10.1016/j.bbrc.2008.10.102

Meng S, Cao J, Feng Q, Peng J, Hu Y (2013) Roles of chlorogenic acid on regulating glucose and lipids metabolism: a review. Evid Based Complement Alternat Med 2013:801457. https://doi.org/10.1155/2013/801457

Mhlongo MI, Piater LA, Madala NE, Labuschagne N (2018) Dubery IA (2018) the chemistry of plant–microbe interactions in the rhizosphere and the potential for metabolomics to reveal signaling related to defense priming and induced systemic resistance. Front Plant Sci 9:112. https://doi.org/10.3389/fpls.2018.00112

Moens M, Perry RN, Starr JL (2009) Meloidogyne species-a diverse group of novel and important plant parasites. In: InRoot-knot nematodes. CABI, Wallingford UK, pp 1–17. https://doi.org/10.1079/9781845934927.0001

Montenegro-Burke JR, Guijas C, Siuzdak G (2020) A tandem mass spectral library of standards. In: Computational methods and data analysis for metabolomics, pp 149–163. https://doi.org/10.1007/978-1-0716-0239-3_9

Morris HR, Paxton T, Dell A, Langhorne J, Berg M, Bordoli RS, Hoyes J, Bateman RH (1996) High sensitivity collisionally-activated decomposition tandem mass spectrometry on a novel quadrupole/orthogonal-acceleration time-of-flight mass spectrometer. Rapid Commun Mass Spectrom 10:889–896. https://doi.org/10.1002/(SICI)1097-0231(19960610)10:8%3C889

Neumann EK, Djambazova KV, Caprioli RM, Spraggins JM (2020) Multimodal imaging mass spectrometry: next generation molecular mapping in biology and medicine. J Am Soc Mass Spectrom 31:2401–2415. https://doi.org/10.1021/jasms.0c00232

Neupert S, Predel R (2005) Mass spectrometric analysis of single identified neurons of an insect. Biochem Biophys Res Commun 323:640–645. https://doi.org/10.1016/j.bbrc.2004.12.086

Nicol JM, Turner SJ, Coyne DL, Nijs LD, Hockland S, Maafi ZT (2011) Current nematode threats to world agriculture. In: Genomics and molecular genetics of plant-nematode interactions, pp 21–43. https://doi.org/10.1007/978-94-007-0434-3_2

Peng D, Jin W, Li J, Xiong W, Pei Y, Wang Y, Li Y, Li B (2017) Adsorption and distribution of edible gliadin nanoparticles at the air/water interface. J Agric Food Chem 65:2454–2460. https://doi.org/10.1021/acs.jafc.6b05757

Petras D, Jarmusch AK, Dorrestein PC (2017) From single cells to our planet—recent advances in using mass spectrometry for spatially resolved metabolomics. Curr Opin Chem Biol 36:24–31. https://doi.org/10.1016/j.cbpa.2016.12.018

Peukert M, Becker M, Matros A, Mock HP (2014) Mass spectrometry-based imaging of metabolites and proteins. In: Plant proteomics: methods and protocols, pp 223–240. https://doi.org/10.1007/978-1-62703-631-3_17

Prell J, Poole P (2006) Metabolic changes of rhizobia in legume nodules. Trends Microbiol 14(161–8):10. https://doi.org/10.1016/j.tim.2006

Puolitaival SM, Burnum KE, Cornett DS, Caprioli RM (2011) Solvent-free matrix dry-coating for MALDI imaging of phospholipids. J Am Soc Mass Spectrom 19(6):882–886. https://doi.org/10.1016/j.jasms.2008.02.013

Quinn RA, Nothias LF, Vining O, Meehan M, Esquenazi E, Dorrestein PC (2017) Molecular networking as a drug discovery, drug metabolism, and precision medicine strategy. Trends Pharmacol Sci 38:143–154. https://doi.org/10.1016/j.tips.2016.10.011

Robichaud G, Garrard KP, Barry JA, Muddiman DC (2013) MSiReader: an open-source interface to view and analyze high resolving power MS imaging files on Matlab platform. J Am Soc Mass Spectrom 24:718–721. https://doi.org/10.1007/s13361-013-0607-z

Rubakhin SS, Churchill JD, Greenough WT, Sweedler JV (2006) Profiling signaling peptides in single mammalian cells using mass spectrometry. Anal Chem 78:7267–7272. https://doi.org/10.1021/ac0607010

Rubakhin SS, Garden RW, Fuller RR, Sweedler JV (2000) Measuring the peptides in individual organelles with mass spectrometry. Nat Biotechnol 18:172–175. https://doi.org/10.1038/72622

Schwamborn K, Caprioli RM (2010) Molecular imaging by mass spectrometry—looking beyond classical histology. Nat Rev Cancer 10:639–646. https://doi.org/10.1038/nrc2917

Soares MS, da Silva DF, Forim MR, Fernandes JB, Vieira PC, Silva DB, Lopes NP, de Carvalho SA, de Souza AA, Machado MA (2015) Quantification and localization of hesperidin and rutin in Citrus sinensis grafted on C. limonia after Xylella fastidiosa infection by HPLC-UV and MALDI imaging mass spectrometry. Phytochemistry 115:161–170. https://doi.org/10.1016/j.phytochem.2015.02.011

Sosnowski P, Zera T, Wilenska B, Szczepanska-Sadowska E, Misicka A (2015) Imaging and identification of endogenous peptides from rat pituitary embedded in egg yolk. Rapid Commun Mass Spectrom 29:327–335. https://doi.org/10.1002/rcm.7112

Spengler B (2015) Mass spectrometry imaging of biomolecular information. Anal Chem 8764-82: 64. https://doi.org/10.1021/ac504543v

Stemmler EA, Cashman CR, Messinger DI, Gardner NP, Dickinson PS, Christie AE (2007) High-mass-resolution direct-tissue MALDI-FTMS reveals broad conservation of three neuropeptides (APSGFLGMRamide, GYRKPPFNGSIFamide and pQDLDHVFLRFamide) across members of seven decapod crustaean infraorders. Peptides 28:2104–2115. https://doi.org/10.1016/j.peptides.2007.08.019

Traxler MF, Kolter R (2012) A massively spectacular view of the chemical lives of microbes. Proc Nat Acad Sci 109:10128–10129. https://doi.org/10.1073/pnas.1207725109

Tuck M, Blanc L, Touti R, Patterson NH, Van Nuffel S, Villette S, Taveau JC, Römpp A, Brunelle A, Lecomte S, Desbenoit N (2020) Multimodal imaging based on vibrational spectroscopies and mass spectrometry imaging applied to biological tissue: a multiscale and multiomics review. Analytical chemistry 93:445–477. https://doi.org/10.1021/acs.analchem.0c04595

Veličković D, Ropartz D, Guillon F, Saulnier L, Rogniaux H (2014) New insights into the structural and spatial variability of cell-wall polysaccharides during wheat grain development, as revealed through MALDI mass spectrometry imaging. J Experimental Botany 65:2079–2091. https://doi.org/10.1093/jxb/eru065

Vinaixa M, Schymanski EL, Neumann S, Navarro M, Salek RM, Yanes O (2016) Mass spectral databases for LC/MS- and GC/MS-based metabolomics: state of the field and future prospects. Trends Anal Chem 78:23–35. https://doi.org/10.1016/j.trac.2015.09.005

Vogt T (2010) Phenylpropanoid biosynthesis. Molecular plant. 3:2–0. https://doi.org/10.1093/mp/ssp106

Wagner A, Fell DA (2001) The small world inside large metabolic networks. Proceedings of the Royal Society of London. Series B: Biol Sci 268:1803–1810. https://doi.org/10.1098/rspb.2001.1711

Wang C, Lower S, Williamson VM (2009) Application of Pluronic gel to the study of root-knot nematode behaviour. Nematology 11:453–464. https://doi.org/10.1163/156854109X447024

Wang M, Carver JJ, Phelan VV, Sanchez LM, Garg N, Peng Y, Nguyen DD, Watrous J, Kapono CA, Luzzatto-Knaan T, Porto C (2016) Sharing and community curation of mass spectrometry data with Global Natural Products Social Molecular Networking. Nature Biotechnol 34:828–837. https://doi.org/10.1038/nbt.3597

Watrous J, Roach P, Alexandrov T, Heath BS, Yang JY, Kersten RD, van der Voort M, Pogliano K, Gross H, Raaijmakers JM, Moore BS (2012) Mass spectral molecular networking of living microbial colonies. Proc Nat Acad Sci 109:E1743–E1752. https://doi.org/10.1073/pnas.1203689109

Yang JY, Phelan VV, Simkovsky R, Watrous JD, Trial RM, Fleming TC, Wenter R, Moore BS, Golden SS, Pogliano K, Dorrestein PC (2012) Primer on agar-based microbial imaging mass spectrometry. J Bacteriol 194:6023–6028. https://doi.org/10.1128/JB.00823-12

Yang L, Wen KS, Ruan X, Zhao YX, Wei F, Wang Q (2018) Response of plant secondary metabolites to environmental factors. Molecules 23:762. https://doi.org/10.3390/molecules23040762

Yang YL, Xu Y, Straight P, Dorrestein PC (2009) Translating metabolic exchange with imaging mass spectrometry. Nat Chem Biol 5:885–887. https://doi.org/10.1038/nchembio.252

Ye H, Gemperline E, Venkateshwaran M, Chen R, Delaux PM, Howes-Podoll M, Ané JM, Li L (2013) MALDI mass spectrometry-assisted molecular imaging of metabolites during nitrogen fixation in the M edicago truncatula–S inorhizobium meliloti symbiosis. Plant J 75:130–145. https://doi.org/10.1111/tpj.12191

Zaima N, Goto-Inoue N, Hayasaka T, Setou M (2010) Application of imaging mass spectrometry for the analysis of Oryza sativa rice. Rapid Commun Mass Spectrom 24:2723–2729. https://doi.org/10.1002/rcm.4693

Ziegler J, Facchini PJ (2008) Alkaloid biosynthesis: metabolism and trafficking. Annu Rev Plant Biol 59:735–769. https://doi.org/10.1146/annurev.arplant.59.032607.092730

Root-Knot Disease Complex: An Interactive Perspective with Microorganisms

9

T. S. Archana, Devendra Kumar, Vipul Kumar,
and Bitaisha Nakishuka Shukuru

Abstract

Root-knot nematodes (RKNs) cause approximately 72% of global crop yield loss and have a vast host range of above 2000 plants. The interaction of nematode with other disease-causing agents increases the disease severity and makes the management strategies difficult. *Meloidogyne*-based disease complexes (MDCs) with plant pathogenic fungi and bacteria are a major constraint in vegetable production. *Meloidogyne* species show close interaction with phytopathogenic fungi in tomatoes. Interaction with fungi, including *Fusarium* spp., *Sclerotium*, *Alternaria dauci*, and *Rhizoctonia* spp., in vegetables, leads to a greater reduction in plant health. They drastically reduced plant growth. Interaction of nematodes with other pathogens is prime necessary for proper disease management. Thus, plants infected with nematodes increase disease severity and influence disease development and etiology.

Keywords

Root-knot nematode · Interaction · Vegetables · Microorganisms

9.1 Introduction

All ecosystems contain the diverse group of creatures known as nematodes; there are estimated to be up to one million global species of nematodes (Mitreva et al. 2005). Some have evolved parasitic lifestyles, while others are free-living (Singh et al.

T. S. Archana (✉) · D. Kumar · V. Kumar · B. N. Shukuru
Department of Plant Pathology, School of Agriculture, Lovely Professional University, Phagwara, Punjab, India
e-mail: archana.26887@lpu.co.in

F. Ahmad, G. Nombela (eds.), *Root-Galling Disease of Vegetable Plants*,
https://doi.org/10.1007/978-981-99-3892-6_9

2021). All vertebrates, including humans, are thought to have experienced parasitism at some point in their history. With over 4100 nematode species known, plants are also parasitized (Decraemer and Hunt 2006). Nearly all crops worldwide contain phytoparasitic nematodes, which lower crop production and quality result in significant losses. Nematode infestations are thought to account for 14% loss of all crops total yield. From an economic perspective, the greatest significant crop-damaging pest nematodes are root-knot and cyst nematode (Jones et al. 2013). They may also affect beneficial plant microbiota and act as virus vectors (Khan et al. 1993; Siddique and Grundler 2018; Jones et al. 2013). Nematodes can result in a number of symptoms, including leaf yellowing, delayed development, and poor crop yields. They are transferred through polluted irrigation water or infected seedlings and seedbeds (Charchar et al. 2008).

Meloidogyne may infect virtually any vascular plant, whether it is in a field, greenhouse, or protected farm. It has species that can be found all over the world. The four primary important nematode species are *Meloidogyne arenaria*, *M. incognita*, *M. javanica*, and *M. hapla* (Wesemael et al. 2011; Coyne et al. 2018; Sikandar et al. 2020). Although there are many host crops, vegetables, soybeans, grains, other solanaceous plants, and tuber crops are the most commercially relevant ones (Trudgill and Blok 2001; Wesemael et al. 2011).

9.2 Root-Knot Nematode Biology

The scientific community ranked the *Meloidogyne* genus as the top most important plant-parasitic nematode in 2013 (Jones et al. 2013; Sikandar et al. 2020; Ali et al. 2017; Ibrahim et al. 2019). *Meloidogyne* includes about 100 species and is one of the most important nematode groups due to its economic significance. In spite of all the challenges provided by obligate nature, study on the nematode Meloidogyne includes all aspects of various survival, evolution, as well as plant responses after invasion (Curtis 2007). Recently, Da Rocha et al. (2021) used extensive transcriptome research to elucidate the Meloidogyne parasitism and regulatory environment. Research based on RKNs has benefited significantly from in-depth knowledge of the model organism *Caenorhabditis elegans*. Despite the additional million years ago divergence, genomic level microsynteny among *Meloidogyne* and *C. elegans* shows that they share developmental and metabolic pathway (Opperman et al. 2008). Worm Base, a sizable database for nematode research created by Harris et al. (2010) for *Caenorhabditis elegans*, now has evidence on various plant parasitic nematodes, including further current sequences of several root knot nematode. Since the 86 Mb and 54 Mb *M. hapla* genomes were sequenced in 2008, another 19 genomic draughts representing six species have been identified (Abad et al. 2008; Opperman et al. 2008), allowing evolutionary and genomic assessments (Lunt et al. 2014; Mitreva et al. 2005).

Despite the fact that the majority of its species reproduce asexually, which is regarded to be an evolutionary dead-end, *Meloidogyne* is well-adapted to shifting environmental conditions (Castagnone-Sereno et al. 2019). Parthenogenesis (apomixis), as seen in various *Meloidogyne*, or mitotic reduction division and further establishment of the chromosome number with second polar nucleus fusion with the pronucleus egg as a part of asexual reproduction (e.g., in *M. chitwoodi* and *M. hapla*). Ironically, *Meloidogyne*, which infects practically all Angiospermae, has the most extensive and diverse host range, which is associated with true asexuality. Asexual reproduction can still epigenetically produce males in harsh environments, but female insemination damages the sperm nucleus (Baniya et al. 2021). Males fertilize the eggs in some *Meloidogyne* species, including *M. megatyla*, *M. pini*, *M. carolinensis*, and *M. microtyla* (Eisenback and Triantaphyllou 2020).

9.2.1 Life Cycle

A gelatinous matrix created by six anal glands houses approximately 500 eggs laid by adult females of RKNs. Glycoproteins in the matrix shield eggs and act as a sensor for temperature and relative humidity specially for growth. Drought stops growth because it reduces the volume of the matrix and causes it to harden. Invading soil microorganisms can be agglutinated by it as an antimicrobial agent. The complete life cycle of RKN is represented in Fig. 9.1. The plant cell degraded to produce a conduit over which egg masses will be kept outside as an expanded root gall. A carbohydrate-binding domain (CBM) that was revealed by Vieira et al. (2011) in the vulval secretion may have this function. It takes 25 to 30 days for an egg to become an adult through several successive molts. Vermiform stage 1 juveniles (J1) undergo their first molt before hatching, becoming juveniles (J2). Second stage also involves acquiring the parasitic stage and creating a feeding site with the host vasculature with sedentarism. The outer cuticles and the non-functional stylet are used to identify the following two stages (J3 and J4). J4, where sexual dimorphism separates the female and male nematodes. Female nematodes that have been dormant for a while restart feeding, convert into a pear-shaped mass and lay egg masses. Various gene expression is associated with this transition. For example, sensory perception genes are upregulated from egg mass to the pre-parasitic mobile phase (ppJ2); stress response genes upregulated generally between J2, J3, and J4; and genes involved in various sensory perception become less expressed leading to sedentary nematode.

Genes involved in lipid metabolism also increased at J3 and J4 in order to get ready for adult phase. For this reason, mature females who are not required to move around repress certain genes. Developmental processes, which include DNA metabolism and membrane transport, dominate gene regulation in the egg. Although there have been conflicting results, several genes have been found by RNA silencing, leading to lower levels of infestation. Having an understanding of the RKN gene expression patterns offers methods for picking target genes logically. Although there have been conflicting results, numerous genes with functions have been found by various RNA silencing mechanism, leading to lesser levels of infestation. Understanding the RKN gene expression patterns offers methods for picking target genes (Iqbal et al. 2020).

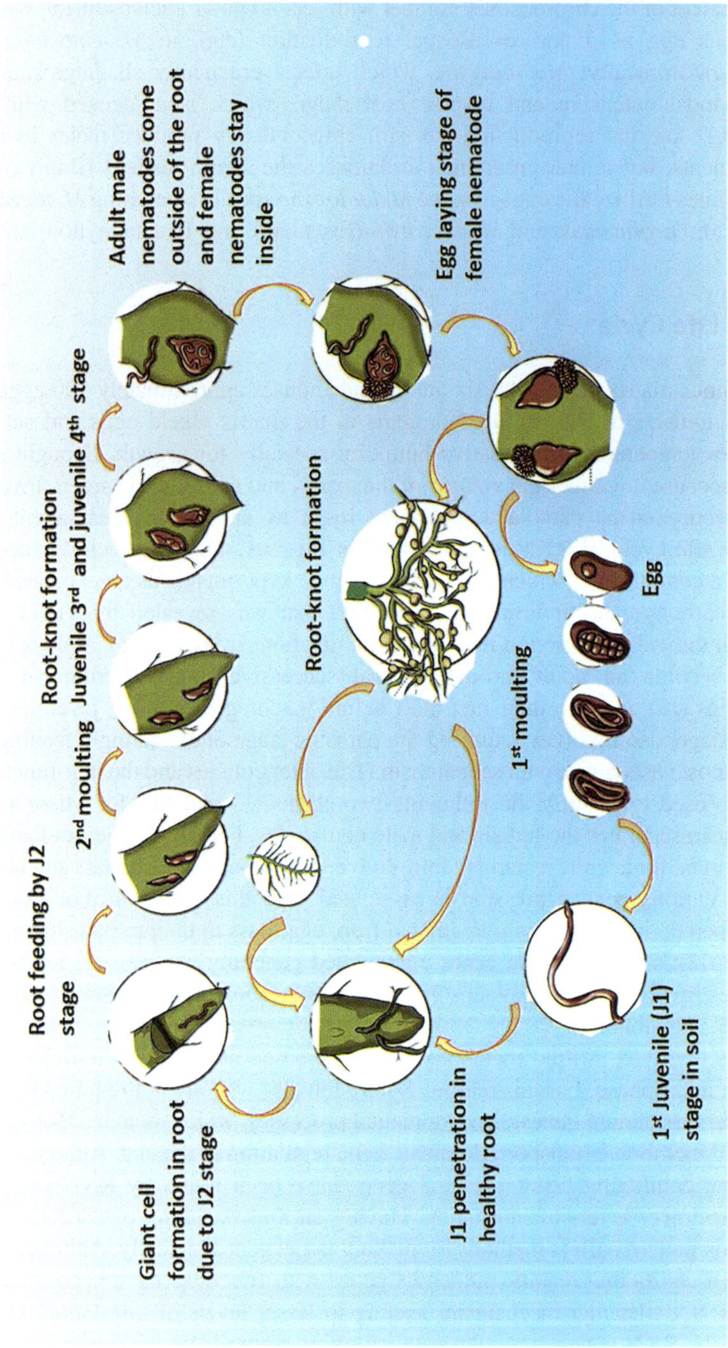

Fig. 9.1 Life cycle of root-knot Nematode

9.3 Nematode Host Assortment and Invasion in Vegetable Crops

Meloidogyne host parasitism highly depends on the species. Some of these that affect various vegetable crops are as follows: hatched J2 swim haphazardly through the soil until they come across a vulnerable root by following chemotactic plant exudates. The species to which they belong and their lipid reserves, which stop invasion when they fall below 65%, determine how long ppJ2 juveniles stay in the soil. In order to enter in to the meristematic region of plant, J2s perforate mechanically at the least resistant spot through which they grasp the root tip of host. The species to which they belong and their lipid reserves, which stop invasion when they fall below 65%, determine how long ppJ2 juveniles stay in the soil (Mitsumasu et al. 2015).

Meloidogyne first migrates toward the root tip to get around the Casparian strip, which acts as a barrier and comprises highly lignified and suberized endodermal cells. They then move upwards the active root growing and tissue differentiation zone, where vascular elements became more visible and adhere to the central cylinder portion (Mende 1997; Holbein et al. 2019). *Meloidogyne* softens the various parts of middle lamella in preparation for their journey by secreting modifying enzymes made in the sub-ventral glands, such as cellulases, various proteins, hemi cellulases, and pectin degrading enzymes (Vieira et al. 2011).

9.4 Interaction of Root-Knot Nematode with Other Microorganisms

A variety of bacterial and fungal pathogens interact with root-knot nematodes, resulting in disease complexes. The physiological changes caused by nematode before the establishment by 2–4 weeks make plant roots more receptive to other pathogens. Galled roots are heavily populated by rotting fungi like *Rhizoctonia solani*, which causes additional damage. Nutrient-rich giant cells serve as substrates for the growth of wilt-causing fungi like *Fusarium*, *Verticillium*, and the bacterium *Pseudomonas solanacearum*. Wilt occurs more frequently and with greater severity when nematodes are present than when absent. A root-knot nematode is thought to be responsible for the breakdown of tobacco's defenses against the *Phytophthora nicotianae* pathogen that causes black shank disease. Similar cases have been reported in numerous other instances. Secondary pathogens are drawn to plants with root-knot nematode infections due to changes in the exudates' quality. The various interaction of nematodes with fungi are given in Table 9.1. Florida tomato fields frequently experience interactions between RKNs and *Fusarium* wilt (*Fusarium oxysporum* f.sp. *lycopersici* race 3, FOL). RKNs and FOL together synergize plant damage in cases where wilted tomato plants with FOL-infected vascular infection are severely galled. In Florida, FOL race 3 predominates, and while furthermost marketable tomato cultivars are resistant to fusarium race 1 and 2, this is not the case for race 3. Due to their increased vulnerability to bacterial spot and blossom-end rot and problems with smaller fruit size, the few resistant cultivars are

Table 9.1 Nematode interaction with plant pathogenic fungi

S. No	*Meloidogyne* species	Pathogen	Crop	Disease	Reference
1.	*Meloidogyne incognita*	*Fusarium oxysporium* f. sp. *Phaseoli*	Bean	Wilt	Carneiro (2010)
2.	*Meloidogyne incognita*	*Fusarium oxysporium* f. sp	Potato	Wilt	El-Shennawy et al. (2012)
3.	*Meloidogyne incognita*	*Rhizoctonia solani*	Green bean	Root rot	Alhazmi and Al-Nadary (2015)
4.	*Meloidogyne incognita*	*Rhizoctonia solani*	Tomato	Damping off	El-Shennawy et al. (2012)
5.	*Meloidogyne incognita*	*Sclerotium rolfsii*	Brinjal	Collar rot	Goswami and Chenula (1974)
6.	*Meloidogyne incognita*	*Phytophthora capsici*	Pepper	Phytophthora blight	Parkunan et al. (2016)
7.	*Meloidogyne javanica*	*Pythium bebaryanum*	Tomato	Dumping-off	Scherlach and Hertweck (2018)
8.	*Meloidogyne* spp	*Fusarium oxysporium* f. sp. *lycopersici*	Tomato	Fusarium wilt	Scherlach and Hertweck (2018)
9.	*Meloidogyne* spp	*Fusarium oxysporium, fusarium solani*	Tomato	Wilt	Hajji-Hedfi et al. (2019)
10.	*Meloidogyne incognita*	*Fusarium oxysporium* f. sp. *conglutinans*	Cauliflower	Fusarium wilt	Rompalli et al. (2016)
11.	*Meloidogyne javanica*	*Fusarium fusarium oxysporium* f. sp. *lycopersici,*	Tomato	Fusarium wilt	Beyan (2019)
12.	*Meloidogyne incognita*	*Fusarium fusarium oxysporium* f. sp. *niveum*	Watermelon	Fusarium wilt	Keinath et al. (2019); Scherlach and Hertweck (2018)
13.	*Meloidogyne incognita*	*Ralstonia solanacearum, Phomopsis vexans*	Eggplant	Phomopsis blight	Khan and Siddiqui (2017)
14.	*Meloidogyne incognita*	*Alternaria dauci, Rhizoctonia solani*	Carrot	Soft rot	Ahmad et al. (2019)

not well-liked by growers (Hutton et al. 2014). Other interactions with pathogens, such as *Pythium* for cucumber and *Fusarium* crown rot for tomatoes, are probably crucial.

Ozdemir et al. (2022) assessed the properties of *Fusarium oxysporum* f. sp. *radicis-lycopersici* Jarvis & Shoemaker and *Meloidogyne incognita* (Ozdemir et al. 2022) on nematode reproduction and wilt severity were examined in tomato hybrids. In January–May 2021, under controlled circumstances, five different combinations of individual, concurrent, and sequential *M. incognita* and FORL inoculations were made to the tomato F1 hybrids Adel, Alberty, Armstrong, Body, Gülizar, and Kaplan. In 60 days, the experiment was finished. Adel, Armstrong, Body, and Gülizar all developed more *M. incognita* galls and egg masses after receiving a simultaneous inoculation. In Alberty and Kaplan, FORL inoculation 10 days after *M. incognita* inoculation (N + 10 FORL) resulted in the highest gall and egg mass numbers.

9.5 Interaction of Root-Knot Nematode with Plant Pathogenic Fungi in Vegetable Crops

The utmost prevalent organisms present in soil and rhizosphere habitats include nematodes and fungi. They perform vital ecosystem services and are instrumental in facilitating nutrient cycling and preserving the stability of food webs. Fungi along with nematodes interact with one another in various ways because they are two of the most prevalent groups of organisms. This chapter explains a comprehensive framework of interactions between fungi and nematodes, focusing on those that affect agricultural ecosystems and vegetable crops. Fungi that live close to nematodes, including fungi that serve as food for nematodes and fungi that consume nematodes and also interact with plant pathogenic fungi and increase the plant disease severity. When pathogens inhabit in soil and plant pests coexist in the soil environment and occupy the same ecological niche, opportunities for interaction between them arise. They can be antagonistic in their rivalry for resources and space, but there is also a chance that they will work together to harm plants, including crops, more severely. For instance, nematode attacks in the rhizosphere can decrease plants' pathogen resistance and make them more vulnerable to infection by soilborne fungal pathogens. As a result of these tripartite interactions, plants become more susceptible to fungal disease and increase the disease severity and yield loss (Lamelas et al. 2020).

In 1892, Atkinson provided the initial description of a nematode and fungi disease complex present in plants when he noted that the presence of root-knot nematodes made cotton's fusarium wilt, which is caused by *Fusarium oxysporum* f. sp. *vasinfectum*, more severe (*Meloidogyne* spp.) (Atkinson 1892). Numerous other collaborative interactions between nematodes and various plant pathogenic fungi have been documented. These cases involve sedentary endoparasitic cysts and root-knot nematodes and the worsening of the disease brought on by *Fusarium* or *Verticillium* wilt fungi. It has been demonstrated that *Meloidogyne* species interact with *Fusarium* wilt to harm several vegetable crops, and cyst nematodes act similarly to worsen wilt diseases. Table 9.1 summarizes recent examples of nematode plant pathogenic fungi disease complexes described in various vegetables.

9.6 Factors Persuading Interactions between RKNs and Plant Pathogenic Fungi

Plant pathogenic nematodes, such as *Meloidogyne*, have the ability to physically harm their host plants by leaving them with minor wounds. Infected plant tissues may be easily accessed by fungus through such injuries. Alternately, few nematodes may cause physiological variations in the plants they eat, causing changes in the fungal pathogen populations surrounding the host plants and increasing their propensity to proliferate and/or become pathogenic. In addition, additional biotic and abiotic elements, such as the genotype of the host plant, the availability of organic matter and nutrients, and other microbes, may influence how nematode pest infections and plant fungal pathogen infections turn out (Ahmad et al. 2019). Depending on whether root-knot nematodes are present in agricultural fields, the species composition of the fungi can change. The most common fungi associated with the presence of *Meloidogyne* species were found to be various species of *Fusarium*, and fungal diversity is crucial in the interactions between host plants and soil microorganisms. Dhami et al. (2022) carried out an experimental study to understand the nature of relative consequences of interaction among *Meloidogyne incognita*, *Fusarium oxysporum*, and tomato leaf curl Palampur virus on disease severity and growth. The findings showed that the growth parameters were reduced to their lowest levels when all three pathogens were inoculated at once. Compared to treatments where RKN was inoculated 10 days after other pathogen, root galling index was more severe in treatments with prior inoculation of RKN or simultaneous inoculation of RKN with another pathogen. When *M. incognita* and *F. oxysporum* f. sp. *melonis* were inoculated simultaneously or sequentially prior or later, the severity of the fusarium wilt was greater than when *F. oxysporum* was used alone.

The effects of the soilborne fungi *Verticillium* spp., *Fusarium oxysporum*, or *Monosporascus* in combination with the *Meloidogyne javanica* against susceptible plant hosts were assessed by Markakis et al. (2021). When *Verticillium dahliae* and *Meloidogyne javanica* were applied separately to split-root plants as opposed to symptoms in whole root plants inoculated with both pathogens, verticillium wilt symptoms in eggplant were significantly worse. When *Fusarium oxysporum* f.sp. *cucumerinum* and *Meloidogyne javanica* were combined in a split-root set-up, the symptoms of root and stem rot and root-knot were more severe than plants when inoculated with a single pathogen. Nematodes and fungi frequently have a synergistic interaction that causes crop loss more remarkable than what would be anticipated from either pathogen acting alone or from the two pathogens affecting additively. For a variety susceptible to the interaction, the outcome could be complete crop failure. Factors like saprophytic ability, a broad host range, and the pathogens' long-term survival compound the issue for the grower; as a result, the productivity of the land for what may be a precious crop is hampered for many years.

9.7 Interaction of RKNs with Plant Pathogenic Bacteria in Vegetable Crops

Potato (*Solanum tuberosum* L.) production is severely harmed and greatly diminished by the soilborne diseases bacterial wilt and RKNs. RKNs and bacterial wilt are both brought on by *Meloidogyne* species and *Ralstonia solanacearum*, respectively. The effects of *Meloidogyne incognita* alone and in combination with the bacterium *Ralstonia solanacearum* were assessed (Markakis et al. 2021). The outcomes demonstrated that when bacteria were added to plants along with nematodes simultaneously, the nematode injury was greatest. The inoculum build-up was greatest, with a higher percent disease incidence and yield loss. *Pseudomonas solanacearum* biotype-3 and *Meloidogyne javanica* had greater combined pathogenic effects on brinjal than either one alone. In contrast to simultaneous inoculation or inoculation of bacteria 4 weeks after the nematode inoculation, the most severe wilt development occurred in plants when inoculated with nematode 2 and 3 weeks before bacterial inoculation. The wilt symptom development was sped up by increased nematode inoculum levels of 50, 100, and 150 egg masses/plant (Sitaramaiah and Sinha 1984). *Meloidogyne* spp., wilt causing *Ralstonia solanacearum*, and *Phomopsis* blight interactions on eggplant growth and the contents of chlorophyll and carotenoids in plants grown were investigated by Khan and Siddiqui (2017). Combined inoculation of these pathogens showed a greater decrease in growth, chlorophyll content, and carotenoid percent than single inoculation. A superior decrease in plant growth was observed when *root knot nematode* was injected 20 days before *R. solanacearum* and *P. vexans* than when *R. solanacearum* and *P. vexans* were injected first. Table 9.2 represents various interactions of RKNs with different plant pathogenic bacteria.

9.8 Nematode Virus Interaction

The first three-step process involved between nematode and virus interaction is the nematode acquires virus particles while feeding on the virus-infected plant roots. Further, nematode vector retains the virus particles at the designated sites; after that, nematode vector retains the virus particles by dissociating from the retention sites. The nematode as vector and virus mode of interaction is very specific. Virus particles are present in the cell sap during the nematode feeding virus particle absorbed at the selective retention sites. In the case of *Xiphinema* spp. virus is associated with the odontophore, esophagus, and esophagus pump; on the other hand, the virus particles are associated with inner surface of the cuticular odontostylet in *Longidorus* species. Different nematode vectors are transmitted, serologically similar viruses, whereas serologically unrelated viruses have common nematode vectors (Taylor 1990).

Table 9.2 Nematode Interaction with Plant Pathogenic Bacteria

S. No	*Meloidogyne* species	Pathogen	Crop	Disease	Reference
1.	*Meloidogyne incognita*	*Agrobacterium tumefaciens*	Tomato	Crown gall	El-Sherif and Elwakil (1991)
2.	*Meloidogyne incognita*	*Clavibacter michiganense*	Tomato	Canker	De Moura et al. (1975)
3.	*Meloidogyne incognita acrita*	*Pseudomonas solanacearum*	Potato	Wilt	Jatala and Martin (1977)
4.	*Meloidogyne incognita acrita*	*Pseudomonas solanacearum*	Tomato	Bacterial wilt	Sowmya et al. (2012)
5.	*Meloidogyne incognita*	*Ralstonia (pseudomonas) solanacearum*	Tomato	Bacterial wilt	Siddiqui and Husain (1991)
6.	*Meloidogyne incognita*	*Protobacterium carotovorum subsp. Carotovorum*	Carrot	Soft rot	Sowmya et al. (2012)

Table 9.3 Nematode interaction with Plant viruses

S. No	*Meloidogyne* species	Pathogen	Crop	Reference
1	*Meloidogyne incognita*	Cucumber mosaic virus	Cucumber	Varshney et al. (2005)
2	*Meloidogyne incognita*	Zucchini yellow mosaic virus	Cucumber	

Another possibility of virus and nematode interaction to the management of nematode disease is by inoculation of the virus. Patel and Patel (1995) reported that they enhance the protein, nitrogen, and total sugar by combining infection of TMV and RKNs. Table 9.3 represents the list of interactions between RKN and the virus.

9.9 Effective Approaches to Study the Plant RKN Interaction

9.9.1 Transcriptomics and Proteomics Study

Plant parasitic nematodes secrete protein effectors that direct the endogenous molecular and physiological pathways of their hosts to their own advantage. The development of unique and profoundly specialized nourishing cells in the roots of the host is an important part of the infection process leading to success. However, there is still a limited understanding of the precise mechanisms contributing to their differentiation. Nevertheless, over the past decade, the techniques of holistic molecular biology, such as transcriptomics and proteomic approach in nematode parasitism biology, have provided detailed information on transcriptional changes in giant cells in the initial stages of differentiation. The interactions among plant and parasitic nematodes occur in a vast molecular network of plant immunity. After initial contact with the host plant's roots, plant-parasite nematodes (PPN) activate basal immune responses. Only a limited number of plant species are analyzed. Therefore, sequencing and proteomic analysis of the next generation is expected to open the possibility of interspecies comparisons for identifying preserved regulated genes and early protein changes in the development of food cells (Vijayashanthi et al. 2020). This "post-genomic" era has introduced powerful approaches to quantify RNA transcription and protein abundance for each gene within the genome—often for a wide range of conditions. Considering the various expression of genes involved in the parasitism in both nematodes and their corresponding host plants is made possible by using microarrays and deep RNA sequencing, which offer novel and extensive insights. For instance, microarray analysis showed that different soybean genes were expressed in the galls that developed on *Glycine max* cultivar William 82 soybean roots during the interaction with *M. incognita*. These modifications comprise the regulation of genes involved in various cell wall remodeling and modification, cell division and mitosis, carbon and energy metabolism, and the downregulation of genes involved in producing defense compounds like salicylic acid and jasmonic acid (Ibrahim et al. 2011).

9.9.2 RKN Effectors

Recent research has revealed several effectors that RKN secretes to promote parasitism by stifling the immune response of its hosts. Plant pathogen-associated molecular pattern (PAMP)-induced immunity is the target of the *Meloidogyne* effector protein MiMsp40 (PTI). In Arabidopsis plants, overexpression of MiMsp40 led to both strong and weak infection leading to susceptible plants by suppressing PTI/ETI signals of immunity, which cause increased susceptibility to nematode infection with increased galls and egg mass after 6 weeks of inoculation. Mc1194, A different protein was identified as an effector that promotes *M. chitwoodi* infection by interacting with the protease and granulin domains of RD21A in *Arabidopsis*, a member of the papain-like cysteine proteases (PLCP) that are involved in programmed cell death. These instances demonstrate how RKN use a variety of effectors to control host plant roots (Nguyen et al. 2018). Increased infestation of nematodes increases the severity of other pathogenic microorganisms and leads to a synergistic effect on disease development and severity.

9.9.3 *Meloidogyne* Parasitism Gene and Their Expression

The most important sedentary endoparasite, Meloidogyne, is obligatory in nature, has a large host range, and significantly contributes to crop losses. The parasitism gene that encodes effector proteins secreted through the stylets of the nematodes modifies the selected plant cell. One of the identified genes is the *Mi8D05* parasitism gene. This shows a specific interaction with the plant aquaporin tonoplast intrinsic protein 2 (TIP 2) and regulates solute-water transport in giant cell to promote parasitic interaction. Pathogenesis-related genes (*PvPR1 and PVPR2*) are involved in *Meloidogyne* parasitism. An effector protein gene named Mel-DOG has been identified, which is a putative Cis regulatory motif associated with expression in the dorsal gland that affects the host spatial-temporal gene expressions. A novel effector protein gene designated as *Mj-nulg1a*, which is expressed specifically within the dorsal gland of *Meloidogyne javanica*, which is a secretary effector protein expressed within the esophageal gland plays an important role in the invasion of host roots and in the formation of feeding sites necessary for the nematodes to complete their life cycle. MeTCTP, an *M. enterolobii* TCTP effector localized in the host cells' cytoplasm, suppresses programmed cell death triggered by the pro-apoptotic protein BAX in host plants that promote parasitism (Vieira and Gleason 2019).

9.10 Conclusion

Nematode infections are widespread throughout the world and result in severe crop losses; standard approaches are insufficient to combat the danger. The severity of the disease and yield loss are exacerbated by the interaction of RKNs with other bacteria. Understanding the fundamentals of nematode interactions and recognizing crucial genes and proteins tangled in the infection process and the plant resistance response are crucial for rising plant lines that are more resistant to nematode infection and interaction. Managing the nematode infection through the use of computational tools, next-generation genome and transcriptome sequencing, and advancements in gene cloning and RNAi are becoming the important areas of research.

References

Abad P, Gouzy J, Aury JM, Castagnone-Sereno P, Danchin EG, Deleury E, Perfus-Barbeoch L, Anthouard V, Artiguenave F, Blok VC, Caillaud MC (2008) Genome sequence of the metazoan plant-parasitic nematode *Meloidogyne incognita*. Nat Biotechnol 26(8):909–915

Ahmad L, Siddiqui ZA, Abd-Allah EF (2019) Effects of interaction of Meloidogyne incognita, Alternaria dauci and Rhizoctonia solani on the growth, chlorophyll, carotenoid and proline contents of carrot in three types of soil. Acta Agric Scand Sect B Plant Soil Sci 69:324–331

Alhazmi A, Al-Nadary SN (2015) Interaction between Meloidogyne incognita and Rhizoctonia solani on green beans. Saudi J Biol Sci 27:570–574

Ali MA, Azeem F, Abbas A, Joyia FA, Li H, Dababat AA (2017) Transgenic strategies for enhancement of nematode resistance in plants. Front Plant Sci 8:750

Atkinson GF (1892) Some diseases of cotton. Ala Agric Exp Stn Bull 41:61–65

Baniya A, Joseph S, Duncan L, Crow W, Mengistu T (2021) The role of Caenorhabditis elegans sex-determination homologs, mi-sdc-1 and mi-tra-1 in Meloidogyne incognita. Eur J Plant Pathol 161(2):439–452

Beyan A (2019) Response of tomato genotypes to Meloidogyne javanica and fusarium oxysporum f.sp. lycopersici co-infestation under glasshouse conditions. Pak J Nematol 37:63–82

Carneiro F (2010) Ramalho, M.; Pereira, M. fusarium oxysporum f. sp. phaseoli and Meloidogyne incognita interaction in common bean. Crop Breed Appl Biotechnol 10:271–274

Castagnone-Sereno P, Mulet K, Danchin EG, Koutsovoulos GD, Karaulic M, Da Rocha M, Bailly-Bechet M, Pratx L, Perfus-Barbeoch L, Abad P (2019) Gene copy number variations as signatures of adaptive evolution in the parthenogenetic, plant-parasitic nematode Meloidogyne incognita. Mol Ecol 28:2559–2572

Charchar MJM, Eisenback JD, Charchar J, Boiteux MEN (2008) Meloidogyne phaseoli n. sp. (Nematoda: Meloidogynidae), a root-knot nematode parasitising bean in Brazil. Nematology 10:525–538

Coyne D, Hallmann J, Timper P (2018) Plant parasitic nematodes in subtropical and tropical agriculture. CabinetMaker

Curtis R (2007) Do phytohormones influence nematode invasion and feeding site establishment? Nematology 9:155–160

Da Rocha M, Bournaud C, Dazenière J, Thorpe P, Bailly-Bechet M, Pellegrin C, Péré A, Grynberg P, Perfus-Barbeoch L, Eves-van den Akker S (2021) Genome expression dynamics reveal the parasitism regulatory landscape of the root-knot nematode Meloidogyne incognita and a promoter motif associated with effector genes. Genes 12:771. https://doi.org/10.3390/genes12050771

De Moura RM, Echandi E, Powell NT (1975) Interaction of Corynebacterium michiganense and Meloidogyne incognita on tomato. Phytopathology 65:1332–1335

Decraemer W, Hunt DJ (2006) Structure and classification. Plant Nematol:3–32

Dhami DS, Kaur S, Sharma A, Dhillon NK, Jain S (2022) Relative consequences of interaction among Meloidogyne incognita, fusarium oxysporum and tomato leaf curl Palampur virus on disease severity and growth of muskmelon. Indian Phytopathol 75:1–13

Eisenback JD, Triantaphyllou HH (2020) Root-knot nematodes: Meloidogyne species and races. In: Manual of agricultural nematology. CRC Press, pp 191–274

El-Shennawy MZ, Khalifa EZ, Ammar MM, Mousa EM, Hafez SL (2012) Biological control of the disease complex on potato caused by root-knot nematode and fusarium wilt fungus. Nematol Medit 40:169–172

El-Sherif AG, Elwakil MA (1991) Interaction between Meloidogyne incognita and agrobacterium tumefaciens or fusarium oxysporum f. sp. lycopersici on tomato. J Nematol 23:239–242

Goswami BK, Chenula VV (1974) Interaction of rootknot nematode Meloidogyne incognita and tobacco mosaic virus in tomato. Ind J Nematol 4:69–80

Hajji-Hedfi LM, Hamdi-Boughalleb N, Horrigue-Raouani N (2019) Fungal diversity in rhizosphere of root-knot nematode infected tomatoes in Tunisia. Symbiosis 79:171–181

Harris TW, Antoshechkin I, Bieri T, Blasiar D, Chan J, Chen WJ, De La Cruz N, Davis P, Duesbury M, Fang R (2010) WormBase: a comprehensive resource for nematode research. Nucleic Acids Res 38:463–D467

Holbein J, Franke RB, Marhavý P, Fujita S, Górecka M, Sobczak M, Geldner N, Schreiber L, Grundler FM, Siddique S (2019) Root endodermal barrier system contributes to defence against plant-parasitic cyst and root-knot nematodes. Plant J 100:221–236

Hutton SF, Scott JW, Vallad GE (2014) Association of the Fusarium wilt race 3 resistance gene, I-3, on chromosome 7 with increased susceptibility to bacterial spot race T4 in tomato. J Am Soc Hortic Sci 139:282–289

Ibrahim HM, Ahmad EM, Martínez-Medina A, Aly MA (2019) Effective approaches to study the plant-root-knot nematode interaction. Plant Physiol Biochem 141:332–342

Ibrahim HMM, Alkharouf NW, Meyer SLF, Aly MAM, Gamal El-Din AEKY, Hussein EHA, Matthews BF (2011) Post-transcriptional gene silencing of root-knot nematode in transformed soybean roots. Exp Parasitol 127:90–99. https://doi.org/10.1016/j.exppara.06:037

Iqbal S, Fosu-Nyarko J, Jones MG (2020) Attempt to silence genes of the RNAi pathways of the root-knot nematode, Meloidogyne incognita results in diverse responses including increase and no change in expression of some genes. Front Plant Sci 11:328

Jatala J, Martin C (1977) Interactions of Meloidogyne incognita acrita and pseudomonas solanacearum on field grown potatoes. Proc Am Phytopath Soc 4:177–178

Jones JT, Haegeman A, Danchin EG, Gaur HS, Helder J, Jones MG, Kikuchi T, Manzanilla-López R, Palomares-Rius JE, Wesemael WM, Perry RN (2013) Top 10 plant-parasitic nematodes in molecular plant pathology. Mol Plant Pathol 14:946–961

Keinath AP, Wechter WP, Rutter WB, Agudelo PA (2019) Cucurbit rootstocks resistant to fusarium oxysporum f. sp. niveum remain resistant when Coinfected by Meloidogyne incognita in the field. Plant Dis 103:1383–1390. [CrossRef]

Khan M, Siddiqui ZA (2017) Interactions of Meloidogyne incognita, Ralstonia solanacearum and Phomopsis vexans on eggplant in sand mix and fly ash mix soils. Sci Hortic 225:177–184

Khan M, Bhattarai A, Kuznia JN, Olson DT (1993) High electron mobility transistor based on a GaN-Al x Ga1 − x N heterojunction. Appl Phys Lett 63(9):1214–1215

Lamelas A, Desgarennes D, López-Lima D, Villain L, Alonso-Sánchez A, Artacho A, Latorre A, Moya A, Carrión G (2020) The bacterial microbiome of Meloidogyne-based disease complex in coffee and tomato. Front Plant Sci 11:136. https://doi.org/10.3389/fpls.2020.00136

Lunt DH, Kumar S, Koutsovoulos G, Blaxter ML (2014) The complex hybrid origins of the root-knot nematodes revealed through comparative genomics. PeerJ 2:e356

Markakis EA, Krasagakis N, Tzortzakakis EA, Tsaniklidis G, Lagogianni CS, Ligoxigakis EK, Pardavella IV, Goumas DE (2021) Investigation of interactions between Meloidogyne javanica and three soilborne fungi against susceptible plant hosts. J Phytopathol 169:98–111

Mende NV (1997) Invasion and migration behaviour of sedentary nematodes. In: Cellular and molecular aspects of plant-nematode interactions. Springer, Dordrecht, pp 51–64

Mitreva M, Blaxter ML, Bird DM, McCarter JP (2005) Comparative genomics of nematodes. Trends Genet 21:573–581

Mitsumasu K, Seto Y, Yoshida S (2015) Apoplastic interactions between plants and plant root intruders. Front Plant Sci 6:617

Nguyen C-N, Perfus-Barbeoch L, Quentin M, Zhao J, Magliano M, Marteu N, Da Rocha M, Nottet N, Abad P, Favery B (2018) A root-knot nematode small glycine and cysteine-rich secreted effector, MiSGCR1, is involved in plant parasitism. New Phytol 217:687–699

Opperman CH, Bird DM, Williamson VM, Rokhsar DS, Burke M, Cohn J, Cromer J, Diener S, Gajan J, Graham S, Houfek TD (2008) Sequence and genetic map of Meloidogyne hapla: a compact nematode genome for plant parasitism. Proc Natl Acad Sci 105:14802–14807

Ozdemir FGG, Arici ŞE, Elekcioğlu İH (2022) Interaction of Meloidogyne incognita (Kofoid & White, 1919) (Nemata: Meloidogynidae) and fusarium oxysporum f. sp. radicis-lycopersici Jarvis & Shoemaker in tomato F1 hybrids with differing levels of resistance to these pathogens. Turkish J Entomol 46:63–73

Parkunan V, Timper P, Ji P (2016) Lack of influence of Meloidogyne incognita on resistance of bell pepper cultivars to Phytophthora capsici. Can J Plant Pathol 38:1–7

Patel KA, Patel BN (1995) Effect of tobacco mosaic virus and root-knot nematode infection on chemical constituents of bidi tobacco. Ind J Mycol Plant Pathol 25:228–230

Rompalli R, Mehendrakar SR, Venkata PK (2016) Evaluation of potential bio-control agents on root-knot nematode Meloidogyne incognita and wilt causing fungus fusarium oxysporum f. sp conglutinans in vitro. African J Biotechnol 15(19):798–805. https://doi.org/10.5897/AJB2015.15235

Scherlach K, Hertweck C (2018) Mediators of mutualistic microbe—microbe interactions. Nat Prod Rep 35:303–308

Siddique S, Grundler FM (2018) Parasitic nematodes manipulate plant development to establish feeding sites. Curr Opin Microbiol 46:102–108

Siddiqui ZA, Husain SI (1991) Studies on the biological control of root-knot nematode. Curr Nematol 2:5–6

Sikandar A, Zhang M, Wang Y, Zhu X, Liu X, Fan H, Duan Y (2020) Nematodes a risk to agriculture. Appl Ecol Environ Res 18:1679–1690

Singh D, Dutta TK, Shivakumara TN, Dash M, Bollinedi H, Rao U (2021) Suberin biopolymer in rice root exodermis reinforces preformed barrier against Meloidogyne graminicola infection. Rice Sci 28:301–312

Sitaramaiah K, Sinha SK (1984) Interaction between Meloidogyne javanica and pseudomonas solanacearum on brinjal. Ind J Nematol 14:1–5

Sowmya DS, Rao MS, Kumar RM, Gavaskar J, Priti K (2012) Bio-management f *Meloidogyne incognita* and *Erwinia carotovora* in carrot (*Daucus carota* L.) using *Pseudomona putida* and *Paecilomyces lilacinus*. Nematol Med 40:189–194

Taylor CE (1990) Nematode interactions with other pathogens. Ann Appl Biol 116:405–416

Trudgill DL, Blok VC (2001) Apomictic, polyphagous root-knot nematodes: exceptionally successful and damaging biotrophic root pathogens. Annu Rev Phytopathol 39:53

Varshney S, Pandey RC, Panday RK, Dwivedi BK, Singh V (2005) Interaction between mungbean yellow mosaic virus and root-knot nematode on growth of mungbean plants. Pak J Nematol 23: 93–98

Vieira MGA, da Silva MA, dos Santos LO, Beppu MM (2011) Natural-based plasticisers and biopolymer films: a review. Eur Polym J 47(3):254–263

Vieira P, Gleason C (2019) Plant-parasitic nematode effectors insights into their diversity and new tools for their identification. Curr Opin Plant Biol 50:37–43

Vijayashanthi S, Shanthi A, Raguchander T, Kavitha PG (2020) Interaction effect of root-knot nematode, Meloidogyne incognita and wilt fungus, fusarium solani in cucumber. J Entomol Zool Stud 8:721–725

Wesemael W, Viaene N, Moens M (2011) Root-knot nematodes (Meloidogyne spp.) in Europe. Nematology 13:3–16

Breeding for Resistance in Vegetables Against *Meloidogyne* Species Causing Root Gall Disease

10

Bitaisha Nakishuka Shukuru, T. S. Archana, Pritha Ghosh, Adesh Kumar, Devendra Kumar, and Vipul Kumar

Abstract

Worldwide, scientists and farmers are struggling to increase productivity and agriculture sustainability to produce more food for people. Vegetable plants are of great importance for human and animal nutrition. The *Meloidogyne* species are one of the most widespread plant parasitic nematodes (PPNs), considered a serious global threat to vegetable crops and causing significant losses. Research scientists reported that using nematode-resistant plant cultivars is a much more significant and environmentally safe alternative than chemical nematicides. Little is known about the effectiveness of breeding for resistance methods to manage *Meloidogyne* populations. Resistant vegetable cultivars are available and have been well documented in their use against *Meloidogyne* species. In this chapter, we have discussed different breeding methods and enumerated some of plant cultivars found to be resistant to root-knot nematodes (RKNs) causing root gall disease. The specific breeding approaches for *Meloidogyne* species-vegetable crop resistance have been used to study the resistance mechanisms among the various varieties of vegetable crops.

Keywords

RKN · Plant host resistance · Breeding approaches · Nonbreeding methods · Genetic engineering

B. N. Shukuru (✉) · T. S. Archana · A. Kumar · D. Kumar · V. Kumar
Department of Plant Pathology, School of Agriculture, Lovely Professional University (LPU), Phagwara, Punjab, India
e-mail: nakishuka.12111882@lpu.in

P. Ghosh
Department of Entomology, School of Agriculture, Lovely Professional University (LPU), Phagwara, Punjab, India

F. Ahmad, G. Nombela (eds.), *Root-Galling Disease of Vegetable Plants*,
https://doi.org/10.1007/978-981-99-3892-6_10

10.1 Introduction

Root-knot nematodes (RKNs) pose severe problems for vegetable plants, including tomato, eggplant, cabbage, carrot, chili, bean, okra, squash, watermelon, broccoli, cauliflower, cucumber, radish, peas, chard, spinach, onion, potato, pointed gourd, and asparagus. Endoparasitic nematodes are found in the plant's tissues, where they spend most of their existence. Among the endoparasitic nematodes, sedentary endoparasites, e.g., *Meloidogyne* species, feed within the host's tissues. The key characteristics for identifying these nematode species include morphological, morphometrical, and anatomical differences using a microscope for image analysis, molecular methods such as fingerprint, and DNA and protein analyses (Shukuru and Archana 2023). In addition, nine main feeding types for PPNs can be identified: plant-feeders, plant-associated nematodes, omnivores, predators, animal parasites, bacterial feeders, hyphal-feeders, nematodes ingesting substrates, and unicellular eukaryotic-feeders. Root-knot nematodes (RKNs) are among them important plant feeders (Fig. 10.1). *Meloidogyne* spp. are among the major genera reported to cause crop losses (Shukuru and Archana 2023). Thereby, the most economically important species directly target plant roots of major production crops and prevent water and nutrient uptake resulting in reduced agronomic performance, overall quality, and yields (Bernard et al. 2017a, b).

RKNs are the most important and destructive among the PPNs (Oka et al. 2000; Chitwood 2003; Mitkowski and Abawi 2011; Singh et al. 2015; Saucet et al. 2016; Edel-Hermann and Lecomte 2019) that commonly include the species *Meloidogyne incognita*, *M. javanica*, *M. arenaria*, *M. hapla*, *M. graminicola*, *M. chitwoodi*, *M. graminis*, *M. naasi*, *M. marylandi*, *M. fallax*, *M. acronea*, and *M. enterolobii* (synonym *M. mayaguensis*) (Onkendi et al. 2014; Suresh et al. 2017; Lima et al. 2018; Shukuru and Archana 2023). The four most important RKNs such as *M. incognita*, *M. hapla*, *M. arenaria*, and *M. javanica* were worldwide. Among them, *M. hapla* is the most common species infesting open fields of temperate climatic regions, while the others are tropical RKN species found in hot tropical or warm climates. *M. incognita* affect more seriously on the carrot plants grown in greenhouses, compared to *M. hapla*, which has a major distribution in open carrot fields (Gugino et al. 2006; Seo et al. 2015).

First observed on cucumbers and reported in 1855 by Berkeley, the Chitwood's work in 1949 defined four species and one subspecies of RKNs (*Meloidogyne incognita acrita*). By mistake, RKNs were considered under the same species, *Heterodera radicicola*. Before, in 1887 and 1892, Goeldi described *Meloidogyne exigua*, species from which Chitwood obtained the name we currently use for RKNs. The name *Meloidogyne* is of Greek origin, meaning "apple-shaped female." Approximately 100 *Meloidogyne* species have been described to date. RKNs are minute, worm-like animals, very common in soil; females are globose and sedentary at maturity, ranging in length from 400 to 1000 μm (Mitkowski and Abawi 2011). As plant parasites, they lead to swellings or nodules on plant roots. For example, in carrots, roots are twisted and deformed, with much forking, and the presence of knobby galls on the outside of the roots is observed. New growth slows as nematode pressure increases above the ground; infected plants pull easily from the ground.

Fig. 10.1 Feeding types of nematodes. Here, we graphically summarize principal feeding habits of plant and soil nematodes, based on food resources utilized

Gall sizes can be remarkably larger with a more serious reduction of the root growths in the plants infected with *M. incognita* than *M. hapla*. In the infection sites of the root tissues, giant cells can be more extensively formed, occupying larger stellar regions with the prominent destruction of adjacent xylem vessels by *M. incognita* than *M. hapla* (Seo et al. 2015). *Meloidogyne* species cause sizable root galls that increase susceptibility to other pathogens. Generally, large galls or knots form throughout the root system; thus, severe infections will result in reduced yields and affect consumer acceptance of vegetable plants (Perry et al. 2009; Beccari et al. 2010; Mitkowski and Abawi 2011). Generally, *M. incognita*, *M. arenaria*, and *M. javanica* require higher temperatures for multiplication and survival than *M. hapla*. For the RKN growth and development, the optimum temperature ranges between 25 and 30 °C.

Nevertheless, M. *hapla* is most prevalent in regions with temperatures around 0 °C to 15 °C or above. The length of the life cycle depends on temperature. It varies from 4 to 6 weeks in summer (with optimum temperature), and it may extend more than 50 days (mostly ranging between 10 and 15 weeks) in the winter season (Abad and Williamson 2010; Mitkowski and Abawi 2011; Gowda et al. 2019). RKN female lays eggs on the root surface (500 to more than 1000 eggs) that rapidly develop into a first-stage juvenile (J_1), residing inside the translucent egg case and molting into a motile J_2 nematode, the only stage capable of initiating infection. Hence, J_{2s} will invade growing root tips after moving to the area of cell elongation and initiating a feeding site by injecting esophageal gland secretions into parasitized plant root cells, leading to dramatic physiological changes (giant-cells formation). RKN juveniles are active, thread-like worms about 0.5 mm long, too small to be seen with the naked eye. Recall that RKNs undergo four juvenile stages, which progress through a molting process each. Thus, spherical-shaped female adults emerge from the J_4 cuticle; occasionally, they develop into males (Karssen 2002; Mitkowski and Abawi 2011). The degree of RGD depends on three main factors RKN density, *Meloidogyne* species and races present, and vegetable host plant species and cultivar (Alves-Santos et al. 2002; Perry et al. 2009; Mitkowski and Abawi 2011). Mainly, physiological races of RKNs occur in *M. incognita*, *M. javanica*, *M. arenaria*, and *M. hapla*. For example, the occurrence of four races for *M. incognita* (race 1, 2, 3 and 4), three races for *M. javanica* (race 1, pepper race 2, groundnut race 3), and one race for *M. arenaria* (race 2) was observed in different parts of India (Gowda et al. 2017). Yield losses due to RKNs can reach over 30% for susceptible vegetable plants (Sikora and Fernandez 2005). In India, economic loss was reported for different crops, including tomato (11–35%), eggplant (10–42%), chili (8–23%), carrot (18.20%), cucumber (6–18%), calabash (21–23%), okra (10–29%), snake gourd (17%), bitter gourd (13–14%), and pumpkin (13%) (Gowda et al. 2017). The interaction with fungi and bacteria aggravates the problem, and the development of a disease complex occurs, increasing the crop losses, which can vary between 40 and 70%; this is because RKNs can break the resistance in vegetable cultivars which are resistant to these soil-borne pathogens (Agrios 2005; Gowda et al. 2017).

10.2 Origins and Mechanisms of Plant Resistance

10.2.1 Plant Resistance Sources

First, it is important to mention that wild plant species, induced mutants, and plant regeneration are potential sources of resistance. No doubt that wild plant species represent the most important source of hypersensitive response (HR) genes, but resenting problems of incompatibility, particularly among the more divergent genotypes. This is due to the association of resistance with various undesirable traits. Therefore, embryo rescue and somatic hybridization techniques can facilitate otherwise difficult gene transfers. In vegetable crops, some mutants induced by irradiation express increased levels of resistance to RKNs. Plant regeneration from cells, tissues, or organs facilitates the selection of somaclonal plant variants with desirable resistance traits arising from single nuclear changes. No research on host plants' induced and constitutive defense mechanisms (Boots and Best 2018) under

RKN attack was already reported (Lee et al. 2021). According to Saucet et al. (2016), one possible means is identifying natural host plant resistance and engineering resistant rootstocks against RKNs. Resistance mechanisms such as HR are highly strong and effective defense reactions (Agrios 2005; Moon et al. 2010; Saucet et al. 2016). First, HR consists of localized cell necrosis at the infection site and is characteristic of single gene resistance (SGR) to nematodes, plant-pathogenic fungi, bacteria, and viruses. Resistance to RKNs occurs in plant hosts soon after germination and then involves induction of HR following the invasion of *Meloidogyne* species.

Furthermore, induced systemic resistance (ISR) and systemic acquired resistance (SAR) are two different phenomena, but both represent active plant defense responses (Choudhary et al. 2007) to nematode attack, including RKNs. ISR is like HR, while SAR is like the inherent immunity of the plant system. SAR requires the signal molecule salicylic acid and is associated with accumulating pathogenesis-related proteins. Like the SAR, a plant can develop defenses against an invader RKN if an infection occurs. In contrast to SAR triggered by salicylic acid accumulation, ISR relies on signal transduction pathways activated by jasmonic acid and ethylene.

10.3 Resistance Mechanisms

10.3.1 Antibiosis, Antixenosis or Non-preference, and Tolerance

The plant employs antixenosis to reduce colonization by nematodes, including *Meloidogyne* spp. For example, plants exhibiting antixenotic resistance should have a reduced initial number of nematodes colonies early in the season. Antibiosis operates after the nematodes has colonized the vegetable plant. The plant's tolerance does not affect the rate of population increase of the target nematodes but does raise the threshold level. Some bacteria and fungi may antagonize nematodes by producing nematicidal/nemastatic compounds (Shukuru and Archana 2023). This mode of action is known as antibiosis. Non-preference and antibiosis are two mechanisms that require a dynamic nematode response or lack of response. Tolerance is more subject to variation because of environmental factors (Singh et al. 2021), including air, water, climate, soil, natural vegetation, and landforms. These three mechanisms of resistance (Fig. 10.2) may also come from plant characteristics or specific genetic traits. Thus, they will affect the behavior of RKNs in soil.

10.3.2 Protease Inhibitors

Protease inhibitors are enzymes that hydrolyze peptide bonds of proteins. In their study, Gawade et al. (2017) reported a significant reduction of RGD index on tomato roots treated with *Cicer arietinum* proteinase inhibitor. This can constitute one of the pathways to designing a control strategy for suppressing RKNs infecting vegetable crops (Gowda et al. 2019).

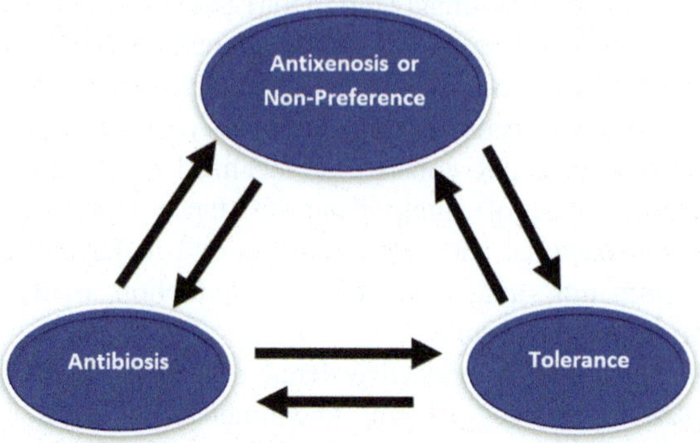

Fig. 10.2 Three interrelated resistance mechanisms present in RKN-resistant/tolerant vegetable crop cultivars

10.3.3 RNAi Strategy

The discovery of RNA interference (RNAi) in the free-living predatory nematode, *Caenorhabditis elegans*, constitutes a pathway of genetically engineered resistance against hidden PPNs. Thus, RNAi is a sequence-specific and homology-dependent gene silencing mechanism in which dsRNA elicits the post-transcriptional silencing of endogenous genes (e.g., nematode parasitism genes) with homologous sequences (Gowda et al. 2019). For instance, host-induced Integrase dsRNA gene silencing was successfully demonstrated against *M. incognita* in tobacco (Yadav et al. 2006). The same mechanism was reported to help manage RKNs in vegetable plants, such as tomato plants carrying *Mi-col-1* and *Lemmi-5* dsRNAs independently (Koulagi and Sirohi 2015; Shivakumara et al. 2017; Banerjee et al. 2018) and other crops (Banerjee et al. 2017).

10.4 Nonbreeding Methods for RKN Control

The coordinated use of multiple strategies to assure stable expected and potential crop yield in vegetable plants is important in managing RKNs. Several cultural methods for RKN control exist and can be applied in respect of the crop nature and timing (e.g., adjusting sowing and harvesting time; adequate rotation with nonhosts; good weed control practices depending on their morphological features—grasses, sedges, and broadleaf, or life cycle—annual, biennial, and perennial). They can include cropping systems, soil admixture, crop rotation, flooding, weed control, adding soil amendments, fallowing, and cultivation. For example, as aquatic animals, nematodes require a water film around soil particles before they can move; thereat, nematode's eggs will not hatch unless there is sufficient moisture in the soil; conditions that are optimum for plant growth can be ideal for the development of RKN. RKNs have a diverse range of natural enemies including fungi, bacteria, predatory nematodes, microarthropods, annelids, protozoa, and other generalist predators, as well as biopesticides, but generally, they are not effective in field as most are being tested in laboratory conditions that are difficult to release in field environments (Stirling 2014; Monteiro et al. 2020; Shukuru and Archana 2023). No doubt that the use of chemicals against RKNs is very effective and most widely practiced in nematode management, but this method has limits due to their effects on ecosystems, biomes, or habitats, including environmental hazards they pose, residues they leave in soil and groundwater environment, and high costs of nematicides (Shukuru and Archana 2023). It was repeatedly reported that the use of resistant plant cultivars for managing RKNs is a significant, effective, and environmentally safe alternative compared to chemical nematicides, which are fortunately being withdrawn from the market. So, it is an opportunity for resistance application (Shukuru et al. 2022).

10.5 Breeding Approaches for Root Gall Disease Resistance

Regardless of the previously discussed control options, using disease-resistant plant material to control nematodes remains the only significantly effective and eco-friendly means. In addition to that, resistant varieties offer the cheapest means of nematode disease control. For Råberg (2014), resistance is the ability to prevent infection or limit parasite replication. Accordingly, Bilichak et al. (2020) reported that the ability to modify a plant's genetic material creates many opportunities for the rapid development of high-quality cultivars with desired characteristics, including nematode disease resistance and an increase in crop yield. Thus, the highly effective and economically reliable method uses disease-resistant vegetable crop varieties to control RKNs.

10.6 Host Plant Resistance Against RKNs

It is important to recall that two relevant primary aspects of plant host resistance against nematodes can be distinguished. The first one is self-protection by the crop, which can be based on the level of plant tolerance to injury caused by the initial infection. The second is rotational aspects in cropping systems conferring protection to the subsequent crops (Rai et al. 2010). Resistance genes in certain vegetable crops effectively control RKNs. For instance, it was reported that the Mi gene confers genetic resistance to RKNs in tomato. Also, many other effective RKN attack-resistance genes have been identified. Among them are *Mi*-2 through *Mi*-8 genes from tomato and *Me* and *N* genes from the pepper. Several genes have not yet been named, and new sources of genetic resistance to RKNs are frequently identified (Mitkowski and Abawi 2011). Thus, the resistance gene *Mi*-1 find in the wild tomato, *Solanum peruvianum*, and now present in many tomato cultivars, confers resistance to three species such as *M. incognita*, *M. javanica*, and *M. arenaria* (Garcia et al. 2007; Newton et al. 2012). However, it was previously reported that the *Mi*-1 resistance gene becomes inactive above 28 °C (Hwang and Williamson 2003; Noling 2019); but other genes like *Mi*-4, *Mi*-5, *Mi*-6, and *Mi*-9 were investigated to be stable at above 28 °C (Jablonska et al. 2007; Newton et al. 2012). Other examples

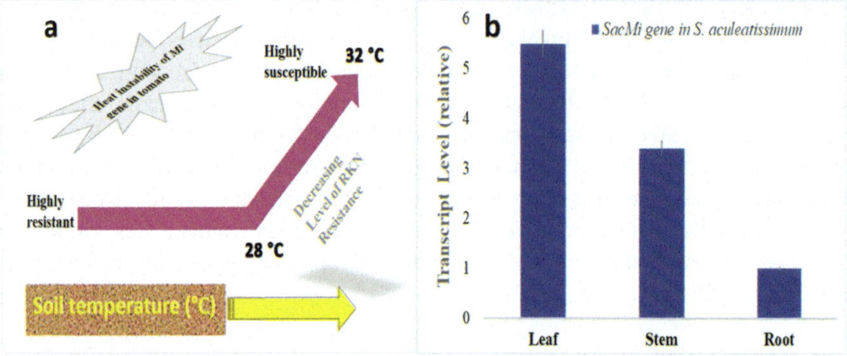

Fig. 10.3 Complete loss of RKN resistance conferred by the Mi gene in tomato with increasing soil T °C (**a**) and expression of *SacMi* gene in Dutch eggplant tissues (*Solanum aculeatissimum*) (**b**)

include *Me*-1 R and *Me*-3 R genes in pepper, and the *Rk R* gene from cowpea (Pegard et al. 2005; Das et al. 2008; Saucet et al. 2016) confers resistance to RKNs. *Solanum aculeatissimum*, a wild relative of eggplant, thanks to the presence of a nucleotide-binding site leucine-rich repeat (NBS-LRR) resistance gene, *SacMi* (Fig. 10.3), possesses resistance to *M. incognita* (Zhou et al. 2018). Accordingly, *S. torvum* resists to RKNs (Gowda et al. 2017).

In addition, it was reported that a maximum number of phenolic compounds, salicylic, chlorogenic, and ascorbic acids present in the vegetable soybean RGD-resistant clones, offer resistance to *M. incognita* (Sato et al. 2019; Ye et al. 2019; Ramzan et al. 2021). Accordingly, to date, researchers and scientists have not yet described the case of resistance in sweet potato plants (Wendimu 2021). As a large Asteraceae or daisy family genus, sweet wormwood (*Artemisia annua*) is repeated to have a compelling nematicidal character (Khan et al. 2019) that intrinsic trait in *Artemisia* can value by offering protection to the plant host and could control *Meloidogyne* species so far. The complex mixtures of volatile compounds such as α-pinene, limonene, 2-methoxy-3-(1-methyl propyl)-pyrazine, methyl salicylate (MeSA), tridecane, and 4,5-di-epi-aristolochene are released by pepper roots. These chemicals allowed the detection of thymol, an active ingredient in pesticide products used against diseases and pests, including RKNs (Kihika et al. 2017). Moreover, in 1977, Harikishore et al. reported the availability of a high degree of resistance in several tubers bearing *Solanum* species, including *Solanum acaule*, *S. acroscopicum*, *S. agrimoniifolium*, *S. ajanhuiri*, *S. boliviense*, *S. brevicaule*, *S. bulbocastanum*, *S. cardiophyllum*, *S. chacoense*, *S. chaucha*, *S. curtilobum*, *S. demissum*, *S. famatinae*, *S. hougasii*, *S. infundibuliforme*, *S. jamesii*, *S. grandarillasii*, *S. kurtzianum*, *S. leptophyes*, *S. lignicaule*, *S. maglia*, *S. microdontum*, *S. multidissectum*, *S. ochranthum*, *S. phureja*, *S. pinnatisectum*, *S. raphanifolium*, *S. recho*, *S. sanitaerosae*, *S. sparsipilum*, *S. spegazzinii*, *S. stenophyllidium*, *S. stenotomum*, *S. stoloniferum*, *S. tuberosum subsp. Andigena*, *S. tuberosum*, *S. vallis mexici*, and *S. vernei*, against RKNs (Prasad 2008).

10.7 Introduction for Resistance

Resistant crop cultivars may be introduced for cultivation in a new area. For instance, *Mi*-1.2, an SGR, dominant locus conferring resistance to the three given RKNs above, is present in many modern tomato (*Solanum lycopersicum*) cultivars, having CC-NBS-LRR gene structure. Mainly, the HR is triggered before the significant initiation of the RKN feeding site (Fuller et al. 2008). When the root gall disease (RGD)-resistant clone becomes available for any vegetable crop, it can benefit a new area where it will be introduced rapidly into susceptible cultivars using available gene transfer. For example, carrot germplasms from Uzbekistan, Poland, and Canada were introduced in California, USA, to offer traits for improved commercial cultivars. The Brasilia carrot cultivar has already shown resistance to *M. javanica*, which was introduced in California (Bryant 2005). In Florida, commercially available resistant varieties to *Meloidogyne* species are currently available, especially for tomato, pepper, southern pea, and sweet potato. The seed packet or label has the code VFN (*Verticillium*, *Fusarium*, Nematodes) (Noling 2019). However, the introduction will depend on environmental factors (genotype × environment interactions: drought, unusual temperature variations: problems of heat instability) prevailing in the new clone agricultural zones (Kuti and Konuru 2005; Altieri et al. 2012; Osei et al. 2018; Sato et al. 2019; Shukuru et al. 2022). Take the case of using tomato varieties which may have to be restricted to spring plantings when cooler soil temperatures prevail in the area (Noling 2019).

10.8 Selection of RKN-Resistant Clones

The selection method refers to resistant vegetable plants that can be obtained from a commercial variety and therefore constitutes the quickest method of developing a resistant vegetable crop cultivar. Many studies on screening for resistance in selected vegetable crop varieties against RKNs reported different positive results (Seid et al. 2017), with now available cultivars resistant to either one, two, or several RKN populations (Table 10.1). For example, sweet potato seedling selection is effective in rapidly breeding RGD-resistant lines (Akunouchi et al. 2013). Reddy et al. (2018) reported that H-88-78-1, considered an advanced tomato breeding line, is the most resistant genotype against *M. incognita* thanks to the presence of *Mi* gene, as reported by the molecular screening with *Mi* gene-linked markers phosphomannose-isomerase (pmi) and *Mi*-2.3 (Gowda et al. 2019). However, Lizardo et al. (2022) currently screened for resistance to *M. incognita* in eight selected tomato germplasm collections and commercially available varieties in the Philippines and found that none of them showed a resistant reaction; this is due to a lack of the *Mi*-1 gene conferring resistance against the RKN species.

Table 10.1 Examples of RKN-resistant varieties and lines of different vegetable crops

Vegetable crop	Resistant varieties/lines	Country	RKN species	References
Sweet potato	Nugget, TUO2, Whatley Loretan, Covington, Evangeline, Jewel	USA	*M. incognita*	Bernard et al. (2017a, b), La Bonte et al. (2008), Yencho et al. (2008)
Sweet potato	W-86, L4–89, BPA4, Sinibastian, Jasper, Jewel, Miracle, Georgia Red, Garcia Yellow, Travis	USA	*Meloidogyne* spp.	Bernard et al. (2017a, b)
Carrot	273, 280, 402, 403, 411,412, 421, 434, 436, 441, 442, 446, 448, 450, 453, 454, 456, 504, 607, 608, 647, 652, 724, 1201(4), 1201(5), 1202(0), 1207(3), 1210(4), 1211(1), 1211(3), 1211(5), 1213(1), 1214(1), 1214(2), 1214(4), 1215(1), 1215(3), 1215(4), 223(1), 223(4), 224(2), 224(4), 228(1), 248(5), 249(1), 249(3), 250(5), 251(1), 251(5), 252(0), 253(3), 254(4), 256(1), 264(1), 265(2), 267(2), 268(1), 402(1), 410(2), 422(2)	South Korea	*M. incognita* race 1	Seo et al. (2014)
Tomato	EC705452, EC699717, EC759288, EC002644, EC035420, EC054644, EC129606-PPEC 006148, LA 2823, LA 3471, H-88-78-1, SL-120, PNR-7, Hisar Lalit, NT-3, NT12, Pusa Hybrid-2, Arka Vardana	India	*Meloidogyne* spp.	Reddy et al. (2018, 2019), Gowda et al. (2017)
Eggplant	Black beauty, Pant Rituraj, Banaras Giant, Rajendra Baigan, Rajendra BaiganII long, IC-90903, IC-127029, IC-122076, KS-224, IC-127040	India	*Meloidogyne* spp.	Gowda et al. (2017)
Tomato	Sanibel	USA	*M. incognita, M. arenaria, M. javanica*	Noling (2019)
Chili pepper	NP-46A, Pusa Jwala, Mohini, Pusa Sadabahar, PSL-3, Surajmukhi, BSS-138, LCA-304,	India	*Meloidogyne* spp.	Gowda et al. (2017)

(continued)

Table 10.1 (continued)

Vegetable crop	Resistant varieties/lines	Country	RKN species	References
	LCA-305, Guchheedar, Hoe-808			
Okra	*Abelmoschus moschatus* genotypes (IC-140970-A, IC-203863) and *A. angulosus* genotypes (IC-470751, IC-203834, IC-203831, IC-203833, 1C-203863)	India	*Meloidogyne* spp.	Gowda et al. (2017)
Cucumber	Long Green		*M. incognita*	Kayani and Mukhtar (2018)
Calabash	PSPL, Hoe-505, Samrat, Bogh-2	India	*Meloidogyne* spp.	Gowda et al. (2017)
Pepper	Carolina Belle, Carolina Wonder	USA	*Meloidogyne* spp.	Noling (2019)
Sweet potato	Evangeline, Covington, Beniharuka, Konamizuki, Hoshikogane, Aikomachi	Japan	*M. incognita*, *Meloidogyne* spp.	Okada et al. (2017)
Cowpea	IT89KD-288	Nigeria	*M. incognita*	Izuogu et al. (2019)
Potato	HC-294	India	*M. incognita*	Prasad (2008)

10.9 Hybridization

The objective is to incorporate RGD-resistance and desirable traits into different vegetable varieties from controlled and free crosses. Artificial crossings characterize this method. It then consists of transferring RGD-resistance from an undesirable variety to a susceptible but otherwise desirable variety (backcross method) and combining RGD-resistance and some other desirable characteristics of one type with the superior attributes of another variety (pedigree method). Hence, RGD-resistant vegetable hybrids varieties generate higher and expected crop yields than open-pollinated varieties. Bhavana et al. (2019) reported the immune response of two tomato genotypes (HAT-310, HAT-311) to *M. incognita*; six crosses have been released with the two previous cultivars considered as sources of resistance (HAT-311 × Swarna Lalima, HAT-296 × HAT-311, EC-596747× HAT- 27311, Swarna Lalima × HAT-310, EC-596743 × HAT-310 and Swarna Lalima × HAT 311) against the same RKN species. RKN resistance has not yet been identified in some vegetable crops, but it has been proven that plant hybrids are good candidates for RKN-resistant rootstocks (Noling 2019). Pluktor and Gloria are among Kenya's most popular cabbage hybrids (Waceke 2007).

10.10 Grafting for RKN-Resistance

The grafting technique is the most efficient method of screening material at the end of the selection. Thus, the production of infectious clones may be necessary. The graft resistance test assumes that if the materials to be tested did not develop the symptoms of RGD, they would be considered resistant to the disease. We focus on the root symptoms of the disease because they are more prominent. The two possibilities are envisaged if the material to be tested is taken as a graft and the RGD-sensitive material as a rootstock or vice versa. Grafting vegetable cultivars onto resistant rootstocks appears to have potential as a practical component of a systems approach for RKN control under several field conditions in many countries worldwide. When market-preferable resistant vegetable crops (yield, size, fruit type, conservation span, color) are unavailable, RKN-susceptible cultivars can be grafted onto RKN-resistant rootstocks (Noling 2019). Considering RKN infestation, growing system, and scion in management, as well as the use of appropriate rootstocks, the grafting method that uses, for example, tomato hybrids (*S. lycopersicum* L.) and interspecific tomato hybrids (*S. lycopersicum* x *S. habrochaites*) is widely utilized worldwide as RGD-resistant rootstocks in grafted tomato production, as it is the case in the USA and India (Gowda et al. 2017; Noling 2019). As for tomato and eggplant (Fig. 10.4), *Solanum torvum* was used as root stock to graft with scions of promising tomato varieties cv. Kashi Aman and Hissar Lalit. As grafted plants were highly compatible, they showed significant resistance by reducing soil RKN populations, reproduction, and gall index (Gowda et al. 2017). In addition, for organic production, Noling (2019) demonstrated that the hybrid rootstocks performed similarly and significantly reduced RGD compared to non-grafted tomato plants.

Fig. 10.4 *Solanum torvum*, a RKN-resistant wild eggplant germplasm (**a**), grafted with a tomato variety, Hissar Lalit (**b**), and another variety Kashi Aman (**c**)

10.11 Somaclonal Variation

Somaclonal variation (SCV) is known as a genetic variation present in plants regenerated from any form of cell/tissue cultures (Krishna et al. 2016) (Fig. 10.5). It can be caused by chromosome aberration and rearrangements, cytoplasmic genetic changes, spontaneous mutation induction, mitotic crossing over, de-amplification and amplification, transposable element activation, DNA methylation and demethylation, altered expression of multigene family, in vitro propagation method used, type and concentration of applied plant growth regulators, or number and duration of subcultures (Bairu et al. 2011; Kumar 2017).

SCVs give rise to the production of novel vegetable genotype variants, as it is observed in potato (largely propagated vegetatively), strawberry, tulip, chili pepper, garlic, soybean, carrot, several cereals, cotton, tea, coffee, banana, cocoa, grapevine, and sugarcane (Bairu et al. 2011; Singh 2013; Dita et al. 2018; Rajan and Singh 2021). Qualitative traits such as flower color, plant height, fruit shape, and flowering habit are highly improved in selected plant variants. SCV can be utilized as a non-conventional breeding method to improve and develop biotic and abiotic stresses-resistant and tolerant varieties (Yusnita et al. 2005), including RKNs. A potato genotype, SVP 53, has achieved new variants with increased yield and quality. Thus, SCV is an effective tool for selecting plant variants, also offering great scope for scientists involved in plant protection and breeding (Rajan and Singh 2021).

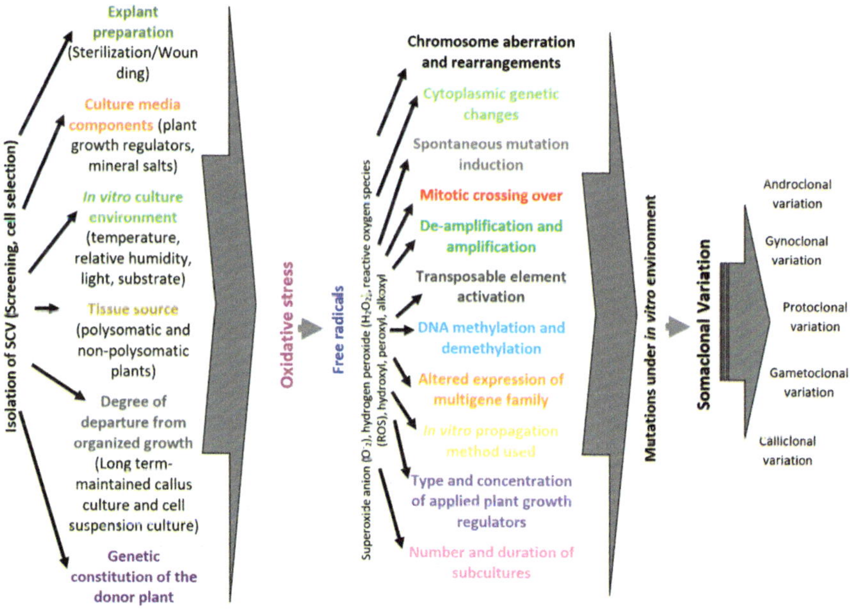

Fig. 10.5 Mechanism of SCV in micropropagated plants including vegetable crops, because of oxidative burst upon in vitro culture

10.12 Biotechnology and Genetic Engineering

The combination of traditional farming breeding methods and some molecular techniques that include genome-editing-based transgenic technology and omics-based analyses, for instance, is very important to develop vegetable crop cultivars with improved resistance to various RKNs (Kim and Yang 2019). Indeed, the genotyping-by-sequencing markers like JB-1, REX-1, pmi12, and *Mi*-23 were evaluated for screening *M. incognita*-resistance in tomato genotypes (Gowda et al. 2019). The Rex-1 CAPS and Rex-1 markers are also used to assay for the *Mi*-1 gene in tomatoes, especially the REX marker for introgressed genes from *Solanum habrochaites* (El Mehrach et al. 2005; Garcia et al. 2007). The codominant SCAR marker for the *Mi*-1.2 gene was found to be located within the *Mi*-1 locus (de Ilarduya et al. 2001; Garcia et al. 2007; Seah et al. 2007). Moreover, the isozyme marker, Aps-1, and DNA marker, Rex-1, were already primarily used in the past for the *Mi* gene (Gowda et al. 2019).

Analyzed by qRT-PCR, the resistant gene *SacMi* can be found in all wild eggplant tissues including leaf, stem, and root, with exactly the highest expression level in leaf tissue (Zhou et al. 2018). Lee et al. (2021) reported differential expression levels of genes involved in proteolysis and biotic stimuli in the uninfected control. The genes related to redox regulation, protease inhibitor, lipid and cell wall metabolism, and proteases were identified as genes conferring defense to plant against *M. incognita*. They concluded that the transcriptional changes in sweet potato genes occur during induced and constitutive defense responses against the RKN infection.

10.13 Conclusion

Vegetable crop varieties as resistance to RKN are available. However, they are not always commercially acceptable. This is due to poor agronomic and marketability characteristics (fruit size, plant productivity, fruit storage). Most resistant clones, as for tomatoes, provide adequate but not absolute protection to plant against RKNs. For example, some races of *M. incognita* can sometimes attack resistant cultivars of different vegetable crops. However, these breeding approaches described in this chapter may play a key role in developing HR in vegetable plants for adequate control of RKNs. Resistant genes such as *Mi*, *Me*, *Rk*, and *SacMi*, identified in different vegetables, are promising in breeding resistance against RKNs.

References

Abad P, Williamson VM (2010) Plant–nematode interaction: a sophisticated dialogue. In: Kader JC, Delseny M (eds) Advances in botanical research, vol 53. Academic Press, Elsevier Science, London, pp 147–192. https://doi.org/10.1016/S0065-2296(10)53005-2

Agrios GN (2005) Plant pathology, 5th ed. Department of Plant Pathology, University of Florida. Elsevier Academic Press, Burlington & California, London

Akunouchi T, Momota Y, Takada A, Nakamura Y, Kumagai T, Nakatani M (2013) Effect of seedling selection on the sweet potato cat resistance test conducted in the field. Breed Res 15(2): 32–35. https://doi.org/10.1270/jsbbr.15.32

Altieri MA, Funes-Monzote FR, Petersen P (2012) Agroecologically efficient agricultural systems for smallholder farmers: contributions to food sovereignty. Agron Sustain Dev 32:1–13. https://doi.org/10.1007/s13593-011-0065-6

Alves-Santos FM, Cordeiro-Rodrigues L, Sayagués JM, Martin-Dominguez R, Garcia-Benavides P, Crespo MC, Diaz-Minguez JM, Eslava AP (2002) Pathogenicity and race characterization of *Meloidogyne* spp. f. Sp. phaseoli isolates from Spain and Greece. Plant Pathol 51:605–611. https://doi.org/10.1046/j.1365-3059.2002.00745.x

Bairu MW, Aremu AO, Van Staden J (2011) Somaclonal variation in plants causes and detection methods. Plant Growth Regul 63:147–173. https://doi.org/10.1007/s10725-010-9554-x

Banerjee S, Banerjee A, Gill SS, Gupta OP, Dahuja A, Jain PK, Sirohi A (2017) RNA interference: a novel source of resistance to combat plant parasitic nematodes. Front Plant Sci 8:834. https://doi.org/10.3389/fpls.2017.00834

Banerjee S, Gill SS, Gawade BH, Jain PK, Subramaniam K, Sirohi A (2018) Host delivered RNAi of two cuticle collagen genes, Mi-col-1 and Lemmi-5 hampers structure and fecundity in Meloidogyne incognita. Front Plant Sci 8:2266. https://doi.org/10.3389/fpls.2017.02266

Beccari G, Covarelli L, Balmas V, Tosi L (2010) First report of Miscanthus × giganteus rhizome rot caused by *fusarium avenaceum*, *Meloidogyne* spp. and *Mucor hiemalis*. Aust Plant Dis Notes 5: 28–29. https://doi.org/10.1071/DN10011

Bernard GC, Egnin M, Bonsi C (2017a) The impact of plant-parasitic nematodes on agriculture and methods of control. In: Shah MM, Mahamood M (eds) Nematology concepts, characteristics and control. London, p 121. https://doi.org/10.5772/intechopen.68958

Bernard GC, Egnin M, Bonsi C, Mortley D, Witola WH, McElhenney W, Samuels S, Land C, Lawrence K (2017b) Evaluation of root-knot nematode resistance in sweetpotato. Afr J Agric Res 12(16):1411–1414. https://doi.org/10.5897/AJAR2017.12169

Bhavana P, Singh AK, Kumar R, Prajapati GK, Thamilarasi K, Manickam R, Maurya S, Choudhary JS (2019) Identification of resistance in tomato against root knot nematode (*Meloidogyne incognita*) and comparison of molecular markers for *Mi* gene. Australas Plant Path 48:93–100. https://doi.org/10.1007/s13313-018-0602-8

Bilichak A, Gaudet D, Laurie J (2020) Emerging genome engineering tools in crop research and breeding. In: Vaschetto L (eds) Cereal genomics. Methods in molecular biology 2072, pp 165–181. https://doi.org/10.1007/978-1-4939-9865-4_14

Boots M, Best A (2018) The evolution of constitutive and induced defences to infectious disease. Proc Biol Sci 285(1883):20180658. https://doi.org/10.1098/rspb.2018.0658

Bryant D (2005) Carrot research probes foreign cultivars. Western Farm Press. https://www.farmprogress.com/carrot-research-probes-foreign-cultivars-0

Chitwood DJ (2003) Research on plant-parasitic nematode biology conducted by the United States Department of Agriculture–Agricultural Research Service. Pest Manag Sci 59(6–7):748–753. https://doi.org/10.1002/ps.684

Choudhary DK, Prakash A, Johri BN (2007) Induced systemic resistance (ISR) in plants: mechanism of action. Indian J Microbiol 47(4):289–297. https://doi.org/10.1007/s12088-007-0054-2

Das S, DeMason DA, Ehlers JD, Close TJ, Roberts PA (2008) Histological characterization of root-knot nematode resistance in cowpea and its relation to reactive oxygen species modulation. J Exp Bot 59(6):1305–1313. https://doi.org/10.1093/jxb/ern036

de Ilarduya OM, Moore AE, Kaloshian I (2001) The tomato Rme1 locus is required for Mi-1-mediated resistance to root-knot nematodes and the potato aphid. Plant J 27(5):417–425. https://doi.org/10.1046/j.1365-313X.2001.01112.x

Dita M, Barquero M, Heck D, Mizubuti ESG, Staver CP (2018) Root gall disease of banana: current knowledge on epidemiology and research needs toward sustainable disease management. Front Plant Sci 9(1468):1–21. https://doi.org/10.3389/fpls.2018.01468

Edel-Hermann V, Lecomte C (2019) Current status of Meloidogyne spp. Formae speciales and races. Phytopathology 109:512–530. https://doi.org/10.1094/PHYTO-08-18-0320-RVW

El Mehrach K, Mejía L, Gharsallah-Couchane S, Salus MS, Martin CT, Hatimi A, Vidavski F, Williamson V, Maxwell DP (2005) PCR-based methods for tagging the Mi-1 locus for resistance to root-knot nematode in begomovirus-resistant tomato germplasm. Acta Hortic 695:263–270. https://doi.org/10.17660/ActaHortic.2005.695.29

Fuller VL, Lilley CJ, Urwin PE (2008) Nematode resistance. New Pathologist 180(1):27–44. https://doi.org/10.1111/j.1469-8137.2008.02508.x

Garcia BE, Mejia L, Salus MS, Martin CT, Seah S, Williamson VM, Maxwell DPA (2007) Co-dominant SCAR marker, Mi23, for detection of the Mi-1.2 gene for resistance to root-knot nematode in tomato germplasm. http://invirlab.plantpath.wisc.edu/GeminivirusResistantTomatoes/Markers/MAS-Protocols/Mi23-SCAR.pdf. Accessed 8 Oct 2022

Gawade BH, Sirohi A, Ganguly AK, Kansal R, Choudhary D, Koulagi R (2017) Effect of chickpea proteinase inhibitor on survival and parasitism of root knot nematode, *Meloidogyne incognita*. J Environ Biol 38:347–352

Gowda MT, Rai AB, Singh B (2017) Root knot nematode: a threat to vegetable production and its management. IIVR- technical bulletin no. 76, ICAR - Indian Institute of Vegetable Research, Varanasi (India)

Gowda MT, Sellaperumal C, Rai AB, Singh B (2019) Root knot nematodes menace in vegetable crops and their management in India: a review. Veg Sci 46(1–2):1–16

Gugino BK, Abawi GS, Ludwig JW (2006) Damage and management of *Meloidogyne hapla* using Oxamyl on carrot in New York. J Nematol 38(4):483–490. https://www.ncbi.nlm.nih.gov/pmc/articles/PMC2586463/

Hwang CF, Williamson VM (2003) Leucine-rich repeat-mediated intramolecular interactions in nematode recognition and cell death signaling by the tomato resistance protein Mi. Plant J Cell Mol Biol 34(5):585–593. https://doi.org/10.1046/j.1365-313x.2003.01749.x

Izuogu NB, Olajide TU, Eifediyi EK, Olajide CM (2019) Effect of root-knot nematode (*Meloidogyne incognita*) on the nodulation of some varieties of cowpea (*Vigna unguiculata* L. Walp). Sci Agric Bohem 50(2):104–109. https://doi.org/10.2478/sab-2019-0015

Jablonska B, Ammiraju JS, Bhattarai KK, Mantelin S, Martinez de Ilarduya O, Roberts PA, Kaloshian I (2007) The Mi-9 gene from Solanum arcanum conferring heat-stable resistance to root-knot nematodes is a homolog of Mi-1. Plant Physiol 143(2):1044–1054. https://doi.org/10.1104/pp.106.089615

Karssen G (2002) The plant-parasitic nematode genus Meloidogyne Goeldi, 1892 (Tylenchida) in Europe. Brill Academic Publishers, Boston, MA

Kayani MZ, Mukhtar T (2018) Reproductivity of Meloidogyne incognita on fifteen cucumber cultivars. Pak J Zool 50(5):1717–1722. https://doi.org/10.17582/journal.pjz/2018.50.5.1717.1722

Khan A, Tariq M, Asif M, Khan F, Ansari T, Siddiqui MA (2019) Integrated management of *Meloidogyne incognita* infecting Vigna radiata L. using biocontrol agent *Purpureocillium lilacinum*. Trends Appl Sci Res 14(2):119–124. https://scialert.net/abstract/?doi=tasr.2019.119.124

Kihika R, Murungi LK, Coyne D, Hassanali A, Teal PE, Torto B (2017) Parasitic nematode *Meloidogyne incognita* interactions with different *Capsicum annum* cultivars reveal the chemical constituents modulating root herbivory. Sci Rep 7:2903. https://doi.org/10.1038/s41598-017-02379-8

Kim YH, Yang JW (2019) Recent research on enhanced resistance to parasitic nematodes in sweetpotato. Plant Biotechnol Rep 13:559–566. https://doi.org/10.1007/s11816-019-00557-w

Koulagi R, Sirohi A (2015) Gene stacking through agrobacterium mediated co-transformation in tomato to engineer resistance against root-knot nematode, *Meloidogyne incognita* and tomato leaf curl virus. Indian J Nematol 45(2):161–168

Krishna H, Alizadeh M, Singh D, Singh U, Chauhan N, Eftekhari M, Sadh RK (2016) Somaclonal variations and their applications in horticultural crops improvement. 3 Biotech 6:54. https://doi.org/10.1007/s13205-016-0389-7

Kumar P (2017) Somaclonal variation: genetic basis and significance in crop improvement. Biotech articles. Available online at: https://www.biotecharticles.com/Agriculture-Article/Somaclonal-Variation-Genetic-basis-and-Significance-in-Crop-Improvement-4207.html. Accessed 9 Oct 2022

Kuti JO, Konuru HB (2005) Effects of genotype and cultivation environment on lycopene content in red-ripe tomatoes. Sci Food Agric 85(12):2021–2026. https://doi.org/10.1002/jsfa.2205

La Bonte DR, Wilson PW, Villordon AQ, Clark CA (2008) Evangeline. Sweet Potato HortSci 43(1):258. https://doi.org/10.21273/HORTSCI.43.1.258

Lee I-H, Kim HS, Nam KJ, Lee K-L, Yang J-W, Kwak S-S, Lee JJ, Shim D, Kim Y-H (2021) The defense response involved in Sweetpotato resistance to root-knot nematode *Meloidogyne incognita*: comparison of root transcriptomes of resistant and susceptible Sweetpotato cultivars with respect to induced and constitutive defense responses. Front Plant Sci 12:671–677. https://doi.org/10.3389/fpls.2021.671677

Lima FSO, Mattos VS, Silva ES, Carvalho, MAS, Teixeira RA, Silva JC, Correa VR (2018) Nematodes affecting potato and sustainable practices for their management. In: Potato - From Incas to all over the world. https://doi.org/10.5772/intechopen.73056

Lizardo RCM, Pinili MS, Diaz MGQ, Cumagun CJR (2022) Screening for resistance in selected tomato varieties against the root-knot nematode *Meloidogyne incognita* in The Philippines using a molecular marker and biochemical analysis. Plan Theory 11(10):1354. https://doi.org/10.3390/plants11101354

Mitkowski NA, Abawi GS (2011) Root-knot nematodes. Plant Health Instructor. https://doi.org/10.1094/PHI-I-2003-0917-01

Monteiro TSA, Pacheco PVM, Gouveia AS, Balbino HM, Grassi de Freitas L (2020) Pochonia. In: Amaresan N, Kumar MS, Annapurna K, Kumar K, Sankaranarayanan A (eds) Beneficial microbes in agro-ecology. Academic Press, pp 669–682. https://doi.org/10.1016/B978-0-12-823414-3.00033-2

Moon HS, Khan Z, Kim SG, Son SH, Kim YH (2010) Biological and structural mechanisms of disease development and resistance in chili pepper infected with the root-knot nematode. Plant Pathol J 26(2):149–153. https://doi.org/10.5423/PPJ.2010.26.2.149

Newton AC, Torrance L, Holden N, Toth IK, Cooke DE, Blok V, Gilroy EM (2012) Climate Change and Defense against Pathogens in Plants. Adv Applied Microbiol 81:89–132. https://doi.org/10.1016/B978-0-12-394382-8.00003-4

Noling JW (2019) Nematode management in tomatoes, peppers, and eggplant. University of Florida's Institute of Food and Agricultural Sciences Extension, USA, pp 16. https://edis.ifas.ufl.edu/publication/NG032

Oka Y, Koltai H, Bar-Eyal M, Mor M, Sharon E, Spiegel Y (2000) New strategies for the control of plant-parasitic nematodes. Pest Manag Sci 56(11):983–988. https://doi.org/10.1002/1526-4998(200011)56:11<983::AID-PS233>3.0.CO;2-X

Okada Y, Kobayashi A, Tabuchi H, Kuranouchi T (2017) Review of major sweet potato pests in Japan, with information on resistance breeding programs. Breed Sci 67(1):73–82. https://doi.org/10.1270/jsbbs.16145

Onkendi EM, Kariuki GM, Marais M, Moleleki LN (2014) The threat of root-knot nematodes (Meloidogyne spp.) in Africa: a review. Plant Pathol 63(4):727–737. https://doi.org/10.1111/ppa.12202

Osei MK, Annor B, Adjebeng-Danquah J, Danquah A, Danquah E, Blay E, Hans Adu-Dapaah H (2018) Genotype × Environment interaction: a prerequisite for tomato variety development. In:

Nyaku ST, Danquah A (eds) Recent advances in tomato breeding and production. https://doi.org/10.5772/intechopen.76011

Pegard A, Brizzard G, Fazari A, Soucaze O, Abad P, Djian-Caporalino C (2005) Histological characterization of resistance to different root-knot nematode species related to phenolics accumulation in *Capsicum annuum*. Phytopathology 95(2):158–165. https://doi.org/10.1094/PHYTO-95-0158

Perry RN, Moens M, Starr FJ (eds) (2009) Root-knot nematodes. CAB International, Wallingford

Prasad KSK (2008) Management of potato nematodes: an overview. J Horticult Sci 3(2):89–106. https://jhs.iihr.res.in/index.php/jhs/article/view/567

Råberg L (2014) How to live with the enemy: understanding tolerance to parasites. PLoS Biol 12(11):e1001989. https://doi.org/10.1371/journal.pbio.1001989

Rai N, Singh S, Rai AB, Singh RK (2010) Resistance response of tomato genotypes against root-knot nematode, *Meloidogyne incognita*. Indian J Nematol 40(2):237–239

Rajan RP, Singh G (2021) A review on application of somaclonal variation in important horticulture crops. Plant Cell Biotechnol Mol Biol 22(35–36):161–175. https://www.ikppress.org/index.php/PCBMB/article/view/6319

Ramzan M, Ahmed RZ, Khanum TA, Akram S, Jabeen S (2021) Survey of root knot nematodes and RMi resistance to *Meloidogyne incognita* in soybean from Khyber Pakhtunkhwa, Pakistan. Eur J Plant Pathol 160:1–13. https://doi.org/10.1007/s10658-019-01740-z

Reddy YS, Sellaperumal C, Prasanna HC, Yadav A, Kashyap SP, Singh S, Rai N, Singh M, Singh B (2018) Screening of tomato genotypes against root-knot nematode and validation of *Mi* 1 gene linked markers. Proc Natl Acad Sci India Sect B: Biol Sci 88(1):65–72. https://doi.org/10.1007/s40011-016-0731-1

Reddy YS, Sellaperumal C, Gowda MT, Mishara P, Tiwari SK, Pandey CD, Pandey S (2019) Characterization of tomato germplasm for root-knot nematode resistance with the help of Mi23 marker. Indian J Nematol 49(2):131–138

Sato K, Kadota Y, Shirasu K (2019) Plant immune responses to parasitic nematodes. Front Plant Sci 10:1165. https://doi.org/10.3389/fpls.2019.01165

Saucet SB, Ghelder CV, Abad P, Duval H, Esmenjaud D (2016) Resistance to root-knot nematodes *Meloidogyne* spp. in woody plants. New Pathologist 211(1):41–56. https://doi.org/10.1111/nph.13933

Seah S, Telleen AC, Williamson VM (2007) Introgressed and endogenous Mi-1 gene clusters in tomato differ by complex rearrangements in flanking sequences and show sequence exchange and diversifying selection among homologues. TAG Theoretical Appl Genet Theoretische und angewandte Genetik 114(7):1289–1302. https://doi.org/10.1007/s00122-007-0519-z

Seid A, Fininsa C, Mekete TM, Decraemer W, Wesemael WML (2017) Resistance screening of breeding lines and commercial tomato cultivars for *Meloidogyne incognita* and *M. javanica* populations (Nematoda) from Ethiopia. Euphytica 213(97):1–15. https://doi.org/10.1007/s10681-017-1886-4

Seo Y, Park J, Kim YS, Park Y, Kim YH (2014) Screening and histopathological characterization of Korean carrot lines for resistance to the root-knot nematode Meloidogyne incognita. Plant Pathol J 30(1):75–81. https://doi.org/10.5423/PPJ.OA.08.2013.0082

Seo Y, Kim YS, Park Y, Kim YH (2015) Comparisons of pathological responses in carrot to root-knot nematodes. Plant Pathol J 31(4):441–445. https://doi.org/10.5423/PPJ.NT.06.2015.0115

Shivakumara TN, Chaudhary S, Kamaraju D, Dutta TK, Papolu PK, Banakar P, Sreevathsa R, Singh B, Manjaiah KM, Rao U (2017) Host-induced silencing of two pharyngeal gland genes conferred transcriptional alteration of cell wall-modifying enzymes of *Meloidogyne incognita* vis-à-vis perturbed nematode infectivity in eggplant. Front Plant Sci 8:473. https://doi.org/10.3389/fpls.2017.00473

Shukuru BN, Archana TS (2023) Biological control of the nematode parasites of crops. In: Kaur T, Sharma S, Sharma A, Singh B (eds) Biological control for plant protection. Nova Science Publishers, New York. [Accepted]

Shukuru BN, Archana TS, Bisimwa EB, Birindwa DR, Sharma S, Kurian JA, Casinga CM (2022) Screening of cultivars against cassava brown streak disease and molecular identification of the phytopathogenic infection-associated viruses. Arch Phytopathol Plant Protect 55(16): 1899–1929. https://doi.org/10.1080/03235408.2022.2123590

Sikora RA, Fernandez E (2005) Nematode parasites of vegetables. In: Luc M, Sikora RA, Bridge J (eds) Plant parasitic nematodes in tropical and subtropical agriculture. Centre for Agriculture and Bioscience International, Oxfordshire, pp 319–392

Singh S (2013) Integrated approach for the management of the root-knot nematode, *Meloidogyne incognita*, on eggplant under field conditions. Nematology 15(6):747–757. https://doi.org/10.1163/15685411-00002715

Singh S, Singh B, Singh AP (2015) Nematodes: a threat to sustainability of agriculture. Procedia Environ Sci 29:215–216. https://doi.org/10.1016/j.proenv.2015.07.270

Singh DP, Singh AK, Singh A (2021) Breeding for resistance to biotic stresses. In: Singh DP, Singh AK, Singh A (eds) Plant breeding and cultivar development. Academic Press, pp 425–464. https://doi.org/10.1016/B978-0-12-817563-7.00010-6

Stirling GR (2014) Biological control of plant-parasitic nematodes, 2nd edn. CAB International, Wallingford

Suresh P, Poornima K, Sivakumar M, Subramania S (2017) Current status of root knot nematodes (*Meloidogyne* spp.) in Tamil Nadu. J Entomol Zool Stud 5(6):610–615

Waceke JW (2007) Plant parasitic nematodes associated with cabbages in Kenya. Afr Crop Sci Conf Proc 8:1071–1074

Wendimu GY (2021) Biology, taxonomy, and management of the root-knot nematode (*Meloidogyne incognita*) in sweet potato. Adv Agric 2021:8820211. https://doi.org/10.1155/2021/8820211

Yadav BC, Veluthambi K, Subramaniam K (2006) Host-generated double stranded RNA induces RNAi in plant-parasitic nematodes and protects the host from infection. Mol Biochem Parasitol 148(2):219–222. https://doi.org/10.1016/j.molbiopara.2006.03.013

Ye W, Robbins RT, Kirkpatrick T (2019) Molecular characterization of root-knot nematodes (*Meloidogyne* spp.) from Arkansas, USA. Sci Rep 9:15680. https://doi.org/10.1038/s41598-019-52118-4

Yencho GC, Pecota KV, Schultheis JR, VanEsbroeck Z-P, Holmes GJ, Little BE, Thornton AC, Truong VD (2008) 'Covington' Sweetpotato. Hort Sci 43(6):911–1914. https://doi.org/10.21273/HORTSCI.43.6.1911

Yusnita Y, Widodo W, Sudarsono S (2005) *In vitro* selection of peanut somatic embryos on medium containing culture filtrate of *Sclerotium rolfsii* and plantlet regeneration. HAYATI J Biosci 12(2):50–56. https://doi.org/10.4308/hjb.12.2.50

Zhou X, Liu J, Bao S, Yang Y, Zhuang Y (2018) Molecular cloning and characterization of a wild eggplant *Solanum aculeatissimum* NBS-LRR gene, involved in plant resistance to *Meloidogyne incognita*. Int J Mol Sci 19(2):583. https://doi.org/10.3390/ijms19020583

An Overview of Predacious Fungi for the Management of Root-Knot Disease in Vegetables

11

Vandana Sahu, Ashwani Kumar Patel, and Shiv Shankar Patel

Abstract

Plant parasitic nematodes (PPN) are ubiquitous in agricultural soils. They damage a range of vegetables as well as other agricultural crops worldwide. Some predaceous fungi, which act as nematode's natural enemies, are one of the best pest management remedies. Some of these microbes create traps, resulting in the eelworms getting trapped and killed. Other predacious fungi behave as parasites inside the nematodes, producing poisons and virulence components that kill the nematodes internally. In order to develop powerful biological control agents against nematodes, it is crucial to understand the underlying principles of microbe-nematode interactions. In addition to focusing on the methods by which predaceous fungi infect worms and the nematode defence against dangerous infections, this book chapter reviews recent developments in our understanding of the interactions between nematodes and predaceous fungi. This chapter comprises important topics for more research and development, including prospective plans for applying our most recent findings to create efficient biocontrol methods for managing root-knot diseases of vegetables.

Keywords

Predacious fungi · Root-knot nematodes · Biological control

V. Sahu (✉) · A. K. Patel · S. S. Patel
Department of Mycology and Plant Pathology, Institute of Agricultural Science, Banaras Hindu University, Varanasi, Uttar Pradesh, India

F. Ahmad, G. Nombela (eds.), *Root-Galling Disease of Vegetable Plants*,
https://doi.org/10.1007/978-981-99-3892-6_11

11.1 Introduction

Vegetables are an excellent source of vitamins and minerals (Neeraj et al. 2017). Therefore, different vegetables are grown to meet the daily requirements of vitamins and minerals. Based on climatic and geographical conditions, various vegetables are grown in India. India, which represents a rich botanical biodiversity, is considered the home of many types of vegetables. Many biotic and abiotic pressures threaten vegetable production. Among biotic stress, pests and plant pathogens adversely affect vegetable production (Chakraborty and Newton 2011). Soil-borne diseases, including plant pathogens, are key bottlenecks hindering vegetable output. Plant parasitic nematodes cause crop losses of up to 21.3%, amounting to INR102,039.79 million (US$1.58 billion) per year (Kumar et al. 2020). An estimated yield loss due to plant parasitic nematodes worldwide was 12.3% ($157 million); US$40.3 million was reported from India (Singh et al. 2015). Plant nematodes are thought to do greater harm than invasive insects, about US$70 billion (Bradshaw et al. 2016).

Among various species of PPN, *Meloidogyne*, *Heterodera*, *Pratylenchus*, *Tylenchorhynchus*., *Ditylenchus*, *Helicotylenchus*, *Hoplolaimus*, *Rotylenchus*, and *Radopholus* are known to cause a reduction in vegetable production to a great extent that varies with crop susceptibility, nematode species, inoculum density, and environment (Taylor and Sasser 1978). Among nematodes, the root-knot nematode (RKN) is considered to be one of the predominant plant parasitic nematodes that cause root-knot disease (galls) in plants, irrespective of their botanical families, because of its polyphagous nature and ability to thrive in harsh situations. Therefore, root-knot disease results in huge losses worldwide (Sasser et al. 1987). This parasitic nematode comprises 98 species identified so far (Jones et al. 2013). It is a sedentary obligate endoparasite prevalent in tropical and subtropical conditions. In 1855, Berkeley was the first Scientist to report an outbreak of clubroot in British greenhouse-grown cucumbers. Chitwood (1949) maintained *Meloidogyne* species in a micro plot. They are more frequent in sandy and sandy loam soils (Kim et al. 2017). It is necessary to manage the nematodes eco-friendly by predators, parasites, and pathogens of nematodes in the ground in the form of biological control agents. Many species of organisms from all spheres of life, including archaea, bacteria, fungi, protists, mammals, and plants, thrive in soil ecology. Fungi are the most versatile and diverse organisms in their morphology, life cycle, and ecology. Large numbers of fungi are known to kill PPN. However, only a few are important and potential sources of biological control, understanding how a pathogen and biological control agent interact in soil or an infectious environment like the rhizosphere is essential for determining how effective a biological control agent will be (Paulitz 2000).

Several microbial pathogens are effective against nematodes. Nematophagous fungi (NPF) are diverse microorganisms that consume nematodes under unfavourable nutritional environments. These nematophagous fungi are nematode parasites since they possess various structural tools in their body and mechanism. According to their nematode-predation features, these NPF are divided into three groups: (1) predatorial/nematode-trapping, (2) ovicidal, and (3) endoparasitic. This chapter summarizes the characteristics of predators, which create modified hyphae known as traps with which they bind and digest nematode larvae through a mechanical/enzymatic process. The predacious fungi can be a biological control agent that can potentially reduce the nematode population as it is widely distributed in soil. Predatory fungus catches and eats tiny or other small animals to obtain some or all of their resources. Within the substrate, predatory fungus grows enormous hyphal systems. Nematodes are attracted to and imprisoned by glue sticks, nets, or clamping rings that the mycelium generated by predacious fungi.

11.2 Symptoms of Root-Knot Diseases

The disease is initially characterized by the formation of endogenously galls/knots on the root system, which is endogenously formed. At the late stage of disease development, symptoms appear above the root systems. They appear in the form of yellowing, stunting, and wilting while under severe conditions leading to the death of the plants because the translocation of nutrients and water uptake to different parts of the plant by galled roots is limited compared to healthy plants. Plants also show nutrient deficiency symptoms due to their condensed capacity to absorb water and nutrients from the soil. If the density of the nematode population increases near the beginning stages of plant development, such plants can die. Some of the vegetable crops affected by the root-knot nematode are represented in Fig.11.1.

11.3 Farmer Views on Nematodes

PPN is a hidden enemy of farmers as they do not produce dramatic symptoms on plants as produced by other plant pathogens such as fungi, bacteria, viruses, viroid, phytoplasma, etc. They are one the most notorious plant pathogens known to cause considerable losses to agricultural crop production worldwide (Jones et al. 2013; Siddique and Grundler 2018). These small soil-borne pathogens, known as nematodes, can harm every part of a plant, including its roots, stems, leaves, blossoms, and seeds.

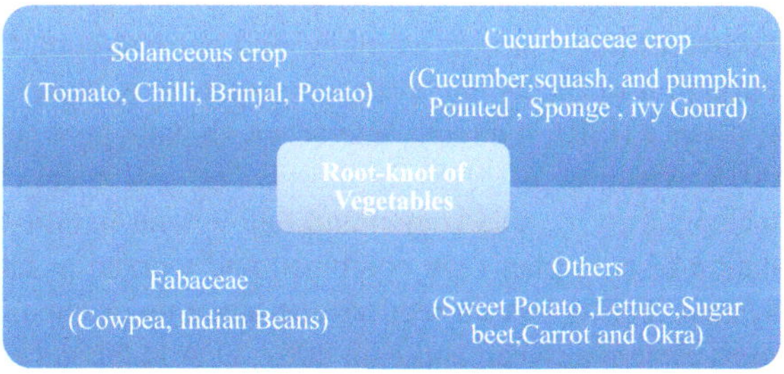

Fig. 11.1 Vegetables affected by root-knot nematodes

11.4 Mechanism of Feeding Nematodes

A protruding stylet is required for feeding from plant cells by plant parasitic nematodes. For feeding, they require a protruding stylet to enter the plant cells. Three to five pharyngeal glands are attached to the stylet and produce effector chemicals that are frequently released, allowing for dissemination, internal moving, and parasitism (Jones et al. 2013; Mejias et al. 2019). Depending on the feeding behaviour of nematodes, they can be classified as endoparasitic or ectoparasitic.

11.5 *Meloidogyne* spp.

From an economic standpoint, this is the most significant obligatory parasite of plant roots, and it may parasitize over 3000 different types of plants. When root-knot nematodes (RKN) are present alongside other diseases, such as *M. arenaria*, *M. hapla*, *M. incognita*, and *M. javanica*, the losses caused by RKN to crops can be estimated to be up to 87–100%. The yield is frequently reduced by 10–20% after the commencement of RKN diseases, and the severity is greater than 75%. RKN scan harms over 50% of greenhouse crops in China and results in yearly economic losses of about 400 million dollars. The root-knot disease of tomato is a widespread disease caused by *Meloidogyne incognita.* Figure 11.2 depicts the root-knot of tomatoes with their symptoms and pathogen.

11.6 Management Aspect

Nematicides can be applied as one of the various techniques for nematode control in agriculture. However, due to European Union Law Legislation (EC No. 1107/2009), which has increased the necessity to deploy effective nematode resistance measures, pesticides are detrimental to human health and pollute the environment (Zhang et al. 2014, 2017). Adopting biocontrol methods as a safer and more effective means of eradicating plant parasitic nematodes is advised. Biological control is using organisms to reduce the population density or impact of a particular pest organism, making it less plentiful or harmful than it would otherwise be. Nematode biological control explicitly refers to the modulation of nematode populations and/or decrease of nematode damage by the activity of antagonists against them, occurring naturally or through manipulating the environment or introducing antagonists. Via antagonistic interactions (such as antibiotic and nutrient competition) or indirect interactions through host plants like systemic acquired resistance (SAR) and induced systemic resistance (ISR), it directly interacts with pathogens (Pal and gardener 2006; Stirling 2018; Xiang et al. 2018). In most cases, biological control agent (BCA); physical techniques, viz., solarization and fallow; and traditional methods such as the alternation of crop plants are combined and worked as an integrated pest management system. This approach reduced the amounts of chemicals and was found to be the most effective (Khan and Kim 2007; D'Addabbo et al. 2019).

Fig. 11.2 Depicting root knot of tomato along with symptoms and *Meloidogyne incognita* (**a**) Root knot of tomato in Pot; (**b**) Symptoms with whole plants; (**c**) Galls on root (**d**) Female of *Meloidogyne inognita;* (**e**) Egg masses and (**f**) Egg of Meloidogyne

11.7 Background of Predacious Fungi

The essence of predacious fungi in managing root-knot of vegetables has been initiated by testing fungi with baits of nematodes. Various predacious fungi have been isolated and identified based on the morphology and molecular-biology characterization (Table 11.1). Linford and Yap (1939) first tried to use fungi as predators to manage RKNs in Hawaii. Then, Linford et al. (1938) demonstrated that there were noticeable decreases in root-knot nematode numbers with the degradation of plant components combined with contaminated soil.

11.8 Nematode-Destroying Fungi

It comprises more than 200 species of taxonomically varied fungi, which may consume living nematodes (eggs, adults, and juveniles) as food. The fungus differs in its saprophytic/parasitic capabilities. The fungal mycelium's development stage at which it may capture nematodes is related to this ability. Ingenious hyphal

Table 11.1 Depicting the phylum and genera of predacious fungi

Phylum	Genera
Zygomycota	*Cystopage, Stylopage, Rhopalomyces*
Ascomycota and anamorph fungi	*Arthrobotrys, Dactylaria, Dactylella, Monacrosporium*
Basidiomycota	*Hohenbuehelia, Hyphoderma, Nematoctonus, Pleurotus*
Chytridiomycota	*Catenaria, Endochytrium, Olpidium, Rhizophydium*
Zygomycota	*Rhopalomyces, Brachymyces, Zoophagus*
Oomycota	*Atkinsiella, Lagenidium, Sommerstorffia, Haptoglossa*
Anamorph fungi	*Rotiferophthora, Harposporium, Haptospora, Pseudomeria, Lecophagus, Cephaliophora, Dwayaangam, Medusamyces, Tolypocladium, Culicinomyces, Tractatus*

Table 11.2 Typical infection structures of some nematophagous fungi

Infection structure	Species	Taxonomic classification
Adhesive nets	*Arthrobotrys oligospora A. conoides A. musiformis A. superba Duddingtonia flagrans*	Ascomycota; Orbiliales
Adhesive branch	*Monacrosporium gephyropagum*	Ascomycota; Orbiliales
Adhesive knobs	*M. ellipsosporum, M. haptotylum*	Ascomycota; Orbiliales
Constricting rings	*A. dactyloides, A. brochopaga*	Ascomycota; Orbiliales
Adhesive knobs and adhesive spores	*Nematoctonus concurrens*	Basidiomycota; Agaricales
Adhesive spores	*N. leiosporus, Drechmeria coniospora, Hirsutella rhossiliensis*	Ascomycota; Hypocreales
Ingested spores	*Harposporium anguillulae*	Ascomycota; Hypocreales
Zoospores	*Catenaria anguillulae, Haptoglossa dickii*	Chytridiomycota; Blastocladiales Oomycota; Haptoglossales
Adhesive hyphae	*Stylopage hadra, Cystopage cladospora*	Zygomycota; Zoopagales
Toxic droplets	*Pleurotus ostreatus*	Basidiomycota; Agaricales
Appressoria	*Pochonia chlamydosporia, Paecilomyces lilacinus*	Ascomycota; Hypocreales

structures, such as hyphae, nubs, branches, or rings to which nematodes attach or are mechanically trapped, have been evolved by scavenger (predatory) fungi (Table 11.2). Various nematode-destroying fungi (NPF) have not yet been found, and 6000–8000 species await recognition (Li et al. 2000; McInnes 2003; Yang et al. 2012). NPF was found in various soils and the rhizosphere (Liu et al. 2009). For

NPF, there are five categories: to effect or kill PPNs, nematode-affecting toxins are secreted by (A) nematode-trapping fungi, (B) endoparasitic fungi, (C) nematode-affecting poisons, (D) fungi parasitizing eggs, and (E) fungi that cause plant resistance and defences (Swe et al. 2011; Maia Filho et al. 2013). The derived metabolites of certain NPF have proven exceptional efficiency in the treatment of parasitic worms (Castañeda-Ramírez et al. 2020; Seong et al. 2021). This overview confers NPF involvement in organic farming and their management strategy for plant-parasitic nematodes.

11.9 Predators

As promising biological agents for controlling PPNs, the predatory fungus is used in experiments as a single fungus often added to organically treated soil. Some commercial preparations have been offered in preliminary testing, but these items were never used, mainly due to uneven performance and quality control issues. But nowadays, various experiments have been successful due to the proper use of the virulent culture of predacious fungi. Royal 350, a related product containing *Arthrobotrys superba* Corda, provided adequate control of the root-knot nematode on tomatoes as long as it was used when nematode numbers were high. A commercial version of *Arthrobotryisr obusta* called "Royal 300" increased yields of the farmed fungus *Agaricus bisporus* while *Rotylenchus myceliophagus* populations declined (Cayrol 1983; Cayrol and Frankowski 1979). In various field soils and nematode-potted cultures, *Monacrosporiumellipso sporium* was commonly seen in conjunction with *Meloidogyne* egg masses (Mankau and Wu 1984). In the field trial, tomato seedlings were transplanted into a substrate with two levels of fungus on wheat grain; the amount of fungus used directly correlated with the increased plant development and *M. incognita* reduction at harvest.

11.10 Endoparasitic Fungi

Only a few endoparasitic fungi's host ranges have been identified, but in general, these fungi were not much more specialized than those that generate traps (Birchfield 1960; Esser 1976; Esser and Ridings 1973). Because of its nematode-attracting solid abilities and the known specificity of conidial adherence to nematodes, *Meriaconiospora* was utilized in biological control studies (Jansson 1982a; Jansson and Nordbring-Hertz 1983; Jansson et al. 1984). In greenhouse pot trials, *M. coniospora* greatly decreased tomato galling caused by *Meloidogyne* spp., a root-knot worm (Jansson et al. 1985). It has been proposed that the endoparasitic fungus *Hirsutella* species could be a valuable organism for the biological control of plant parasitic nematodes (Stiirhan and Schneider 1980). Nematodes primarily provide the sustenance of the endoparasitic fungi, producing little mycelial development outside the host. Few attempts have been made to use endoparasitic fungi for nematode control due to the challenges involved in growing them. *Nematoctonus concurrens* conidia counts in sterile sand were lowered by *Dreschler* and *N. haptocladus*.

11.11 Parasitic Fungi on Eggs of Nematodes

Under various climatic and soil environmental circumstances, *Paecilomyces lilacinus* effectiveness and adaptability in suppressing several harmful nematodes have been examined (Candanedo et al. 1983; Davide and Zorilla 1983; Noe and Sasser 1984; de Sisler et al. 1985; Roman and Rodriguez-Marcano 1985; Cabanillas and Barker 1989). According to various accounts, *Paecilomyces lilacinus* and *P. variotii* can be detected in the eggs of *Meloidogyne arenaria* and *M. incognita* in North America and Peru and the cysts of *Globodera* and *Heterodera* (Dowsett and Reid 1977, 1979; Friman et al. 1985). It is widespread in many plant rhizospheres and generates leucinostatin and lilacin antibiotics (Samson 1974). In Peru, *P. lilacinus*was found to be parasitizing RKN, *M. incognita* egg masses (Jatala et al. 1979). In adult *Meloidogyne* females, penetration typically occurs through the anus or vulva. The fungus was discovered to have destroyed 80 to 90% of the nematode eggs it had infected. According to Jatala et al. (1980), *P. lilacinus* parasitized the egg of *M. incognita* and has the ability to control *M. incognita* on potatoes in the field. In tomato and okra, *P. lilacinus* was infected, and *M. incognita* was much under control (Noe and Sasser 1984). In the eggs of *M. arenaria*, *P. lilacinus* was identified in substantial numbers (Morgan-Jones et al. 1984). Cabanillas et al. (1988) observed no galls forming in tomato roots transplanted with nematode eggs inoculated with *P. lilacinus*.

11.12 Mechanism of Antagonism

A heterogonous group, predacious soil fungi, is a natural adversary of parasitic nematodes. Carbon, nitrogen, other vital components, and nematode biomass are crucial to these predacious fungi (Siddiqui and Mahmood 1996). While certain nematode parasites are required, commonly, they are facultative saprophytes (Lopez-Llorca et al. 2007). The NPF is being researched for possible use as a biotic deterrent to root-knot nematode (RKN). The capability towards the outbreak of nematode in various phases, including young, adult, and eggs, has been demonstrated by additional 200 species of taxonomically diverse fungi (Nordbring-Hertz et al. 2006). The two unique defences against fungal invasion are provided by nematode morphology. The initial barrier is the eggshell, which is made up of three layers: an inner lipoprotein layer, an outer vitelline layer comprised mostly of proteins, and an outer chitin layer. Eggshells are found in root-knot and cyst nematodes. The second barrier is the cuticle. This barrier thickness varies widely depending on the nematode species (Morton et al. 2004). The three primary fungi infection methods for nematodes are parasitism, harmful substances, and enzymes.

11.13 Parasitism

Fungi that live on or inside their host organisms and gain sustenance from them are parasitic nematode fungi. The nematode-trapping, endoparasitic, and egg- and female-parasitic fungi can all be classified under this category (Abd-Elgawad and Askary 2018). They developed specialized structures in their mycelium to capture nematodes, nematode-trapping fungus transition into their parasitic stage. The fungi can enter the nematode through constructions, which perform as 2D or 3D constrictor rings and sticky nets and exploit it as a new basis for nutrition (Lopez-Llorca et al. 2007). Nematodes cuticle is damaged by the traps made by the mycelium of the fungus. The hyphae spread throughout the interior of the worm body and create a penetration peg, and at the last stage, the hyphae project through the nematode's shell (Soares et al. 2018). In greenhouse environments, the *M. javanica* has been controlled by the fungus *Arthrobotrys oligospora*, which traps nematodes by creating a specific penetration tube to pierce their cuticle (Mostafanezhad et al. 2014).

Endoparasitic organisms are different; instead of developing specialized structures to infect worms, they produce spores (conidia or zoospores). Most of this group are obligatory nematode parasites that grow entirely inside the nematode (Lopez-Llorca et al. 2007). When fungus generates their spores, the nematode becomes infected, as with *Harposporium* spp., or they adhere toward the cuticle and then inoculate contents within nematode, with *Drechmeri aconiospora* (Morton et al. 2004). Zoospore-producing fungi, such as *Pythium caudatum*, lead to the encrustation of openings, viz., its mouth, anus, and vulva of nematode, as the spores are lured to the secretions and swim towards them. When zoospores germinate, they become immobile, form a hyphal penetration tube, and enter the nematode through the body opening (Kim 2015). Fungal hyphae develop specialized appressoria, compressed, expanded mycelial ends, cling toward shells, and then ease diffusion to eggs as they go towards the nematode egg (Nordbring-Hertz et al. 2006). These infested shells expand and swell as per the diffusion endures. The NPF may devour their contents to obtain nutrition and energy to keep growing (Kim 2015). Using the fungus *Trichoderma harzianum* as a usual nematode control showed a noticeable decrease in *M. incognita* in tomatoes, illustrating this category (Feyisa and Lencho 2015).

11.14 Toxic Compounds

Certain chemical compounds produced by some NPF species are poisonous to worms which paralyze the nematode (Satou et al. 2008). However, most research on NPF has been on endoparasitic and predatory moulds (Soares et al. 2018). Mostly, fungi belonging to basidiomycetes produce toxins. In addition, certain fungi also yielded complexes that stand poisonous to nematode but not to fungi, as they cannot infect the worm (Soares et al. 2018). NPF possess various chemical compounds, viz., simple fatty acids, other natural acids, pyrones, lactones, benzo quinones, anthraquinones, furans, alkaloids, cyclodepsi peptides, peptaibiotics, and hybrid structures like lactam-bearing macro lactones (Degenkolb and Vilcinskas 2016a). The publications of Degenkolb and Vilcinskas (2016a, b) provide an outstanding survey of arsenal weapons producing lethal chemical-producing fungi and their metabolites.

11.15 Enzymes

Specific enzymes are shared by almost all five families of nematophagous fungi, which are crucial for nematode killing and digestion (Braga and de Araújo 2013; Soares et al. 2018). These large molecules can catalyze reactions in living things. As a result, enzyme activity speeds up the responses. Nematodes are shielded from the acts of natural predators by physical barriers. One of these obstacles is the cuticle of immature nematode (Lee 1967; Ekino et al. 2017). Proteins are present in great abundance throughout their makeup. NF have mechanical and enzymatic methods to get over this obstacle. Proteases, a mainly neutral serine protease, and an alkaline serine protease remain macromolecules involved in cuticle absorption. The protease enzymes hydrolyzed the peptide bonds of cuticular proteins (Liang et al. 2010). *Lecanicilliump salliotae* (also known as *Verticillium psalliotae*) produces an alkaline serine protease that causes cuticles to break down in periods and immobilizes nematode (Yang et al. 2005). Neutral serine protease generated by *Arthrobotrys oligospora* is involved in nematode pathogenicity (Zhao et al. 2004). By developing serine proteases, *Arthrobotrys oligospora* is useful for in vitro regulation of *Haemonchuscontortus* and *Caenorhabditis elegans* (Junwei et al. 2013; Yang et al. 2022). As a result, they play a critical part in the fungus infection process. PPN eggs have chitin and protein-rich eggshells. Exochitinase and endochitinase are two types of chitinases that catalyze the hydrolysis of glycosidic connections about the N acetylglucosamine (EC 3.2.1.14) (Tikhonov et al. 2002). Hence, chitinase is the major mycological enzyme involved in infection and shell destruction (Khan et al. 2004). Chitinases generated by NF, *Monacrosporium thaumasium*, exhibited nematicidal activity against the worm *Panagrellus redivivus* (Soares et al. 2014). Enzymes have shown nematicidal effects when chitinase is used alone, without fungi, and in conjunction with the physical processes of NF infection and digestion (Soares et al. 2012; Braga et al. 2015). This makes it possible for novel PPN control strategies to be developed.

11.16 Special Attack Strategies

Some nematophagous species create unique strategies that they use to combat nematodes. The tool is comparable to those employed by nematophagous fungi to damage nematode cuticles before completing the attack on the nematode. Strategies come in various shapes, including spears, swords, and rackets with thorns (Soares et al. 2018). The processes for employing these strategies in an attack can be broken down into three stages: (A) the nematode is being pressed as the hyphae grow in its direction. (B) The formation of the penetration peg uses to break through the nematode cuticle. (C) The nematode body will be completely covered with nematophagous fungi via hyphae (Luo et al. 2004, 2006, 2007).

11.17 Biological Control and Commercial Product

Due to the escalating expenses associated with pesticide testing and certification, developing novel nematicides has practically stopped, forcing the development of additional non-chemical management techniques. Realistic nematode management in the future will increasingly rely on biological control. Hence, over the past few years, natural regulation has changed.

11.17.1 Biological Control of *Meloidogyne* Species Using Fungi

Several fungi naturally function against *Meloidogyne* (Viaene and Abawi 1998; Duponnois et al. 1998; Stirling et al. 1998; Stirling and Smith 1998; Kumar and Singh 2006; Thakur and Devi 2007; Liu et al. 2008; Collange et al. 2011). Cuticle penetration, nematode immobilization, invasion, and digesting comprise the typical infection process (Huang et al. 2004). The cuticles of nematodes and the walls of their eggs are crucial in the fungal invasion. Chitin, collagen, and fibers comprise most of the cuticle, which may be a precursor to the nematophagous fungus that infects worms (Huang et al. 2004). The contact between a fungus and a host organism or prey needs penetration of the body's outer shell regardless of the fungus' mode of activity, whether predation or parasitism. This penetration occurs before the fungus colonizes the body's internal tissues through digestion, allowing it to achieve its nutritional requirements (Gaspard et al. 1990; Hajieghrari et al. 2008).

11.17.2 Toxins of Microbes

Natural repellents, nematostatics, and nematicides which stop nematode eggs or larvae from maturing, or a combination of these, can be found in some substances (sensusstricto). Several reports suggest that some fungi-made enzymes, such as 6-pentyl-pyrone produced by *Trichoderma harzianum*, may harm nematodes, including RKN (Sarhy-Bagnon et al. 2000).

11.17.3 Production of Biological Control Agents

Production of predatory fungi can be done in either solid-state fermentation methodology (SSF) or liquid-state fermentation methodology (LSF). The desired final composition—liquid or wettable powder—determines the approach to use. It also relies on the workload and the price of production.

11.17.4 Production by Liquid-State Fermentation

Large, agitated, temperature-controlled, and aerated tanks are used for liquid media fermentation, which is ideal for growing some fungi as such *Fusarium venenatum*, which is used to generate Qorn®, can be produced by LSF using two microbes, *P. penetrans* and *B. thuringiensis*.

11.17.5 Production by Solid-State Fermentation

Solid-state fermentation is usually understood to be the growth of microbes (ideally filiform fungus) on a compact surface without fluid movement (Hesseltine 1987; Mitchell et al. 2002; Barrios-González 2012). It uses these microorganism's growth and metabolism to break down solid substrates and create biopesticides (biomass and secondary metabolites) (biomass and secondary metabolites). Microbes develop outside and inside the compact ground without any liquid flow. A natural substrate that can absorb nutrients contained in a melted condition in a solution may be employed to generate the porous matrix.

11.17.6 Formulation of Biological Control Agents

Combining active components, viz., spores as inert surfactants, is used to maintain the survival and virulence of the employed strain. Commercial items must also display the proper form (liquid or powder) for their intended use in the field. The strains viability and ability to germinate must be preserved during formulation, and it must be assisted in maintaining its severity toward associated disease. Carriers may remain inorganic, for example, talc or zeolites (Chaube et al. 2003; Küçük and Kivanç 2005). The BCA must also be stabilized for storage and use circumstances and protected from the sun's UV radiation. One method for producing biocontrol organisms is to embed liquid/solid biomass in polymers like alginate and carrageen (Cho and Lee 1999). It has been discovered that adding fungal mycelia to alginate pellets can efficiently transport biocontrol fungus (Papavizas et al. 1987; Küçük and Kivanç 2005). Cell entrapment is frequently employed in the biotechnology sector to speed up the synthesis of bioproducts, lower cell mortality, and improve cell recovery. Lewis and Papavizas (1983, 1985) developed alginate bits comprising fruiting bodies of several fungus and yeast cells (Serp et al. 2000). Compared to conidial suspensions, such preparations have various advantages, such as the ability to store pellets dry.

11.18 Role of Predatory Nematodes in Inducing the Activity of Predacious Fungi in the Management of Plant Parasitic Nematodes

Nematodes with predatory behaviour encompass nematode management near the beginning of the twentieth century. However, research on their potential has just started in addition to serving as BCAs against plant-parasitic nematodes. They stimulate the activity of predacious fungi employed as bionematicides by parasitism to control nematodes (Abd-Elgawad and Askary 2018; Sarker et al. 2020; Comans-Pérez et al. 2021; Girardi et al. 2022). The literature provides many examples, including *M. gaugleri*, which is efficient against *M. incognita* and *Heterodera oryzae*. The *Odontopharynx longicaudata* shows effectiveness against both *M. incognita* and *M. javanica* (Khan and Kim 2007).

11.19 Outlook

It must be understood that the biological nematicides require favourable BCA and the utilization of their genes and metabolites that minimize the special effects of nematodes, including RKN, and induce beneficial reactions in the developing plant resistance. Furthermore, even though many fungi and bacteria's by-products may improve plants' resistance to nematode assault, they are not typically regarded as bionematicides when utilized to reinforce or promote plant growth (Wilson and Jackson 2013). Therefore more research is needed, particularly on the environmental science, ecosystem, interactions with other farming inputs, and modes of accomplishment of these fungal and bacterial biocontrol agents.

11.20 Conclusion

This eco-friendly management approach of root-knot nematode (RKN) in vegetables can substitute chemical control. To protect the environment from chemicals, the predacious fungus may be utilized to treat pests like root-knot nematodes. The paramount strategy to guarantee the value of this biological control agent is to isolate the original strain because they are then adapted to pest management and the environment. These strain formulations may be exemplary microbial conservation with high virulence against pests.

References

Abd-Elgawad MMM, Askary TH (2018) Fungal and bacterial nematicides in integrated nematode management strategies. Egyptian J Biol Pest Control 28(1). https://doi.org/10.1186/s41938-018-0080-x

Barrios-González J (2012) Solid-state fermentation: physiology of solid medium, its molecular basis and applications. Process Biochem 47(2):175–185. https://doi.org/10.1016/j.procbio.2011.11.016

Birchfield W (1960) A new species of *Catenaria* parasitic on nematodes of sugarcane. Mycopathologia 13:331–338. https://doi.org/10.1007/978-1-4615-4815-7_15

Bradshaw CJ, Leroy B, Bellard C, Roiz D, Albert C, Fournier A et al (2016) Massive yet grossly underestimated global costs of invasive insects. Nat Commun 7:1–8. https://doi.org/10.1038/ncomms12986

Braga FR, de Araújo JV (2013) Nematophagous fungi for biological control of gastrointestinal nematodes in domestic animals. Appl Microbiol Biotechnol 98(1):71–82. https://doi.org/10.1007/s00253-013-5366-z

Braga FR, Soares FEF, Giuberti TZ, Lopes CG, Lacerda T, de Ayupe TH, Araújo JV (2015) Nematocidal activity of extracellular enzymes produced by the nematophagous fungus *Duddingtonia flagrans* on cyathostomin infective larvae. Vet Parasitol 212(3–4):214–218. https://doi.org/10.1016/j.vetpar.2015.08.018

Cabanillas E, Barker KR (1989) Impact of *Paecilomyces lilacinus* inoculum level and application time on control of *Meloidogyne incognita* on tomato. J Nematol 21(1):115–120

Cabanillas E, Barker KR, Daykin ME (1988) Histology of the interactions of *Paecilomyces lilacinus* with *Meloidogyne incognita* on tomato. J Nematol 20(3):362–365

Candanedo E, Lara J, Jatala P, Gonzales F (1983) Control biologico del nematode *Meloidogyne incognita* con elhongo *Paecilomyceslilacinus* (Abstr.), XXIII Meeting APS Caribbean Div

Castañeda-Ramírez GS, Torres-Acosta JFJ, Sánchez JE, Mendozade- Gives P, González-Cortázar M, Zamilpa A et al (2020) The possible biotechnological use of edible mushroom bioproducts for controlling plant and animal parasitic nematodes. Biomed Res Int 2020:1–12

Cayrol JC (1983) Luttebiologiquecontre les *Meloidogyne* au moyen*d'Arthrobotrysirregularis*. Rev Nematol 6:265

Cayrol JC, Frankowski JP (1979) Une methode de luttebiologiquecontre les nematodes a gaUes des racinesappartenant an genre *Meloidogyne* Pepinieristes Horticultures Maraichers. Revue Horticole 193:15–23

Chakraborty S, Newton AC (2011) Climate change, plant diseases and food security: an overview. Plant Pathol 60(1):2–14. https://doi.org/10.1111/j.1365-3059.2010.02411.x

Chaube HS, Mishra S, Varshney DS, Singh SU (2003) Biocontrol of plant pathogens by fungal antagonists historical background, present status and future prospects. Ann Rev Pl Pathol 2:1–42

Chitwood BG (1949) Root-knot nematodes – part I. A revision of the genus *Meloidogyne Göldi*, 1887. Proceedings of the Helminthological Society of Washington 16:90–104

Cho C-F, Lee W-C (1999) Formulation of a biocontrol agent by entrapping biomass of *Trichoderma viride* in gluten matrix. J Biosci Bioeng 87(6):822–824. https://doi.org/10.1016/s1389-1723(99)80161-3

Collange B, Navarrete M, Peyre G, Mateille T, Tchamitchian M (2011) Root-knot nematode (Meloidogyne) management in vegetable crop production: the challenge of an agronomic system analysis. Crop Prot 30(10):1251–1262. https://doi.org/10.1016/j.cropro.2011.04.016

Comans-Pérez RJ, Sánchez JE, Al-Ani LKT, González-Cortázar M, Castañeda-Ramírez GS, Gives PM et al (2021) Biological control of sheep nematode *Haemonchuscontortus* using edible mushrooms. Biol Control 152:104420. https://doi.org/10.1016/j.biocontrol.2020.104420

D'Addabbo, Laquale, Perniola, Candido (2019) Biostimulants for plant growth promotion and sustainable management of phytoparasitic nematodes in vegetable crops. Agronomy 9(10):616. https://doi.org/10.3390/agronomy9100616

Davide RG, Zorilla RA (1983) Evaluation of a fungus *Paecilomyceslilacinus* (Thorn.) Samson. for the biological control of the potato cyst nematode *Globodera rostochiensis* Woll. as compared with some nematicides. Philos Agric 66:397–404

de Sisler GM, Silvestri L, Acita JO (1985) Utilizacion de *Paecilomyces lilacinus*para el control de *Nacobbus aberrans* (Nematoda, Nacobbidae) en campo. Fitopathologia 20:17–20

Degenkolb T, Vilcinskas A (2016a) Metabolites from nematophagous fungi and nematicidal natural products from fungi as alternatives for biological control. Part II: metabolites from nematophagous basidiomycetes and nonnematophagous fungi. Appl Microbiol Biotechnol 100:3813–3824. https://doi.org/10.1007/s00253-015-7233-6

Degenkolb T, Vilcinskas A (2016b) Metabolites from nematophagous fungi and nematicidal natural products from fungi as an alternative for biological control. Part I: metabolites from nematophagous ascomycetes. Appl Microbiol Biotechnol 100:3799–3812. https://doi.org/10.1007/s00253-015-7234-5

Dowsett JA, Reid J (1977) Light microscope observations on the trapping of nematodes by *Dactylaria candida*. Can J Bot 55:2963–2970

Dowsett JA, Reid J (1979) Observations on the trapping of nematodes by *Dactylaria scaphoides*using optical, transmission and scanning electron microscope techniques. Mycologia 71:379–391. https://doi.org/10.2307/3759158

Duponnois R, Ba AM, MateilleT. (1998) Effects of some rhizosphere bacteria for the biocontrol of nematodes of the genus *Meloidogyne* with *Arthrobotrys oligospora*. Fund Appl Nematol 21(2): 157–163

Ekino T, Yoshiga T, Takeuchi-Kaneko Y, Kanzaki N (2017) Transmission electron microscopic observation of body cuticle structures of phoretic and parasitic stages of Parasitaphelenchinae nematodes. PLoS One 12(6):e0179465. https://doi.org/10.1371/journal.pone.0179465

Esser RP (1976) *Haptoglossa heterospora* Drechsler, a fungus parasite of Florida nematodes, Florida Dept of Agriculture and Consumer Service, No 23, 2 pp

Esser RP, Ridings WH (1973) Pathogenicity of selected nematodes by *Catenaria anguillulae*. Soil Crop Sci Florida 33:60–64

Feyisa B, Lencho A (2015) Evaluation of some botanicals and *Trichoderma harzianum* for the management of tomato root-knot nematode *(Meloidogyne incognita* (Kofoid and White) Chit Wood). Adv Crop Sci Technol 04(01). https://doi.org/10.4172/2329-8863.1000201

Friman E, Olsson S, Nordbring-Hertz B (1985) Heavy trap formation by *Arthrobotrys oligosporain liquid culture*. FEMS Microbiol Lett 31(1):17–21. https://doi.org/10.1111/j.1574-6968.1985.tb01126.x

Gaspard JT, Jaffee BA, Ferris H (1990) Association of *Verticillium chlamydosporium* and *Paecilomyces lilacinus* with root-knot nematode infested soil. J Nematol 22(2):207–213

Girardi NS, Sosa AL, Etcheverry MG, Passone MA (2022) In vitro characterization bioassays of the nematophagous fungus *Purpureocillium lilacinum:* evaluation on growth, extracellular enzymes, mycotoxins and survival in the surrounding agroecosystem of tomato. Fungal Biol 126:300–307. https://doi.org/10.1016/j.funbio.2022.02.001

Hajieghrari B, Torabi-Giglou M, Mohammadi MR, Davari M (2008) Biological potential of some Iranian *Trichoderma* isolates in control of soil borne plant pathogenic fungi. Afr J Biotechnol 7(8):967–972

Hesseltine CW (1987) Solid state fermentation: an overview. Int Biodeterior 23:79–89. https://doi.org/10.1080/07388550590925383

Huang X, Zhao N, Zhang K (2004) Extracellular enzymes serving as virulence factors in nematophagous fungi involved in infection of the host. Res Microbiol 155(10):811–816. https://doi.org/10.1016/j.resmic.2004.07.003

Jansson HB (1982a) Attraction of nematodes to endoparasitic nematophagous fungi. Trans Br Mycol Soc 79:25 29

Jansson H-B, Nordbring-Hertz B (1983) The endoparasitic nematophagous fungus *Meria coniospora* infects nematodes specifically at the chemosensory organs. Microbiology 129(4): 1121–1126. https://doi.org/10.1099/00221287-129-4-1121

Jansson HB, von Hofsten A, von Mecklenburg C (1984) Life cycle of the endoparasitic nematophagous fungus *Meria coniospora*: a light and electron microscopic study. Antonie Van Leeuwenhoek 50(4):321–327

Jansson HB, Jeyaprakash A, Zuckerman BM (1985) Control of root-knot nematodes on tomato by the endoparasitic fungus *Meriaconiospora*. J Nematol 17:327–329

Jatala P, Kaltenbach R, Bocangel M (1979) Biological control of *Meloidogyne incognita acrita*and *Globodera pallida* on potatoes. J Nematol II:303

Jatala P, Kaltenbach R, Bocangel M, Devax AJ, Compos R (1980) Field application of *Paecilomyces lilacinus*for controlling *Meloidogyne incognita* on potatoes. J Nematol 12:226–227

Jones JT, Haegeman A, Danchin EG, Gaur HS, Helder J, Jones MG et al (2013) Top 10 plant-parasitic nematodes in molecular plant pathology. Mol Plant Pathol 14:946–961. https://doi.org/10.1111/mpp.12057

Junwei W, Qingling M, Jun Q, Weisheng W, Shuangqing C, Jianxun L et al (2013) The recombinant serine protease XAoz1 of *Arthrobotrys oligospora* exhibits potent nematicidal activity against *Caenorhabditis elegans* and *Haemonchuscontortus*. FEMS Microbiol Lett 344:53–59

Khan Z, Kim YH (2007) A review on the role of predatory soil nematodes in the biological control of plant parasitic nematodes. Appl Soil Ecol 35:370–379. https://doi.org/10.1016/j.apsoil.2006.07.007

Khan A, Williams KL, Nevalainen HK (2004) Effects of *Paecilomyces lilacinus* protease and chitinase on the eggshell structures and hatching of *Meloidogyne javanica* juveniles. Biol Control 31:346–352. https://doi.org/10.1016/j.biocontrol.2004.07.011

Kim YH (2015) Predatory nematodes as biocontrol agents ofphytonematodes. In: Askary TK, Martinelly PRP (eds) Biocontrol agents of phytonematodes. CABI, pp 393–420. https://doi.org/10.1079/9781780643755.0393

Kim E, Seo Y, Kim YS, Park Y, Kim YH (2017) Effects of soil textures on infectivity of root-knot nematodes on carrot. Plant Pathol J 33(1):66. https://doi.org/10.5423/PPJ.OA.07.2016.0155

Küçük C, Kivanç M (2005) Effect of formulation on the viability of biocontrol agent, *Trichoderma harzianum*conidia. Afr J Biotechnol 4(5):483–486

Kumar D, Singh KP (2006) Assessment of predacity and efficacy of *Arthrobotrys dactyloides*for biological control of root-knot disease of tomato. J Phytopathol 154:1–5. https://doi.org/10.1111/j.1439-0434.2005.01047.x

Kumar V, Khan MR, Walia RK (2020) Crop loss estimations due to plant-parasitic nematodes in major crops in India. Natl Acad Sci Lett 43(5):409–412. https://doi.org/10.1007/s40009-020-00895-2

Lee DL (1967) The structure and composition of the helminth cuticle. Adv Parasitol 4:187–254. https://doi.org/10.1016/s0065-308x(08)60450-9

Lewis JA, Papavizas GC (1983) Production of chlamydospores and conidia by *Trichoderma* spp in liquid and solid growth media. Soil Biol Biochem 15(3):351–357. https://doi.org/10.1016/0038-0717(83)90083-4

Lewis JA, Papavizas GC (1985) Characteristics of alginate pellets formulated with *Trichoderma* and *Gliocladium* and their effect on the proliferation of the fungi in soil. Plant Pathol 34(4):571–577. https://doi.org/10.1111/j.1365-3059.1985.tb01409.x

Li TF, Zhang KQ, Liu X (2000) Taxonomy of nematophagous fungi. Chinese Scientific and Technological Publications, Beijing

Liang L, Meng Z, Ye F, Yang J, Liu S, Sun Y et al (2010) The crystal structures of two cuticle-degrading proteases from nematophagous fungi and their contribution to infection against nematodes. FASEB J 24:1391–1400. https://doi.org/10.1096/fj.09-136408

Linford MB, Yap F (1939) Root-knot nematode injury restricted by a fungus. Phytopathology 29:596–609

Linford MB, Yap F, Oliveira JM (1938) Reduction of soil populations of the root-knot nematode during decomposition of organic matter. Soil Sci 45:127–141

Liu T, Wang L, Duan YX, Wang X (2008) Nematicidal activity of culture filtrate of *Beauveria bassiana*against *Meloidogyne hapla*. World J Microbiol Biotechnol 24:113–118. https://doi.org/10.1007/s11274-007-9446-z

Liu X, Xiang M, Che Y (2009) The living strategy of nematophagous fungi. Mycoscience 50:20–25. https://doi.org/10.1007/S10267-008-0451-3

Lopez-Llorca LV, Maciá-Vicente JG, Jansson H-B (2007) Mode of action and interactions of nematophagous fungi. In: Integrated management and biocontrol of vegetable and grain crops nematodes, pp 51–76. https://doi.org/10.1007/978-1-4020-6063-2_3

Luo H, Mo MH, Huang XW, Li X, Zhang KQ (2004) *Coprinus comatus*: a basidiomycete fungus forms novel spiny structures and infects nematodes. Mycologia 96:1218

Luo H, Li X, Li G, Pan Y, Zhang K (2006) Acanthocytes of *Stropharia rugoso* annulata function as a nematode-attacking device. Appl Environ Microbiol 72:2982–2987. https://doi.org/10.1128/AEM.72.4.2982-2987.2006

Luo H, Liu Y, Fang L, Li X, Tang N, Zhang K (2007) *Coprinus comatus* damages nematode cuticles mechanically with spiny balls and produces potent toxins toimmobilize nematodes. Appl Environ Microbiol 73:3916. https://doi.org/10.1128/AEM.02770-06

Maia Filho FDS, Vieira JN, Berne MEA, Stoll FE, Nascente PDS, Pötter L et al (2013) Fungal ovicidal activity on *Toxocaracanis* eggs. Revista Iberoamericana de Micología 30:226–230. https://doi.org/10.1016/j.riam.2012.12.009

Mankau R, Wu (1984) Biological control of *Meloidogyne incognita* on tomato with the predaceous fungus *Monacrosporiumellipsosporum*, First Int. Congo Nematol Canada (Abstr), 116

McInnes SJ (2003) A predatory fungus (*Hyphomycetes: Lecophagus*) attacking Rotifera and Tardigrada in maritime Antarctic lakes. Polar Biol 26:79–82. https://doi.org/10.1007/s00300-002-0449-9

Mejias J, Truong NM, Abad P, Favery B, Quentin M (2019) Plant proteins and processes targeted by parasitic nematode effectors. Front Plant Sci 10:970. https://doi.org/10.3389/fpls.2019.00970

Mitchell DA, Berovic M, Krieger N (2002) Overview of solid state bioprocessing. Biotechnol Ann Rev 8:183–225. https://doi.org/10.1016/S1387-2656(02)08009-2

Morgan-Jones G, White JF, Rodriguez-Kabana R (1984) Phytonematode pathology: ultrastructural studies II. Parasitism of *Meloidogyne arenaris* egg and larvae by *Paecilomyces lilacinus*. Nematropica 14:57–71

Morton O, Kerry B, Hirsch P (2004) Infection of plant-parasitic nematodes by nematophagous fungi – a review of the application of molecular biology to understand infection processes and to improve biological control. Nematology 6(2):161–170. https://doi.org/10.1163/1568541041218004

Mostafanezhad H, Sahebani N, NourinejhadZarghani S (2014) Control of root-knot nematode (*Meloidogyne javanica*) with combination of *Arthrobotrys oligospora* and salicylic acid and study of some plant defense responses. Biocontrol Sci Tech 24:203–215. https://doi.org/10.1080/09583157.2013.855166

Neeraj, Bisht V, Johar V (2017) Bael (*Aegle marmelos*) extraordinary species of India: a review. Int J Curr Microbiol App Sci 6(3):1870–1887. https://doi.org/10.20546/ijcmas.2017.603.21

Noe JP, Sasser JN (1984) Efficacy of *Paecilomyces lilacinus* in reducing yield losses due to *Meloidogyne incognita*, First Int. Cong. Nematol, Canada (Abstr.) 116

Nordbring-Hertz B, Jansson H-B, Tunlid A (2006) Nematophagous fungi. In: Encyclopedia of life sciences. Wiley, Chichester

Pal KK, Gardener BMS (2006) Biological control of plant pathogens. Plant Health Instr 2:1–25. https://doi.org/10.1094/PHI-A-2006-1117-02

Papavizas GC, Fravel DR, Lewis JA (1987) Proliferation of *Taleromyces flavus* in soil and survival in alginate pellets. Phytopathology 77:131–136. https://doi.org/10.1094/Phyto-77-131

Paulitz TC (2000) Population dynamics of biocontrol agents and pathogens in soils and rhizospheres. Eur J Plant Pathol 106(5):401–413. https://doi.org/10.1023/A:1008733927515

Roman J, Rodriguez-Marcano A (1985) Effect of the fungus *Paecilomyceslilacinus*on the larval population and root-knot formation of *Meloidogyne incognita* in tomato. J Agric Univ Puerto Rico 69:159–167. https://doi.org/10.46429/jaupr.v69i2.7339

Samson RA (1974) *Paecilomyces* and some allied hyphomycetes. Stud Mycol 6:1

Sarhy-Bagnon V, Lozano P, Saucedo-Castaneda G, Roussos S (2000) Production of 6-pentyl-alpha-pyrone by *Trichoderma harzianum*in liquid and solid state cultures. Process Biochem 36: 103–109. https://doi.org/10.1590/S1517-838220080004000022

Sarker MS, Mohiuddin KM, Al-Ani LKT, Hassan MN, Akter R, Hossain MS et al (2020) Effect of bio-nematicide and bau-biofungicide against root-knot (*Meloidogyne Spp.*) of soybean. Malaysian J Sustain Agric 4:44–48. https://doi.org/10.26480/mjsa.02.2020.44.48

Sasser JN, Freckman DW, Veech JA, Dickson DW (1987) Vistas on nematology. Soc Nematol 7:14

Satou T, Kaneko K, Li W, Koike K (2008) The toxin produced by *Pleurotusostreatus* reduces the head size of nematodes. Biol Pharmaceut Bull 31:574–576. https://doi.org/10.1248/bpb.31.574

Seong J, Shin J, Kim K, Cho B-K (2021) Microbial production of nematicidal agents for controlling plant-parasitic nematodes. Process Biochem 108:69–79. https://doi.org/10.1016/j.procbio.2021.06.006

Serp D, Cantana E, Heinzen C, Von Stockar U, Marison IW (2000) Characterization of an encapsulation device for the production of monodisperse alginate beads for cell immobilization. Biotechnol Bioeng 70(1):41–53

Siddique S, Grundler FM (2018) Parasitic nematodes manipulate plant development to establish feeding sites. Curr Opin Microbiol 46:102–108. https://doi.org/10.1016/j.mib.2018.09.004

Siddiqui ZA, Mahmood I (1996) Biological control of plant parasitic nematodes by fungi: a review. Bioresour Technol 58:229–239. https://doi.org/10.1016/S0960-8524(96)00122-8

Singh S, Singh B, Singh AP (2015) Nematodes: a threat to sustainability of agriculture. Procedia Environ Sci 29:215–216. https://doi.org/10.1016/j.proenv.2015.07.270

Soares FE, Braga FR, Araújo JV, dos Santos Lima W, Mozer LR, Queiróz JH (2012) In vitro activity of a serine protease from *Monacrosporium thaumasium* fungus against first-stage larvae of *Angiostrongylus vasorum*. Parasitol Res 110:2423–2427. https://doi.org/10.1007/s00436-011-2781-x

Soares FEF, Queiroz JH, Araújo JV, Gouveia A, Queiroz PV, Hiura E et al (2014) Nematocidal action of chitinases produced by the fungus *Monacrosporium thaumasium* under laboratorial conditions. Biocontrol Sci Tech 25:337–344

Soares FEF, Sufiate BL, de Queiroz JH (2018) Nematophagous fungi: far beyond the endoparasite, predator and ovicidal groups. Agric Nat Resour 52:1–8. https://doi.org/10.1016/j.anres.2018.05.010

Stiirhan D, Schneider R (1980) *Hirsutella heteroderae,* a new nematode-parasitic fungus. Phytopatholische Zeitschrifl 99:105–115

Stirling, G. R. (2018). Biological control of plant-parasitic nematodes, Diseases of nematodes, eds G. O. Poinar and H.-B. Jansson (Boca Raton: CRC Press), 103–150

Stirling GR, Smith LJ (1998) Field tests of formulated products containing either *Verticillium chlamydosporium* or *Arthrobotrys dactyloides* for biological control of root knot nematodes. Biol Control 11:229–237

Stirling GR, Licastro KA, West LM, Smith LJ (1998) Development of commercially acceptable formulations of the nematophagous fungus *Verticillium chlamydosporium*. Biol Control 11: 217–223. https://doi.org/10.1006/bcon.1997.0602

Swe A, Li J, Zhang KQ, Pointing SB, Jeewon R, Hyde KD (2011) Nematode-trapping fungi. Curr Res Environ Appl Mycol 1:1–26. https://doi.org/10.1128/microbiolspec

Taylor AL, Sasser JN (1978) Biology, identification and control of root-knot nematodes. North Carolina State University Graphics, p 111

Thakur NSA, Devi G (2007) Management of *Meloidogyne incognita* attacking okra by nematophagous fungi, *Arthrobotrys oligospora* and *Paecilomyces lilacinus*. Agric Sci Dig 27: 50–52

Tikhonov VE, Lopez-Llorca LV, Salinas J, Jansson HB (2002) Purification and characterization of chitinases from the nematophagous fungi *Verticillium chlamydosporium* and *V. suchlasporium* fungal. Genet Biol 35:67. https://doi.org/10.1006/fgbi.2001.1312

Viaene NM, Abawi GS (1998) Management of *Meloidogyne hapla* on lettuce in organic soil with Sudan grass as acover crop. Plant Dis 82:945–952. https://doi.org/10.1094/PDIS.1998.82.8.945

Wilson MJ, Jackson TA (2013) Progress in the commercialization of bionematicides. BioCont 58: 715–722. https://doi.org/10.1007/s10526-013-9511-5

Xiang N, Lawrence KS, Donald PA (2018) Biological control potential of plant growth-promoting rhizobacteria suppression of *Meloidogyne incognita* on cotton and *Heterodera glycines* on soybean: a review. J Phytopathol 166:449–458. https://doi.org/10.1111/jph.12712

Yang JK, Huang XW, Tian BY, Wang M, Niu QH, Zhang KQ (2005) Isolation and characterization of a serine protease from the nematophagous fungus, *Lecanicillium psalliotae*, displaying nematicidal activity. Biotechnol Lett 27:1123–1128. https://doi.org/10.1007/s10529-005-8461-0

Yang E, Xu LL, Yang Y, Zhang XY, Xiang MC, Wang CS et al (2012) Origin and evolution of carnivorism in the Ascomycota (fungi). *Proc Natl Acad Sci U S A* 109:10960–10965. https://doi.org/10.1073/pnas.1120915109

Yang L, Li X, Bai N, Yang X, Zhang K-Q, Yang J (2022) Transcriptomic analysis reveals that rho GTPases regulate trap development and lifestyle transition of the nematode-trapping fungus *Arthrobotrys oligospora*. Microbiol Spectrum 10. https://doi.org/10.1128/spectrum.01759-21

Zhang S, Gan Y, Xu B, Xue Y (2014) The parasitic and lethal effects of *Trichoderma longibrachiatum* against *Heterodera avenae*. Biol Control 72:1–8. https://doi.org/10.1016/j.biocontrol.2014.01.009

Zhang S, Gan Y, Ji W, Xu B, Hou B, Liu J (2017) Mechanisms and characterization of *Trichoderma longibrachiatum* T6 in suppressing nematodes (*Heterodera avenae*) in wheat. Front Plant Sci 8: 1491. https://doi.org/10.3389/fpls.2017.01491

Zhao ML, Mo MH, Zhang KQ (2004) Characterization of a neutral serine protease and its full-length cDNA from the nematodetrapping fungus *Arthrobotrys oligospora*. Mycologia 96:16–22. https://doi.org/10.1080/15572536.2005.11832991

12 Biofertilizer of Organic Origin for Management of Root Galling Disease of Vegetables

Sushmita Sharma, Rishil Gupta, Faryad Khan, Sachin Upadhayaya, and Faheem Ahmad

Abstract

Increasing global food demand necessitates the intensification of crop production in modern agriculture, which entails the significant use of synthetic fertilizers to increase crop output. Plant parasitic nematodes pose a grave danger to agricultural production. Root-knot nematode, *Meloidogyne* spp., has been deemed a limiting factor in the production of the majority of crops, including vegetables. Hence, root-knot nematodes *Meloidogyne* spp. treatment is an obligatory challenge. Present methods of soil management are primarily dependent on inorganic chemical fertilizers, which pose a significant risk to both human health and the environment. The proliferation of biofertilizers in modern agriculture can be attributed to the fact that these substances are beneficial to the environment, economical, and simple to use. Because of the potential role they could play in ensuring food safety and maintaining sustainable crop production, the use of beneficial microorganisms as a source of biofertilizer has assumed a position of utmost importance in the agricultural sector. The environmentally friendly methods inspire various applications of plant growth-promoting rhizobacteria

S. Sharma
Department of Entomology, Agriculture and Forestry University, Rampur, Chitwan, Nepal

R. Gupta · F. Ahmad (✉)
Department of Botany, Aligarh Muslim University, Aligarh, India
e mail: faheem.bt@amu.ac.in

F. Khan
Department of Botany, Aligarh Muslim University, Aligarh, India

Department of Botany, Krishna Devi Balika P.G. College, Farrukhabad, India

S. Upadhayaya
Department of Microbiology and Parasitology, Agriculture and Forestry University, Rampur, Chitwan, Nepal

F. Ahmad, G. Nombela (eds.), *Root-Galling Disease of Vegetable Plants*,
https://doi.org/10.1007/978-981-99-3892-6_12

(PGPRs), endo- and ectomycorrhizal fungi, cyanobacteria, and many other useful microscopic organisms, which led to improved nutrient uptake, plant growth, and plant tolerance to both biotic and abiotic stress. In comparison to the detrimental effects of chemical fertilizers, biofertilizers can directly or indirectly contribute to food security. The direct mechanism of biofertilizers is phytostimulus and nutrient mobility, while the indirect mechanism is biocontrol activity. Direct mechanisms include phytohormone synthesis and phosphate, potassium, zinc solubilization, etc. In contrast, indirect means include the synthesis of HCN, siderophores, antibiotics, etc. These possible biological fertilizers play a vital role in the production and sustainability of soil, as well as in the protection of the environment, serving as inputs for farmers that are both environmentally friendly and cost-effective.

Keywords

Biofertilizers · Eco-friendly · *Meloidogyne* spp. · Sustainable food security

12.1 Introduction

Biofertilizers are natural fertilizers that include living microorganisms helping to increase the availability and mobility of nutrients from the soil, which may be of fungal, bacterial, or algal origin (Mitter et al. 2021). The interaction of plant-associated microorganisms and soil improves soil fertility and promotes sustainable agriculture. Biofertilizers can be used as a substitute for synthetic fertilizers to solve various problems like pollution and fertilizer residues caused by chemical fertilizers (Kumar and Kumar 2019). These fertilizers could be used as an alternative source of synthetic fertilizers, which are eventually beneficial in reducing the detrimental effects of chemical fertilizers on the environment and human health (Odoh et al. 2020). Biofertilizers are applied either by root dipping, seed treatment, or soil application. Biofertilizers directly affect plants through phytostimulation and help in nutrient mobility (Mahmud et al. 2021).

Root-knot nematodes are small round colorless worms measuring about 0.5 mm to 0.75 mm, mostly belonging to the *Meloidogyne* family (Janati et al. 2018). The symptoms of root-knot nematode are seen as the development of galls in the plant parts, and the damage is seen more often when plants are under different abiotic stress, such as water stress, temperature stress, etc. Root-knot nematode in vegetables causes profound loss in the production of vegetable crops (Janati et al. 2018), causing about 50% damage in solanaceous crops and up to 30% loss in cucurbitaceous and root crops, and mostly, infestation is seen in broad leaf vegetables (Gowda et al. 2007). Janati et al. (2018) recorded more than 80% of infections due to *Meloidogyne javanica* in vegetables grown under greenhouse conditions causing yellowing of leaves and stunting as visible above-ground part symptoms. Izuogu et al. (2019) reported a negative correlation between root nodule formation in cowpea and root-knot nematodes. Nematode-suppressing plants such as marigold can be used in the field for crop rotation for many years, which does not eliminate the nematode but reduce the population in the area (Lopez-Perez et al. 2010).

12.2 Biofertilizers: Why Are They Advantageous in Sustainable Agriculture for Managing Root-Knot Nematode?

Farmers are still haphazardly using chemical fertilizers, although soil contamination, health hazards, and toxicity have been increasing. There is a need to search for alternatives to chemical and synthetic fertilizers before destroying the soil. Biofertilizers contribute to environmental health and can be used to some extent to nullify the dangerous effects of chemical fertilizers and improve the condition of unhealthy soil to transform it into healthy and sustainable soil structure and soil performance. Biofertilizers are low-cost renewal sources of plant nutrients to improve crop productivity and fertility. The activities and interaction of microorganisms with soil help maintain the soil ecosystem's structure and increase crop yield. Biofertilizers are arranged in a coordinated complex ecosystem that influences living and non-living components of the soil (Odoh et al. 2020). Once these biofertilizers are inoculated in the seed or applied in the soil, they provide nutrition and help the soil ecosystem for a more extended period by fixing nitrogen, promoting growth stimulants, etc. (Malusa et al. 2012). Biofertilizers can mobilize nutritive elements by nitrogen fixation and mobilizing and enhancing the uptake of various elements from the soil. Biofertilizers indirectly improve the health and vigor of the plant, which eventually provides resistance to plants against the nematode attack. Fungal and bacterial biofertilizers are closely associated with plants and improve plant immunity. Algal biofertilizers can be replace the traditional use of synthetic fertilizers as they make biological nitrogen, phosphorus, and potassium available to plants (Ghosh et al. 2022). Biochar and charcoal-based fertilizers can be used for sustainable soil productivity as it is found stable in the soil and helps in the availability of inoculants for a longer period by increasing the bacterial population in the soil (Wolna-Maruwka et al. 2021). Inoculation of valuable bacteria and fungus in combination with biochar and charcoal increases seed viability and germination and helps in an overall increase in productivity and fertility of the soil for a longer period. Biofertilizers produce indole acetic acid, gibberellin, biotin, vitamin B, etc., which catalyzes crops' growth and yield. Seaweeds act as the suppressive agent for the penetration of nematodes in the plants singly or mixed with the nematicides by reducing the population of nematodes and the number of nematode eggs per root (Afia and El-Nuby 2016). The various sources and types of biofertilizers are described below.

12.2.1 Bacterial Biofertilizers

Many rhizospheric bacteria, such as plant growth promoting rhizobacteria (PGPR), can be applied as biofertilizers which help increase the yield of vegetables and improve soil fertility (García-Fraile et al. 2015). Mostly used bacterial biofertilizers for improving vegetable production are nitrogen-fixing bacteria, phosphorus solubilizing bacteria, and potassium-solubilizing bacteria (Kumar and Kumar 2019). These bacteria enhance crop production by nitrogen fixation, synthesis of phytohormones, degradation of organic compounds, etc. Azotobacter is the most used biofertilizer and is commonly found in arable soil as free-living bacteria. In greenhouse test and invitro cultivation, *B. laterosporus* caused high nematode mortality in solanaceous vegetables, while *B. megaterium* was supposed to reduce the nematode population to half of the potatoes (El-Hadad et al. 2011). Plant growth promoting rhizobacteria *Burkholderia vietnamiensis* B418 application singly or combined with the nematicide was effective against the *Meloidogyne* in watermelon. The use of *B. vietnamiensis* B418 singly was most effective compared to a combined application (Liu et al. 2022). The rhizosphere colony decreased the number of galls and second juvenile stage by about 75% and 85%, respectively, in the field of tomatoes (Alfianny et al. 2017). About 80% of juvenile stage 2 root-knot nematode was controlled using *Pseudomonas*, as Sun et al. (2021) reported. There are enlisted some bacterial biofertilizers and their targeted root-knot nematodes as hosts (Table 12.1).

12.2.2 Fungal Biofertilizers

This biofertilizer includes fungal agents that can be used in seeds as a seed treatment, sprayed on the plant surface, or applied on the soil surface (Odoh et al. 2020). Plant growth-stimulating fungi, mycorrhiza fungi, phosphorus/potassium solubilizing fungi, and enzyme-producing fungi are primarily used as fungal biofertilizers in crop production (Odoh et al. 2020). The amount of fungus in the rhizosphere, the rate of development of eggs in egg mass, and the size of the gall affect the effectiveness of fungal biofertilizers in the crops. Fungal biofertilizer influences different biochemical development in plants and suppresses biotic and abiotic stress. Fungal biofertilizer competes with phytopathogens by making colonies in the rhizosphere and preventing pathogens from affecting the plants. *Trichoderma brevicompactum* was found to suppress the production of the egg of nematode *M. incognita* by about 85%, and *Trichoderma asperellum* was supposed to suppress the second stage of the juvenile by about 80% (Affokpon et al. 2011). Different mycorrhizal fungi (*Rhizophagus aggregatus*, *Funneliformis mosseae*, *Gigaspora gigantean*) mass-reared, sterilized, and used to control *Meloidogyne incognita* resulted in the obstruction of egg hatching percentage by 80.71% (Alamri et al. 2022). Mycoparasitism and entomopathogenicity are the two basic mechanisms that fungal biological control agents use to kill agricultural pests and insects. Mycoparasitism is the process by which fungi feed on other fungi. Mycoparasitism is the relationship between a fungal

Table 12.1 List of bacterial biofertilizers and their targeted root-knot nematode

Bacterial biofertilizer	Root-knot nematode and their host	References
Bacillus laterosporus	*M. incognita* and vegetables	Hadad et al. (2011)
B. megaterium	*M. incognita* and potato	Hadad et al. (2011)
Burkholderia vietnamiensis B418	*M. incognita* and watermelon	Liu et al. (2022)
Pseudomonas fluorescens, P. lilacinus, and P. guilliermondii	*M. incognita* and tomato	Sun et al. (2021)
Bacillus firmus	*M. incognita* and tomatoes	Terefe et al. (2009)
B. paralicheniformis FMCH001, B. subtilis FMCH002	*Meloidogyne javanica* and tomatoes	Díaz-Manzano et al. (2023)
B. velezensis BZR 86	*M. incognita* and tomato, cucumber	Migunova et al. (2021)
Pseudomonas spp.	*M. incognita* and tomato crop	Ahmed et al. (2023)
B. Licheniformis IRh9, P. megaterium IRh10, P. Putida IRh15	*M. incognita* and tomato crop	Gowda et al. (2023)
Pseudomonas fluorescens, P. putida, Bacillus amyloliquefaciens, B. megaterium	*Meloidogyne incognita and* bottle gourd	Rani et al. (2022)
Bacillus altitudinisis KMS-6, Bacillus cereus KMT-5, B. megaterium KMT-8, Bacillus subtilis	*Meloidogyne javanica, M. incognita* and eggplant, tomato	Adiwena et al. (2023), Antil et al. (2022a, b)
B. subtilis	*Meloidogyne incognita* and pistachio	Pourkhaloei et al. (2022)

parasite and the fungal host. Entomopathogenic fungi, on the other hand, are parasitic fungi that can destroy infections. *Gliocladium* and *Trichoderma* are two types of fungi that manage fungal infections in plants by engaging in a process known as mycoparasitism. There are enlisted some fungal biofertilizers and their targeted root-knot nematodes as hosts (Table 12.2).

12.2.3 Algal Biofertilizers

Algal fertilizer includes mainly blue-green algae and Azolla. They are the favorable channel that translates solar energy into different gases, which finally into important chemical substances help in biomass production and yield of the crop. Green microalgae and some cyanobacteria species, such as *Chlorella Vulgaris*, are commonly used as biofertilizers in soil and biofertilizer studies (Ammar et al. 2022). Algal-based biofertilizer provides better nutrients as compared to Farm Yard Manure and other chemical fertilizers since algal biofertilizer has high organic content and moisture-retaining capacity (Baweja et al. 2019). Algal biofertilizers transform solar energy and other atmospheric gases into useful chemical products by generating large-scale biomass and helping in carbon dioxide sequestration (Ghosh et al. 2022).

Table 12.2 List of fungal biofertilizers and their targeted Root- Knot Nematode

Fungal biofertilizer	Root-knot Nematode and their host	References
Syncephalastrum racemosum	*Meloidogyne incognita* and cucumber	Huang et al. (2014)
Trichoderma longibrachiatum, Trichoderma viride	*Meloidogyne incognita* and cucumber	Zhang et al. (2015)
Trichoderma brevicompactum, Trichoderma asperellum	*Meloidogyne* spp. and vegetables	Affokpon et al. (2011)
Trichoderma harzianum	*Meloidogyne javanica* and tomato	Nafady et al. (2022)
Trichoderma sp.	*Meloidogyne* spp. and tomato	Kiriga et al. (2018)
Trichoderma virens	*Meloidogyne incognita* and chickpea	Khan et al. (2022)
Trichoderma harzianum	*Meloidogyne incognita* and tomato	d'Errico et al. (2022)
Trichoderma harzianum MZ025966	*Meloidogyne javanica* and tomato	Nafady et al. (2022)
Trichoderma album	*Meloidogyne incognita* and tomato	Khalil et al. (2022)
Trichoderma harzianum AMUTH-1 + Pseudomonas putida AMUPP-1	*Meloidogyne graminicola* and rice	Haque and Khan (2022)
Trichoderma asperellum	*Meloidogyne* spp. and tomato	Expósito et al. (2022)
Trichoderma harzianum + Purpureocillium lilacinum	*M. javanica* and soybean	Soares et al. (2021)

Besides this, presence of algae in the soil results in less runoff of nitrogen, phosphorus, and other organic matter (Raouf et al. 2012). There are many algae species, and the algae extracts have nematicidal properties. Dry powder of *Sargassum swartzii* was found to repress the root-knot nematode in solanaceous crops (Afia and El-Nuby 2016). Some algal biofertilizers and their targeted root-knot nematodes as hosts are enlisted (Table 12.3).

12.2.4 Biochar-Based Biofertilizers

Biochar is a modern technology that is applied in the soil for the sequestration of carbon from the atmosphere improving soil nutrient retention and crop productivity. Biochar-type biofertilizer is used as inoculant carriers and helps in the soil's stable availability of the inoculated substances. Because of its one-of-a-kind physicochemical qualities, such as its high carbon content and capacity to fix metals, the application of biochar in soil remediation may prove advantageous. They increase the production of crops by improving crop growth parameters and enhancing the soil's physical and chemical properties (Kumar et al. 2022). Biochar inoculated

Table 12.3 List of algal biofertilizers and targeted root-knot nematode

Algal biofertilizers	Beneficial role	Root-knot nematode	References
Phacelocarpus tristichus, *Turbinaria ornata*	Suppression of root gall and development stage of nematode in tomato	*Meloidogyne incognita*	Ibrahim et al. (2021)
Ulva lactuca, *Jania rubens*, *Laurencia obtusa*, and *Sargassum vulgare*	Reduction in the number of galls in banana	*Meloidogyne* spp.	El-Ansary and Hamouda (2014)
Spirulina and *amphora*	Reduced root-knot nematode numbers in cucumber	*Meloidogyne incognita*	El-Eslamboly et al. (2019)
Ulva fasciata Delile (UF) (green algae), *Corallina mediterranea*, *Corallina officinalis* (red algae)	Enhanced the tomato defense genes	*Meloidogyne incognita*	Ghareeb et al. (2019)
Ascophyllum nodosum	Reduce RKN performance	*Meloidogyne* spp.	Williams et al. (2021)
Chlorella vulgaris	Decreased in mature females, egg masses and root galls in cowpea	*M. incognita*	Abo-Korah et al. (2022)
Spirulina platensis	Inhibited the count of the RKN in banana	*Meloidogyne incognita*	Hamouda et al. (2019)

with *Rhizobium* and *Bacillus* sp. was effective in increasing seed germination and seed viability for the treated seeds (Kumar et al. 2017). Khan et al. (2021) reported that biochar treated with urea nitrogen reduced the loss of nitrogen from different soil types. The co-application of chemically induced nitrogen-loaded biochar and biofertilizer was effective in crop growth and enhanced the crop growth parameters. Biochar remains in the soil longer, increasing organic matter, decreasing nutrient loss, and immobilizing toxic compounds (Dahal et al. 2016). Biochar not only improves biomass production but also decreases the harmful gases released from the soil to reduce the climate change impacts by reducing greenhouse gas emissions. In another piece of research, Huang et al. (2017) and colleagues demonstrated that biochar reduces the susceptibility of rice plants to infections caused by root-knot nematodes.

12.3 Antagonistic Role of Biofertilizers to Root Galling Disease of Vegetables

Biofertilizers enhance crop growth and output while being environment friendly. They interact with the soil's natural microbiota in both synergistic and antagonistic ways, and they take part in a variety of ecologically important processes (Fig.12.1) Migunova et al. (2021) investigated a variety of bacterial strains for RKND management (Root-knot nematode diseases). The application of *B. velezensis* BZR 86 greatly reduced the development of root-knot disease on tomato and cucumber plants and significantly boosted cucumber plant growth and biomass in proportion to bacterial concentration. They showed that the strain *B. velezensis* BZR 86 is a rich source of new, creative products for sustainable agricultural systems. It can be used as a biofertilizer and as an additional tool to manage the root-knot disease on horticultural crops in an ecologically safe manner. *Bacillus* spp. is another group of bacterial agents identified as one of the most promising nematode antagonists. These nematode antagonists, such as *B. cereus* and *B. megaterium*, have been found to be crucial in effectively managing root-knot nematodes and improving crop production (El-Wakeel et al. 2020). As a result, *Bacillus* spp. function as biofertilizers, in addition to their roles as hormones and enzymes that promote plant growth (discussed earlier), which all work together to boost plant growth and

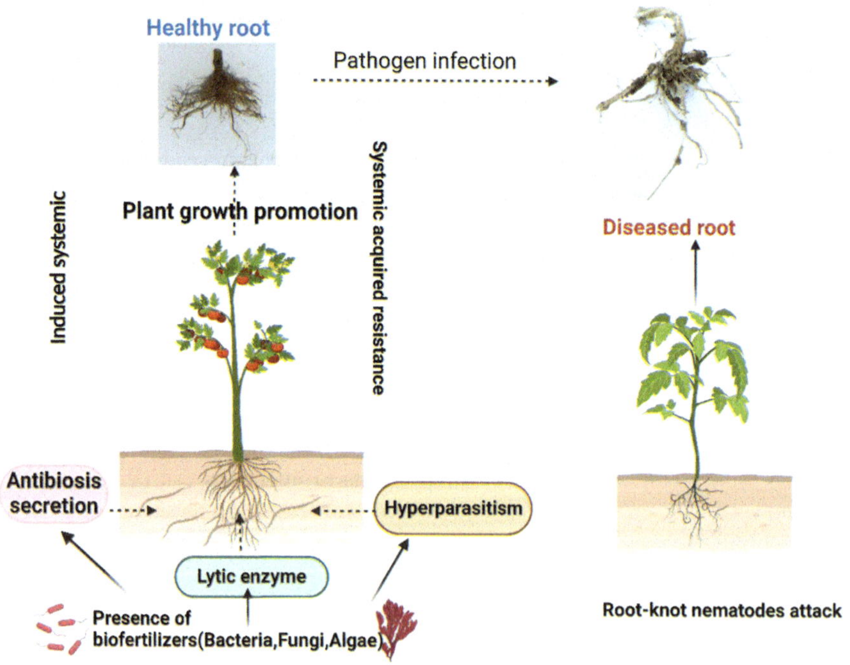

Fig. 12.1 Mechanism of antagonistic biofertilizers (Bacteria, Fungi, Algae) in controlling root-knot diseases

yield. It is important to note that the release of phytotoxic substances, bacterial metabolites, enzymes, and growth regulators may play a role in the selective action of bacterial strains as a bioherbicide (Li and Kremer 2006).

Fungi belonging to the genera *Trichoderma* and *Fusarium* are known to be able to counteract the effects of *Meloidogyne* species. It has been established that *Trichoderma* species can invade plants' root surfaces. This has been connected to its ability to reduce the severity of the disease known as root-knot nematode and the competition it faces from other pests (Mukhtar et al. 2021). *Trichoderma* sp. could reduce the number of *Meloidogyne* sp. second-stage juveniles (J2) and eggs in tomato roots. *Trichoderma* is an organism that lives in the rhizosphere and spreads to the surface of plant roots. Their antimicrobial activity effect is mostly on fungi but also affects the RKN life cycle (Kiriga et al. 2018).

Biofertilizer became an alternative because it is better for the environment and human health. One of the fungal cultures explored for this purpose is *Trichoderma*. *Trichoderma* can synthesize volatile compounds, and its capacity to solubilize phosphates, making them available to the plant, has complicated its use as a biofertilizer. Farmers also utilize it as a biofertilizer since it promotes the plant's uptake of macro and micronutrients. *Trichoderma* works as a biofungicide through a variety of processes, including mycoparasitism, antibiosis, competitive advantage in the rhizosphere, and priming of the crop's defense mechanisms. The effectiveness of different concentrations of *Trichoderma virens* against *Meloidogyne incognita* in vitro was examined by Khan et al. (2022). Additionally, the use of *T. virens* in combination with *M. incognita* was examined in pot-grown chickpea plants. It was discovered that this combination was substantially more efficient at preventing root galling disease and enhanced the growth and physiological characteristics of the plants. It is well known that biofertilizers, particularly *Trichoderma*, can create toxins and antibiotics such as viridian, fusaric acid, lilacin, oxalic acid, trichoderin, trichodermol A, harzianolide, and penicillic acid, all of which inhibit the formation of RKNs (Devi and Bora 2018). Mycorrhizal fungi provide resistance to the plant root and soil against different pathogens (Odoh et al. 2020). Biofertilizers secrete fungistatic and antibiotic-like substances, minimizing the effects of harmful fertilizers, bacteria, and nematodes. Biochar biofertilizer induces resistance in the crop systematically against fungus and bacteria. Use of commercial biofertilizer inoculated with NPK and *Bacillus* spp. showed effective management of *Meloidogyne javanica* (Osman et al. 2021). Application of *Serratia* spp. in combination with urea fertilizer was found effective against root-knot nematode, causing higher mortality of second-stage juveniles in greenhouse conditions (Ketabchi et al. 2016).

According to Abdulrahman and Yüksel (2019), the gall of root-knot nematode was reduced, causing a decrease in the number of galls by using *Paenibacillus polymyxa* followed by mixing *T. harzianum* and *T. viride*. The effect of biofertilizer mixed with compost manure and cattle manure when applied to the greenhouse effect twice (1 week before and after root-knot nematode inoculation) was increased significantly. Inoculation of *P. polymyxa* recorded the highest reduction of hatched juveniles and females, resulting in biological control of *M. incognita* (El-Hadad et al.

2011). *Azotobacter* is used to suppress the growth of saprophytic and pathogenic microorganisms near the root system of plants. Fungi such as *Trichoderma*, Mycorrhiza, and other endophytic fungus induce chemicals in the crops, which increase the resistance of plants against nematodes. Two different strains of *Bacillus* spp., namely, BMH and INV, suppressed root-knot nematode more effectively when applied in combination than when two strains were applied differently individually (Cruz-Magalhães et al. 2022). The combined application of *Amphora* by spraying on the plant surface and by soil drenching was effective in nematode control by enhancing the plants' resistance against nematodes. Combined application of *Amphora* by spraying and soil drenching provided effective results against nematodes by increasing resistance against nematodes and reducing the reproduction rate of nematodes (El-Eslamboly et al. 2019). Alfianny et al. (2017) reported that there are different rhizosphere bacteria association species capable of eliminating the root-knot nematode in the rhizosphere.

12.4 Production and Formulation of Biofertilizers

Biofertilizers, also called bio-inoculants, are environmentally safe and easy-use fertilizers containing living or dormant microorganisms in suitable carrier materials. They are formed by the fermentation and are easily available to plants. The major composition of biofertilizers is bacteria, fungus, algae, etc., creating a symbiotic relationship with plants. The biofertilizer is prepared by the selecting an efficient microbial strain in the suitable nutrient medium by formulating it in a solid or liquid base. The different factors influencing the application of biofertilizers are the specificity of the strain of microorganisms, soil properties, field and laboratory conditions, etc. For mass production of biofertilizers, the following steps are involved (Figs. 12.2 and 12.3):

- Mother cultures are selected based on the performance in the greenhouse and at the field level. The pure culture is grown in the respective medium in the lab. A loopful of inoculum is transferred 250 ml conical flask containing a liquid medium. The conical flask was kept in a rotary shaker for 3–5 days. The mother cultures are further multiplied in larger flasks.

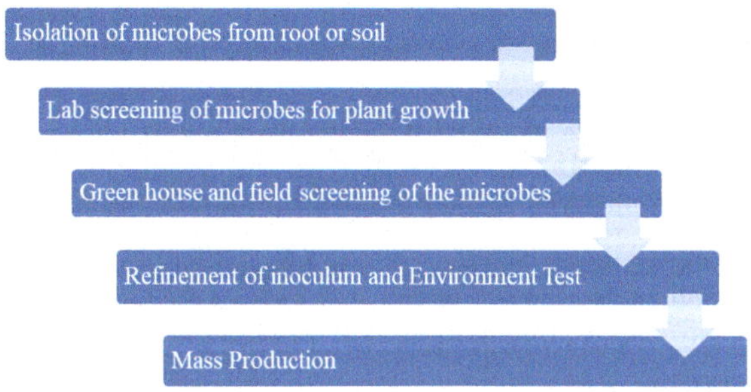

Fig. 12.2 Production and formulation of biofertilizers

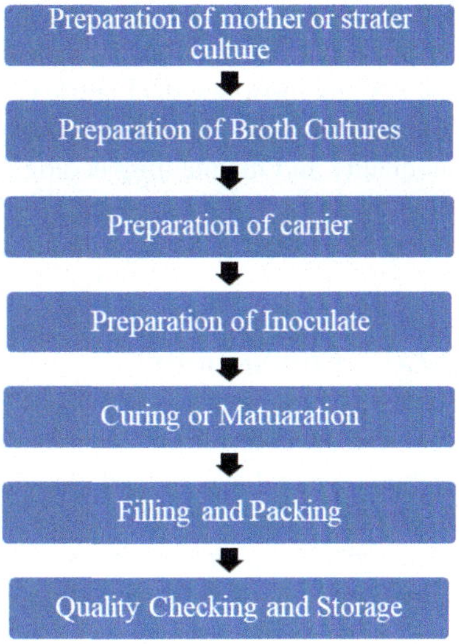

Fig. 12.3 Steps of production of biofertilizers

- Distribute an equal quantity of liquid medium in big conical flasks. Sterilize it in an autoclave for half an hour at 15 lb. pressure. Each flask is inoculated with the mother culture in a ratio of 1:5. The flask is kept in a rotary shaker for about 120 hours until the population reaches 10^9 cells per ml.
- The carrier should have high organic matter and high moisture capacity of 150–200% by weight and provide a nutritive medium for growth. Peat is mostly used as a carrier which is crushed and powdered to 200–300 mesh.
- The sterilized peat is mixed with a high-count broth culture. About 1 part by broth weight is required for two parts of the dry carrier. Final moisture varies from 40 to 50%.
- Curing should be done at room temperature (28 degrees centigrade) for 5–10 days.
- After curing, the sieved powder is filled in a polythene bag and packed by sealing.
- Quality checking should be done, and storage should be done at a temperature of 15 degrees centigrade, not exceeding 30 degrees centigrade, for 6 months.

A combination of more than two biofertilizer strains, including fungal and bacterial, can be used effectively to manage gall nematodes which act by complementing each other with a synergetic effect (Pirttilä et al. 2021).

12.5 Future Outlook

Due to the excessive use of chemical fertilizers, there is a depletion of soil health and an increase in resistance to disease, pests, and nematodes against the control measures. Before the destruction of soil fertility and productivity, alternatives to chemical fertilizers are to be introduced, which minimize the effect of chemical fertilizers and improve the soil structure. Nematode infestations can be managed sustainably by using biofertilizers of organic origin. The application of biofertilizers may be the most practical approach to controlling root-knot nematode infestations. The effectiveness of the association of biofertilizers towards improving environmental quality and maintaining ecological balance will finally be realized. Biofertilizers provide an opportunity to reduce climate change to reduce the impacts of climate change and sustainable agriculture adaptability.

12.6 Conclusions

The use of biofertilizers for managing different diseases and nematodes is a new technology that has gained popularity for its eco-friendly and safe exploitation. Bacterial, fungal, algal, and biochar are major biofertilizers commonly used in nematode management. These biofertilizers improve soil and plant health by improving plants' soil structure and resistance ability against nematodes. The crop associated with biofertilizers can be used in sustainable agriculture cultivation, ecological stability, and enhancing the immunity of plants against insects and pests. Hence, the use of biofertilizers can be very fruitful in the management of nematodes in a sustainable way in vegetable production.

References

Abdulrahman AM, Yüksel A (2019) Effect of some agroecosystems management applications on plant flowering and root-knot nematodes activates using some soil improvements in greenhouses. https://doi.org/10.5829/idosi.ijmr.2019.148.157

Abo-Korah M (2022) Integrated management of root-knot nematode, *Meloidogyne Incognita* infecting cowpea plants. Menoufia J Plant Protect 7:89–95. https://doi.org/10.21608/mjapam.2022.247520

Adiwena M, Murtilaksono A, Egra S, Hoesain M, Asyiah IN, Pradana AP, Izatika ZN (2023) The effects of micronutrient-enriched media on the efficacy of *Bacillus subtilis* as biological control agent against *Meloidogyne incognita*. Biodiversitas J Bio Divers 24

Affokpon A, Coyne DL, Htay CC, Agbèdè RD, Lawouin L, Coosemans J (2011) Biocontrol potential of native *Trichoderma* isolates against root-knot nematodes in West African vegetable production systems. Soil Bio Biochem 43:600–608. https://doi.org/10.1016/j.soilbio.2010.11.029

Afia AI, El-Nuby AS (2016) Seaweeds as a managing approach for root-knot nematode disease. Egyptian J Agronematol 15:95–108

Ahmed N, Ghramh HA, Shakeel Q, Ashraf W, Abbas HT, Binyamin R, Masroor A, Raheel M, Khan Z (2023) Evaluation of Rhizospheric-Pseudomonas spp. for the management of Meloidogyne incognita in tomato. J King Saud Univ Sci 35:102395. https://doi.org/10.1016/j.jksus.2022.102395

Alamri S, Nafady NA, El-Sagheer AM, El-Aal MA, Mostafa YS, Hashem M, Hassan EA (2022) Current utility of arbuscular mycorrhizal fungi and hydroxyapatite nanoparticles in suppression of tomato root-knot nematode. Agronomy 12:671. https://doi.org/10.3390/agronomy12030671

Alfianny R, Aryantha IN, Syamsudin TS (2017) Role of indigenous rhizosphere bacteria in suppressing root-knot nematode and improve plant growth tomato. J Plant Pathol 16:25–32. https://doi.org/10.3923/ppj.2017.25.32

Ammar EE, Aioub AA, Elesawy AE, Karkour AM, Mouhamed MS, Amer AA, El-Shershaby NA (2022) Algae as bio-fertilizers: between current situation and future prospective. Saudi J Biol Sci 29:3083–3096. https://doi.org/10.1016/j.sjbs.2022.03.020

Antil S, Kumar R, Pathak DV, Kumar A, Panwar A, Kumari A, Kumar V (2022a) Potential of *Bacillus altitudinis* KMS-6 as a biocontrol agent of *Meloidogyne javanica*. J Pest Sci 1:1–0. https://doi.org/10.1007/s10340-021-01469-x

Antil S, Kumar R, Pathak DV, Kumar A, Panwar A, Kumari A (2022b) Plant growth-promoting rhizobacteria-*Bacillus cereus* KMT-5 and *B. megaterium* KMT-8 effectively suppressed *Meloidogyne javanica* infection. Appl Soil Eco 174:104419. https://doi.org/10.1016/j.apsoil.2022.104419

Baweja P, Kumar S, Kumar G (2019) Organic fertilizer from algae: a novel approach towards sustainable agriculture. Biofertilizers for sustainable agriculture and environment. Springer, pp 353–370. https://doi.org/10.1007/978-3-030-18933-4_16

Cruz-Magalhães V, Guimarães RA, da Silva JC, de Faria AF, Pedroso MP, Campos VP, Marbach PA, de Medeiros FH, De Souza JT (2022) The combination of two *Bacillus* strains suppresses *Meloidogyne incognita* and fungal pathogens, but does not enhance plant growth. Pest Manag Sci 78:722–732. https://doi.org/10.1002/ps.6685

d'Errico G, Greco N, Vinale F, Marra R, Stillittano V, Davino SW, Woo SL, D'Addabbo T (2022) Synergistic effects of *Trichoderma harzianum*, 1, 3 Dichloropropene and organic matter in controlling the root-knot nematode *Meloidogyne incognita* on tomato. Plan Theory 11:2890. https://doi.org/10.3390/plants11212890

Dahal N, Bajracharya RM, Merz J (2016) Prospects of bicar as soil amendment in Nepal hill farming systems. J Food Agric Environ 17:92–103. https://doi.org/10.3126/aej.v17i0.19865

Devi G, Bora LC (2018) Effect of some biocontrol agents against root-knot nematode (*Meloidogyne incognita* race2). Int J Environ Agric Biotech 3:265260. https://doi.org/10.22161/ijeab/3.5.23

Díaz-Manzano FE, Amora D, Martínez-Gómez Á, Molbak L, Escobar C (2023) Biocontrol of *Meloidogyne spp.* in *Solanum lycopersicum* using a dual combination of *Bacillus* strains. Front Plant Sci 4:5205. https://doi.org/10.3389/fpls.2022.1077062

El-Ansary MS, Hamouda RA (2014) Biocontrol of root-knot nematode infected banana plants by some marine algae. Russ J Mar Biol 40:140–146. https://doi.org/10.1134/S1063074014020047

El-Eslamboly AA, Abd El-Wanis MM, Amin AW (2019) Algal application as a biological control method of root-knot nematode *Meloidogyne incognita* on cucumber under protected culture conditions and its impact on yield and fruit quality. Egypt J Biol Pest Control 29:1–9. https://doi.org/10.1186/s41938-019-0122-z

El-Hadad ME, Mustafa MI, Selim SM, El-Tayeb TS, Mahgoob AE, Aziz NH (2011) The nematicidal effect of some bacterial biofertilizers on *Meloidogyne incognita* in sandy soil. Braz J Microbiol 42:105–113. https://doi.org/10.1590/S1517-83822011000100014

El-Wakeel AM, El-Nagdi MA, Soliman MG, Elkelany US (2020) Dwal manage-ment of root-knot nematode Meloidogyne incognita and weeds infested *solanum lycopersicum* with protoplast fusants of *Bacillus cereus* and *Pseudomonas aeruginosa*. Plant Archives 20:9133–9143

Expósito A, García S, Giné A, Escudero N, Herranz S, Pocurull M, Lacunza A, Sorribas FJ (2022) Effect of molasses application alone or combined with *Trichoderma asperellum* T-34 on *Meloidogyne spp.* management and soil microbial activity in organic production systems. Agronomy 12:1508. https://doi.org/10.3390/agronomy12071508

García-Fraile P, Menéndez E, Rivas R (2015) Role of bacterial biofertilizers in agriculture and forestry. AIMS Bioeng 2:183–205. https://doi.org/10.3934/bioeng.2015.3.183

Ghareeb RY, Adss IA, Bayoumi SR, El-Habashy DE (2019) The nematicidal potentiality of some algal extracts and their role in enhancement the tomato defense genes against root knot-nematodes. Egyptian J Biol Pest Control 29:1–0. https://doi.org/10.1186/s41938-019-0153-5

Ghosh D, Ghorai P, Debnath S, Indrama T, Kondi V, Tiwari ON (2022) Algal biofertilizer towards green sustainable agriculture. In: New and future developments in microbial biotechnology and bioengineering. Elsevier, pp 27–45. https://doi.org/10.1016/B978-0-323-85579-2.00019-8

Gowda MT, Rai A B, Singh B (2007) Root knot nematode: a threat to vegetable production and its management https://doi.org/10.1002/ps.6685

Gowda T, Prasanna R, Kundu A, Rana V, Chawla G (2023) Characterization and evaluation of native rhizobacteria isolated from *Meloidogyne incognita*-infected tomato (*Solanum lycopersicum*). Indian J Agric Sci 93:175–180

Hamouda R, Al-Saman M, El-Ansary M (2019) Effect of *Saccharomyces cerevisiae* and *Spirulina platensis* on suppressing root-knot nematode, *Meloidogyne incognita* infecting banana plants under greenhouse conditions. Egyptian J Agronematol 18:90–102. https://doi.org/10.21608/ejaj.2019.52593

Haque Z, Khan MR (2022) Integrated management of rice root-knot nematode, Meloidogyne graminicola with *Pseudomonas putida* and fluopyram. Indian J Nematol 52:66–75. https://doi.org/10.5958/0974-4444.2022.00010.5

Huang WK, Sun JH, Cui JK, Wang GF, Kong LA, Peng H, Chen SL, Peng DL (2014) Efficacy evaluation of fungus *Syncephalastrum racemosum* and nematicide avermectin against the root-knot nematode *Meloidogyne incognita* on cucumber. PLoS One 9:e89717. https://doi.org/10.1371/journal.pone.0089717

Huang Z, Lu Q, Wang J, Chen X, Mao X, He Z (2017) Inhibition of the bioavailability of heavy metals in sewage sludge biochar by adding two stabilizers. PLoS One 12:e0183617. https://doi.org/10.1371/journal.pone.0183617

Ibrahim DS, Ghareeb RY, Abou El Atta DA, Azouz HA (2021) Nematicidal properties of methanolic extracts of two marine algae against tomato root-knot nematode, *Meloidogyne incognita*. J Plant Prot Pathol 12:137–144. https://doi.org/10.21608/JPPP.2021.154414

Izuogu NB, Olajide TU, Eifediyi EK, Olajide CM (2019) Effect of root-knot nematode (*Meloidogyne incognita*) on the nodulation of some varieties of cowpea (*Vigna unguiculata L. Walp)*. Sci Agric Bohem 50:104–109. https://doi.org/10.2478/sab-2019-0015

Janati S, Houari A, Wifaya A, Essarioui A, Mimouni A, Hormatallah A, Sbaghi M, Dababat AA, Mokrini F (2018) Occurrence of the root-knot nematode species in vegetable crops in Souss region of Morocco. J Plant Pathol 34:308. https://doi.org/10.5423/PPJ.OA.02.2018.0017

Ketabchi S, Charehgani H, Majzoob S (2016) Impact of rhizosphere antagonistic bacteria and urea fertilizer on root knot nematode (*Meloidogyne incognita*) under green house condition. J Anim Plant Sci 26

Khalil HA, Yahya EB, Jummaat F, Adnan AS, Olaiya NG, Rizal S, Abdullah CK, Pasquini D, Thomas S (2022) Biopolymers based aerogels: A review on revolutionary solutions for smart therapeutics delivery. Prog Mater Sci 31:101014. https://doi.org/10.1016/j.pmatsci.2022.101014

Khan Z, Rahman MH, Haider G, Amir R, Ikram RM, Ahmad S, Schofield HK, Riaz B, Iqbal R, Fahad S, Datta R (2021) Chemical and biological enhancement effects of biochar on wheat growth and yield under arid field conditions. Sustainability 13:5890. https://doi.org/10.3390/su13115890

Khan A, Bani Mfarrej MF, Nadeem H, Ahamad L, Hashem M, Alamri S, Gupta R, Ahmad F (2022) Trichoderma virens mitigates the root-knot disease progression in the chickpea plant. Acta Agric Scand, Section B-Soil Plant Sci 72:775–787. https://doi.org/10.1080/09064710.2022.2080107

Kiriga AW, Haukeland S, Kariuki GM, Coyne DL, Beek NV (2018) Effect of *Trichoderma* spp. and *Purpureocillium lilacinum* on *Meloidogyne javanica* in commercial pineapple production in Kenya. Biol Control 119:27–32. https://doi.org/10.1016/j.biocontrol.2018.01.005

Kumar M, Kumar K (2019) Role of bio-fertilizers in vegetables production: a review. J Pharmacogn Phytochem 8:328–334

Kumar A, Usmani Z, Kumar V (2017) Biochar and flyash inoculated with plant growth promoting rhizobacteria act as potential biofertilizer for luxuriant growth and yield of tomato plant. J Environ Manag 190:20–27. https://doi.org/10.1016/j.jenvman.2016.11.060

Kumar A, Kumar V, Bruno LB, Rajkumar M (2022) Synergism of industrial and agricultural waste as a suitable carrier material for developing potential biofertilizer for sustainable agricultural production of eggplant. Horticulturae 8:444. https://doi.org/10.3390/horticulturae8050444

Li J, Kremer RJ (2006) Growth response of weed and crop seedlings to deleterious rhizobacteria. Biol Control 39:58–65. https://doi.org/10.1016/j.biocontrol.2006.04.016

Liu M, Philp J, Wang Y, Hu J, Wei Y, Li J, Ryder M, Toh R, Zhou Y, Denton MD, Wu Y (2022) Plant growth-promoting rhizobacteria *Burkholderia vietnamiensis* B418 inhibits root-knot nematode on watermelon by modifying the rhizosphere microbial community. Sci Rep 12:1–3. https://doi.org/10.1038/s41598-022-12472-2

Lopez-Perez JA, Roubtsova T, de Cara GM, Ploeg A (2010) The potential of five winter-grown crops to reduce root-knot nematode damage and increase yield of tomato. J Nematol 42:120

Mahmud AA, Upadhyay SK, Srivastava AK, Bhojiya AA (2021) Biofertilizers: a nexus between soil fertility and crop productivity under abiotic stress. Curr Res Environ Sustain 3:100063. https://doi.org/10.1016/j.crsust.2021.100063

Malusá E, Sas-Paszt L, Ciesielska J (2012) Technologies for beneficial microorganisms Inocula used as biofertilizers. Sci World J 491206:1. https://doi.org/10.1100/2012/491206

Migunova VD, Tomashevich NS, Konrat AN, Lychagina SV, Dubyaga VM, D'Addabbo T, Sasanelli N, Asaturova AM (2021) Selection of bacterial strains for control of root-knot disease caused by *Meloidogyne incognita*. Microorganisms 9:1698. https://doi.org/10.3390/microorganisms9081698

Mitter EK, Tosi M, Obregón D, Dunfield KE, Germida JJ (2021) Rethinking crop nutrition in times of modern microbiology: innovative biofertilizer technologies. Front Sustain Food Syst 5. https://doi.org/10.3389/fsufs.2021.606815

Mukhtar T, Tariq-Khan M, Aslam MN (2021) Bioefficacy of *Trichoderma* species against javanese root-knot nematode, *Meloidogyne javanica*, in green gram. Gesunde Pflanzen 73:265–272. https://doi.org/10.1007/s10343-021-00544-8

Nafady NA, Sultan R, El-Zawahry AM, Mostafa YS, Alamri S, Mostafa RG, Hashem M, Hassan EA (2022) Effective and promising strategy in management of tomato root-knot nematodes by Trichoderma harzianum and arbuscular mycorrhizae. Agronomy 12:315. https://doi.org/10.3390/agronomy12020315

Odoh CK, Eze CN, Obi CJ, Anyah F, Egbe K, Unah UV, Akpi UK, Adobu US (2020) Fungal biofertilizers for sustainable agricultural productivity. Agriculturally important fungi for sustainable agriculture: volume 1: perspective for diversity and crop productivity. Springer, pp 199–225. https://doi.org/10.1007/978-3-030-45971-0_9

Osman HA, Ameen HH, Mohamed M, El-Sayed GM, Dawood MG, Elkelany US (2021) Bio-fertilizers' protocol for controlling root knot nematode *Meloidogyne javanica* infecting peanut fields. Egypt J Biol Pest Control 311–319. https://doi.org/10.1186/s41938-021-00471-w

Pirttilä AM, Mohammad Parast Tabas H, Baruah N, Koskimäki JJ (2021) Biofertilizers and biocontrol agents for agriculture: how to identify and develop new potent microbial strains and traits. Microorganisms 9:817. https://doi.org/10.3390/microorganisms9040817

Pourkhaloei A, Saberi Riseh R, Moradi M, Vatankhah M, Loit E (2022) Biological control of root-knot nematode *Meloidogyne incognita* in pistachio using bacterial biocontrol agents. Pistachio Health J 5:62–75

Rani P, Singh M, Prashad H, Sharma M (2022) Evaluation of bacterial formulations as potential biocontrol agents against the southern root-knot nematode, *Meloidogyne incognita*. Egypt J Biol Pest Control 32:29. https://doi.org/10.1186/s41938-022-00529-3

Raouf AN, Al-Homaidan AA, Ibraheem IB (2012) Agricultural importance of algae. Afr J Biotechnol 11:11648–11658. https://doi.org/10.5897/AJB11.3983

Soares PL, Nascimento DD (2021) Integrated nematode management of root lesion and root-knot nematodes in soybean in Brazil. In: Integrated Nematode Management: State-of-the-art and visions for the future. CABI, Wallingford UK, pp 103–110. https://doi.org/10.1079/9781789247541.0015

Sun X, Zhang R, Ding M, Liu Y, Li L (2021) Biocontrol of the root-knot nematode *Meloidogyne incognita* by a nematicidal bacterium *Pseudomonas simiae* MB751 with cyclic dipeptide. Pest Manag Sci 77:4365–4374. https://doi.org/10.1002/ps.6470

Terefe M, Tefera T, Sakhuja PK (2009) Effect of a formulation of *Bacillus firmus* on root-knot nematode *Meloidogyne incognita* infestation and the growth of tomato plants in the greenhouse and nursery. J Invertebr Pathol 100:94–99. https://doi.org/10.1016/j.jip.2008.11.004

Williams TI, Edgington S, Owen A, Gange AC (2021) Evaluating the use of seaweed extracts against root knot nematodes: a meta-analytic approach. Appl Soil Ecol 168:104170. https://doi.org/10.1016/j.apsoil.2021.104170

Wolna-Maruwka A, Piechota T, Niewiadomska A, Kamiński A, Kayzer D, Grzyb A, Pilarska AA (2021) The effect of biochar-based organic amendments on the structure of soil bacterial community and yield of maize (*Zea mays* L.). Agronomy 11:1286. https://doi.org/10.3390/agronomy11071286

Zhang S, Gan Y, Xu B (2015) Biocontrol potential of a native species of *Trichoderma longibrachiatum* against *Meloidogyne incognita*. Appl Soil Ecol 94:21–29. https://doi.org/10.1016/j.apsoil.2015.04.010

Prospects for the Use of Metabolomics Engineering in Exploring and Harnessing Chemical Signaling in Root Galls

13

Faryad Khan, Ekta Pandey, Saba Fatima, Arshad Khan, Saeeda Zaima Zeb, and Faheem Ahmad

Abstract

It is fascinating to note that plant parasitic nematodes (PPNs) have been shown to destroy crops all over the globe extensively. Synthetic nematicides or chemicals are used to stop their spread, but their prolonged use has adversely harmed human, animal, and plant populations. In natural habitats, interactions between host plant roots, various growth-promoting microbes, and plant parasitic nematodes (e.g., cyst nematodes and root-knot nematodes) are frequent. While each of these interactions between the host plant and the plant parasitic nematode, or PGPMs, influences each other's biological activity via various chemical signals such as secondary metabolites, phytohormones, enzymes, etc. Many metabolomics strategies, including gas and liquid chromatography, mass spectrometry, nuclear magnetic resonance, electrospray ionization, mass spectrometry imaging, and Fourier transform infrared spectroscopy, are being used to investigate these chemical and metabolic interactions. Metabolomics offers qualitative and quantitative techniques for analyzing the different defense and resistance mechanism approached by PGPMs and host against diverse pathogen and PPNs. This chapter studies the modern metabolomics approach to identify the metabolites synthesized and released during the host plant roots and gall-inducing nematode interactions and their role in different chemical signaling pathways.

F. Khan (✉)
Department of Botany, Aligarh Muslim University, Aligarh 202002, India

Department of Botany, Krishna Devi Balika P.G. College, Farrukhabad, India

E. Pandey · S. Fatima · A. Khan · S. Z. Zeb · F. Ahmad
Department of Botany, Aligarh Muslim University, Aligarh 202002, India

F. Ahmad, G. Nombela (eds.), *Root-Galling Disease of Vegetable Plants*,
https://doi.org/10.1007/978-981-99-3892-6_13

Keywords

Metabolites · Metabolomics · Phytohormones · Plant parasitic nematodes · Rhizosphere

13.1 Introduction

In the past few years, metabolomics has become one of the most important scientific breakthroughs. It has enabled researchers to accurately profile metabolites in microbes, plants, and animals (Ryan and Robards 2006; Heyman and Dubery 2016; Zeng et al. 2020). The term "metabolomics" was coined by Fiehn et al. in 2001. They defined it as "a complete and quantitative analysis of all metabolites in a biological system." It makes a profile of small molecules that emerge from cellular metabolism and can directly show the results of complicated systems of biochemical reactions. In the field of metabolomics, metabolites are profiled and characterized, and their relative abundance is evaluated using analytical techniques like chromatography, mass spectrometry (MS), nuclear magnetic resonance (NMR) and IR spectroscopy, and Fourier transform (FT) spectroscopy. This gives us information about many different parts of how cells work (Liu and Locasale 2017). Metabolomics looks at all the small molecule parts and how they change in individual cells, cellular components, tissue types, or organs. It is frequently utilized to study plants and microbial systems. Presently, metabolomics is a growing field in the omics sciences that emphasizes high-throughput snapshots of metabolomes (Shafi et al. 2021). The plant kingdom is thought to include five million potential metabolites. In addition to structural variability, the metabolome cannot be completely covered due to geographical and seasonal differences, as well as wide concentration ranges. Thus, combining knowledge gathered with various extraction procedures and analytical instruments such as GC-MS, LC-MS, NMR, or FT-IR (Weckwerth 2003, 2011) has consistently been advised.

Roots of plants often interact with microorganisms in their natural habitats. The exudates secreted by plant roots in the rhizosphere are a common way to communicate between plants and microorganisms. The chemical nature of these root exudates affects the microbial populations in the rhizosphere, which is the zone surrounding the roots (Sasse et al. 2018). Complex chemical communication is used by plant roots to interact with microbes in the rhizosphere. Plant-microbe (including pathogen) interactions have been elucidated with the use of metabolomics. Plant parasitic nematodes (PPNs) are a significant agroeconomic problem due in part to the lack of efficient countermeasures and their intricate relationship with their host. There are more than 4000 known PPN species (Decraemer and Hunt 2006; Nicol et al. 2011); most of them feed on roots, but some also feed foliage (Fuller et al. 2008). Although there are many different types of sedentary PPN, the root-knot nematodes (*Meloidogyne* spp.) and cyst nematodes are responsible for most economic losses (*Heterodera* spp. and *Globodera* spp.) (Fuller et al. 2008; Nicol et al. 2011). Infections caused by PPNs result in average postharvest losses of 12.3% and an annual economic loss of 157 billion dollars (Singh et al. 2015). Root-knot nematodes (RKNs) of the genus *Meloidogyne* and cyst nematodes of the genera *Heterodera* and *Globodera* are among the 10 most destructive plant nematodes (Jones et al. 2013). *M. incognita* and *M. javanica* are regarded as the most quickly spreading pests and diseases in the globe (Bebber et al. 2014). RKNs cause the formation of galls or knots, which are syncytial feeding structures in the host roots. Tiny root areas inside nematode-induced syncytia grow rapidly. Each of these galls is made up of several large cells (Jones and Payne 1978). Strong sink tissues are formed by the feeding structures, which are hypothesized to be metabolically active (Hofmann et al. 2010). Nothing is understood currently regarding metabolic changes that occur during syncytium formation. Although there are many parasitic plant species with economic importance, the largest hazard to agricultural crops globally comes from the root gall-forming nematodes of the family Heteroderidae.

Many metabolomics investigations have been carried out recently with the goal of expanding our knowledge of plant-nematode interactions (Ali et al. 2015). Metabolomics has proven to be an effective tool for elucidating the specificity of plant-RKN relationships. Severe changes in the primary metabolism of plants are probable because of the high nutritional and energy demands of pathogen and the dramatic reconfiguration of infected plant cells. In addition, nematodes might create novel metabolic pathways in the host plants by stimulating the manufacture of certain substances required to their food (Hofmann et al. 2010). Metabolomes reveal the metabolic profiling of root galls and how the changes occur in the cellular pathways of giant cells. In this chapter, we put more emphasis on metabolomics and considered various metabolic pathways and signaling pathways, which are directly and indirectly involved in the development of root galls.

13.2 Rhizospheric Biology of Host–Pathogen Interaction

The rhizosphere is a center for diverse and interesting microorganism interaction as well as among the most complicated ecosystems on earth, and it provides habitat for a dense population, diversified collection of intensively metabolizing soil microorganisms like bacteria, mycorrhizal fungi, protists, herbivore insects, nematodes, invertebrates, etc., each of which interacts with one another in sophisticated trophic trading networks (Mhlongo et al. 2018; Khanna et al. 2021; Li et al. 2021). Rhizosphere microbes may be helpful or hazardous to the host plant development (Nihorimbere et al. 2011). The harmful microorganisms, like soil-borne pathogenic organisms and parasites, limit growth of the plant, start causing yield decline, and degrade agricultural output that have been intensively investigated for past decade (Ab Rahman et al. 2018). Besides this, beneficial microorganisms (including mutualistic microbes) may stimulate plant development by improving food availability, generating phytohormones, and raising resistant to plant parasitic nematodes and biotic or abiotic barriers (Rolli et al. 2015; Yin et al. 2021). The typically limited organic content in soil stimulates a struggle between microorganisms resulting in the development of unique interconnections between them. Hence, organisms generated several specialized signaling serving their community to improve fitness in continuously adjusting soil circumstances. The interaction of plants with soil microbes is mostly driven by signaling molecules that takes place at the root area.

Roots serve a core part of soil ecosystem in promoting biochemical processes like anchoring and enhancement of water and nutrients transport, and they also exude an array of compounds called root exudates supporting activities which include lubricating, defensive, and other physiological roles. For example, sugars, organic acids, amino acids, polyphenols, flavonoids, hormones, mucilage, enzymes, alkaloids, vitamins, and terpenoids are a few metabolites that are secreted by roots for connecting efficiently with microbes present in rhizosphere (Kaur and Sodhi 2022). Basically, such chemical substances substitute as a source of nitrogen and carbon supply to plants and include promoting multiplication of helpful microorganisms as well as suppressing soil-borne pathogens.

The chemical warfare in rhizosphere via exudation released by root culminates into both negative and positive responses to each other. For understanding, chemotaxis, the positive biochemical signals generated via the roots towards plant growth-promoting microbes (PGPM). These encourage the expression of growth elicitors and promote cross communication among plants and rhizospheric microbiota. On its other hand, the unfavorable associations prompted many antimicrobial compounds, toxicants, and nematicide compounds (Knights et al. 2021; Khanna et al. 2021). Additionally, competing on resources, allelopathy, chemical invasion, and pathogen are also the major factors causing negative interactions. Surprisingly, root exudates communicate all these routes creating complexity for diverse reactions. Although a few little root exudates work like phytotoxins, some are crucial to changing soil physical attributes, microbial populations, and symbiotic. Fascinatingly, most root secreted exudates are important to supporting crucial defensive mechanisms in plants and minimizing the vulnerability to pathogenicity (Chagas et al. 2018). However, some defensive pathways are also activated by VOCs (volatile compound) that actively work against plant prey like plant parasitic nematode.

Plant parasitic nematodes (PPNs) are the biggest obstacles worldwide, crossing significant costs of agricultural destruction. They are recognized to be the persistent pathogens to cause severe deterioration of crops. Approximately, 4000 species of PPNs have been recognized worldwide, entirely obligate parasites contributing to significant harm in agriculture (Zinovieva 2014; Khanna et al. 2021). Mainly, research has been predominantly focused on two groups of PPNs, respectively, (CNs) cyst nematodes (*Heterodera* sp. and *Globodera* sp.) and (RKNs) root-knot nematodes (*Meloidogyne* sp.) due to economic aspect. PPNs infect the plants mostly in juvenile stage following egg hatching and shattered root infrastructure followed by filching plant nutrients and expose plant to certain other harmful pathogenic attacks. Numerous studies have emphasized the beneficial microorganism's capability to avoid or mitigate the PPNs as biological controller. The availability of beneficial microorganisms and their own metabolites are sufficient to stably reduce or restrict the plant parasitic nematodes development. Figure 13.1 represents

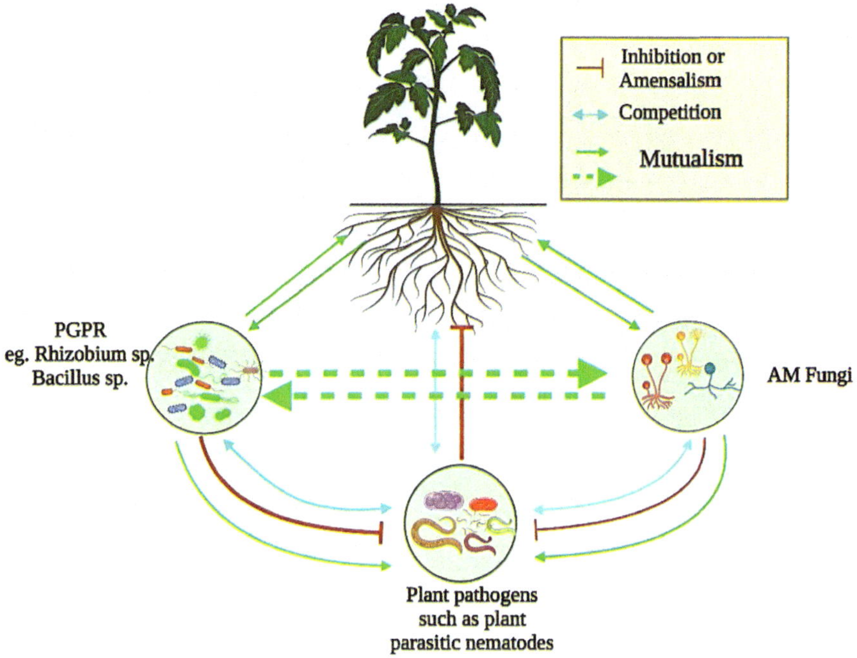

Fig. 13.1 Interactions and different relationship between PGPR, AM fungi, plant parasitic nematodes, and host plant

interactions and different relationships between PGPR, AM fungi, plant parasitic nematodes, and host plant.

13.3 Metabolites and Metabolomics

Plants may create thousands of different metabolites that work as natural chemicals, attracting pollinators, avoiding herbivores, defending against microbial diseases, and protecting against environmental stresses. Primary and specialized (secondary) metabolisms are the two broad categories into which plant metabolites may be classified (Pott et al. 2019; Castro-Moretti et al. 2020). While the plant's primary metabolites contain substances essential to its development and reproduction, specialized or secondary metabolites include substances required to the plant for withstand plant parasitic pathogens, abiotic and biotic pressures (Fig. 13.2). The metabolites of fundamental metabolic biochemical pathways including EMP-pathway, the TCA cycle, and the (PPP) pentose-phosphate pathway also act as the basic components of secondary metabolic pathways, highlighting the intrinsic link between these classes of metabolism (Tsugawa 2018). For example, amino acids have a role in the absorption of nitrogen as well as serving as intermediates for a variety of specialized chemicals, such as pigments and phytohormones.

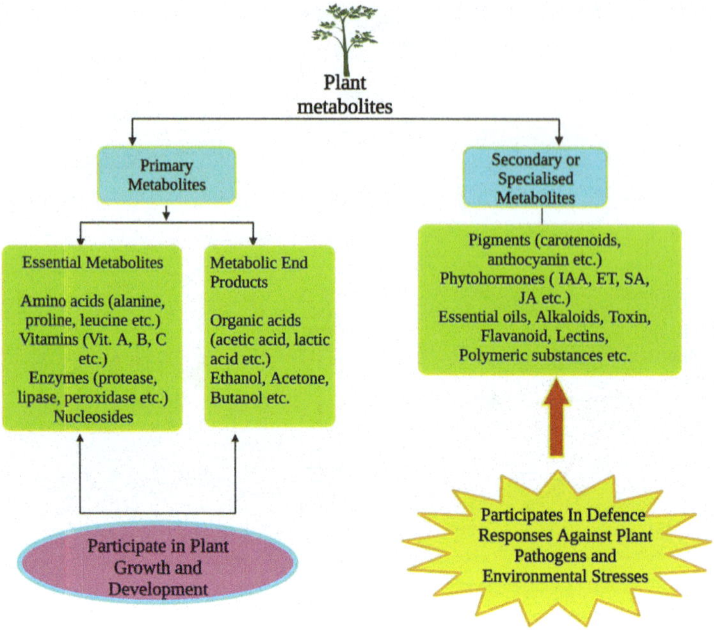

Fig. 13.2 The chart shows the classification of plant metabolites and their role in plant development and defense system

Phytoanticipins and phytoalexins are two categories of biocidal secondary (specialized) metabolites synthesized by plants to defend themselves against pathogenic and pests attack (Ren et al. 2018; Desmedt et al. 2020). Besides this, phytoalexins could be naturally present in an inactivated storage form (such as a glycoside) from which they are released in response to the perception of a pest or pathogen.

One postgenomic method for examining the microorganisms in the rhizosphere is metabolomics. The field of metabolomics makes use of analytical methods like GC and LC (gas and liquid chromatography), mass spectrometry (MS), NMR (nuclear magnetic resonance), infrared spectroscopy (IR spectroscopy), and Fourier transform (FT) spectroscopy to recognize, profile, and evaluate the comparative abundance of metabolites at a specific time. The focus of metabolomics is the fingerprinting, analyzing, and profiling of metabolites. The detection of each metabolite in a sample, regardless of identity, is the process of fingerprinting. These techniques, NMR, FT-IR spectroscopy, and electrospray ionization (ESI)-MS, are frequently used to screen a biological system to determine if different metabolites are present in the control and testing material. Thus, a less expensive first technique is provided before more expensive metabolic profiling. Detection, classification, and, if applicable, identifying metabolites inside an extract by using chromatographic separation techniques (such as GC or liquid chromatography (LC)) in association with MS techniques are all part of metabolic profiling. Some popular technologies applied in metabolomics are listed and briefly discussed below.

13.3.1 Nuclear Magnetic Resonance (NMR)

Nuclear magnetic resonance (^{1}H-NMR) reveals the organization of hydrogen atoms in a molecule, while ^{13}C-NMR reveals the order of carbon atoms in a molecule. NMR is typically used in metabolomics to analyze polar substances. It is less accurate (micromolar range) than MS-based approaches, but more robust in terms of identification and repeatability (van Dam and Bouwmeester 2016). Nuclear magnetic resonance (NMR) is an objective and unbiased analytical method that accurately identifies the molecular makeup of substances. The analysis of metabolites using NMR, both quantitative and qualitative, has been used extensively.

Theoretically, ^{1}H-NMR may provide an individual signal for each chemically different hydrogen nucleus, enabling the operator to link together the structure of a substance. In contrast to other metabolomic techniques, ^{1}H-NMR may be thought of as nonbiased since a significant portion of the biological compounds (metabolites) contains hydrogen (Bharti and Roy 2012). So, despite having a relative sensitivity that is just half those of MS-based techniques, NMR is becoming the mainstream technology for metabolic profiling. However, NMR responsiveness does vary depending on substance class, with resolving and spectrum crowding also having an influence. The possibilities of NMR-based metabolomics are substantially enhanced by LC-NMR. By combining LC and NMR, the complicated sample may be significantly simplified using recent (HPLC and UPLC) column chromatography methods (De Koning et al. 1998; Liu et al. 2021).

13.3.2 FTIR Spectroscopy and Electrospray Ionization (ESI)-MS

An organized and continually evolving analytical method, FTIR spectroscopy allows for the non-destructive, high-throughput (thousands of samples per day), and incredibly quick (seconds per sample) analysis of a vast range of different sample forms. The fundamental idea behind this technique is that whenever an infrared beam probes a sample, functional groups inside the sample absorbed the light and vibrate in one of many present ways, such as stretching, bending, and deformation vibrations. As these peaks of absorptions and vibrations are directly connected to biochemical species, the resulting infrared spectrum may be thought of as the infrared—or even metabolic—fingerprint of any biochemical compound (Ellis et al. 2002; Allwood et al. 2008). Specified IR wavelengths cause different kinds of biochemical substances to interact. The spectral frames for the following types of compounds may be found in the mid-IR region (4000 cm^{-1}–600 cm^{-1}): fatty acids (2800 cm^{-1}–3050 cm^{-1}), amides (1650 cm^{-1}–1800 cm^{-1}), mixed area (1200 cm^{-1}–1450 cm^{-1}), and polysaccharides (1050 cm^{-1}–1150 cm^{-1}) (Allwood et al. 2008).

FT-IR is comparatively less costly than mass spectrometric or other spectroscopic methods, and it is well suited to becoming a quick first-round screening technique.

The most popular technique to ionize molecules prior to a mass spectrometer is electron spray ionization (ESI), which may be used with both LC and GC technology. The analytes are either deprotonated (M−, negative mode) or protonated (M+, positive mode), depending on the potential across the ESI nozzle (van Dam and Bouwmeester 2016). Analyzing a similar sample in M+ and M− mode broadens the range of metabolites that are recorded because molecules vary in their tendency to receive or release a proton.

13.3.3 Mass Spectrometry Imaging

The spatiotemporal arrangement of several biological molecules in tissues may be measured using the emerging method known as mass spectrometry imaging (MSI). Since it may directly connect molecular alterations and histology, mass spectrometry imaging (MSI) is a prospective approach for pathogenic analysis and the exploration of causes. The study of components, metabolites, peptides, and amino acids is made possible by MSI's wide mass range, simple sample preparation, and lack of need for radioisotope or fluorescence labeling (Miura et al. 2012; Boughton et al. 2016). MSI is primarily divided into the following three categories based on the ionization mode (probe), which must be used in vacuum for secondary ion mass spectrometry (SIMS), desorption electrospray ionization (DESI), and matrix-assisted laser desorption ionization (MALDI).

Highly energetic primary ions, such as Ar+, Ga+, and In+, are used in SIMS to hit the sample surface. The primary ions penetrate the surface of the sample and generates a cascade of clashes with the molecules and atoms. As secondary ion kinetic energies rise above the energy at which they bond to the substrate, they are discharged off the surface. This normally happens at a depth of 10 Å and is size independent. Usually, SIMS ionizes and desorbs components and tiny molecules. Large-scale surface fragmentation causes the practical mass range to be constrained to m/z 1000 (Stevie et al. 1994; King 2003). There are drawbacks to developing an analytical platform for MSI, even though it is a cutting-edge technology that allows us to detect the distribution of exogenous or endogenous compounds in tissue. The invention and application of a new matrix are demanded in MALDI-MSI. Contrarily, MALDI only has a spatial resolution of 20 m, but DESI-MSI has a spatial resolution of around 200 m (Liu et al. 2021).

13.3.4 Gas Chromatography–Mass Spectrometry

The most commonly used technique for global metabolites profiling at the moment is GC-MS. Good separation ability, simple to use, and low cost are all benefits of GC-MS. It has a standard metabolites spectrum database that allows for the rapid and accurate qualitative analysis of metabolites and can analyze hundreds of components at once. Even though GC-MS has noticeable limitations: Although GC-MS is appropriate for non-thermosensitive and highly volatile molecules, it is not as appropriate for less volatile compounds and may change in certain compounds based to the efforts required for derivatives. Thermolabile metabolites are not captured by GC-MS, which is fundamentally biased towards non-volatile high-molecular weight metabolites and in favor of those that are volatile up to 250C (such as esters, alcohols, and monoterpenes) (Ellis et al. 2002; Coulier et al. 2006; Jeckel et al. 2022).

However, the best method for analyzing volatile chemicals is still high-resolution gas chromatography (HRGC). When combined with high-resolution mass spectrometry (HRMS), comprehensive two-dimensional gas chromatography (GC x GC) offers better peak capacity for target analyzation as compared to one-dimensional GC.

13.3.5 Liquid Chromatography–Mass Spectrometry

Liquid chromatography-mass spectrometry (LC-MS) is used in metabolomics to analyze and identify molecules similarly to GC-MS, but it overcomes the limitations of GC-MS. LC-MS is an essential technology in metabolomics research because it is ideal for metabolites with lesser volatility and low thermal durability. Hence, LC-MS-based analysis technique is practical for metabolomic analysis samples, particularly when using RP (reverse-phase) separation technology (Allwood and Goodacre 2010; Gika et al. 2019). Without any prior preparation, sections of the materials may be inserted straight into the chromatographic column. Middle- and low-polarity molecules may be analyzed using this reversed-phase gradient elution separation, while more polar substances (such amino acids and carbohydrates) can be identified using hydrophilic exchange chromatography (HILIC) (Jandera 2011).

Even though LC-MS has been used in several research, there are still some issues with metabolomics. As an example, an increased salt concentrations in the solvent impairs the ionization efficiency of ESI and influences the effectiveness and reproducibility of quantitative analysis. The matrix effect is a massive problem with LC-MS/MS analysis (Liu et al. 2021); hence it is crucial to eliminate or minimize this impact.

13.4 Metabolomics as a Tool for Signaling in Root Galls

More than $80 billion economic losses are reportedly caused each year by plant parasite nematodes, which seriously harm and reduce agricultural yields in a variety of crops worldwide. Several nematicides have now been prohibited or are being phased out due to health and environmental concerns in Europe and other countries of the globe (Atolani and Fabiyi 2020). To prevent damage to crops, we must concentrate on sustainable and alternative nematode management techniques. Plant roots produce and expel a diverse array of bioactive specialized metabolites, most of

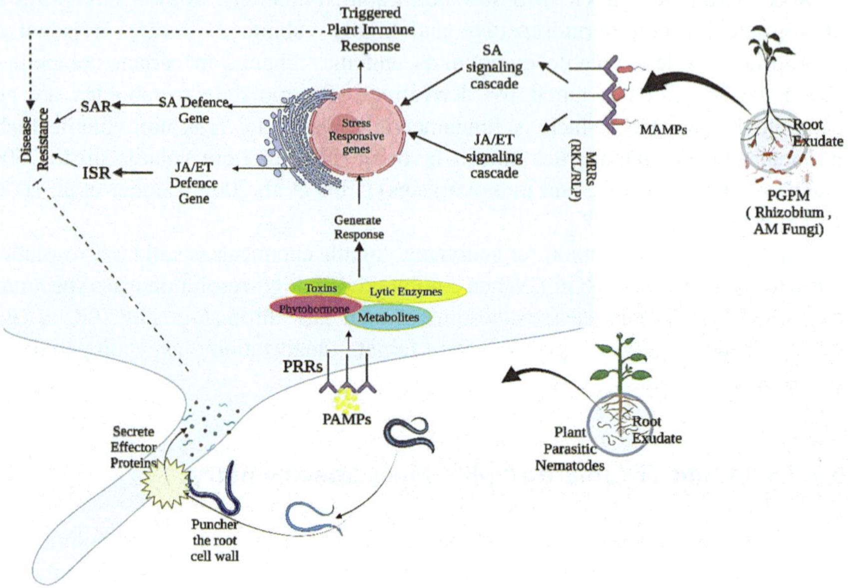

Fig. 13.3 Chemical signaling between PGPM to host plant, host plant to plant parasitic nematode, and plant-parasite nematode to host plant. PAMP/MAMP—pathogen/microbe-associated molecular pattern, PRR—pathogen recognition receptors, SA—salicylic acid, JA—jasmonic acid, ET—ethylene, SAR—systemic acquired resistance, ISR—induced systemic resistance

which are recognized as defensive chemicals. Root metabolites have nematode-attracting, nematode-repelling, nematode-stimulating, nematode-inhibiting properties. Hence, thorough knowledge of the root-mediated interaction between PGPM to plant parasitic nematodes, host plant to plant parasitic nematodes, and plant parasitic nematodes to host plant may help with effective pest nematode management (Figs. 13.1 and 13.3).

13.4.1 PGPM to Plant Parasitic Nematodes

Plant growth-promoting microbes (PGPM), a common class of microbe microbes, have a great deal of potential for use as biocontrolling agents against soil-borne pathogens like root-knot nematode. They provide the host plants with many essential functions such as the discharge of different phytohormones such as Indole-3-acetic acid (IAA), ethylene (ET), Abscisic acid (ABA), gibberellins (GA), brassinosteroids (BRs), salicylic acid (SA), and jasmonic acid (JA); enzymes like 1-aminocyclopropane-1-carboxylate(ACC)-deaminase, glucanases, chitinases, etc.; phosphate solubilization; nitrogen fixation; siderophore synthesis; and defense against many pathogenic microbes, especially PPNs.

The activity of PGPR is connected to secondary metabolite synthesis, defense-related genes expression, primary metabolite modifications, and cell wall reconfiguration (Mhlongo et al. 2018). Phytohormones are very well-known plant metabolites that play a role in various plant-defensive responses or plant priming stages. As an example, jasmonic acid and ethylene are key hormones in induced systematic resistance (ISR), while SA is the key hormone in the development of systemic acquired resistance (SAR) (Fig. 13.3) (Denancé et al. 2013; Uhrig et al. 2013).

By profiling the metabolism of soybean roots treated with *Bacillus simplex* and infested with soybean cyst nematodes, Kang et al. (2020) aimed to detect metabolic variations that could explain nematode resistance. He draws the conclusion that soybean roots treated with *B. simplex* had lower concentrations of sucrose, fructose, glucose, and maltose than control soybean roots, which reduced the nematode's food supply. Besides that, *B. simplex* treatment increased the levels of lactic acid, gluconic acid, melibiose, noradrenaline, and phytosphingosine in soybean roots, enhancing their nematocidal effect (Kang et al. 2020). The goal of Khanna et al. (2019)'s study was to identify metabolic changes that could explain nematode tolerance by evaluating the metabolism of tomato plant roots that have been treated with *Pseudomonas aeruginosa* and *Burkholderia gladioli* and infested with *Meloidogyne incognita*. He draws the conclusion that tomato roots treated with *P. aeruginosa* and *B. gladioli* show increase in the levels of phenolic compounds, flavonoids, anthocyanins, osmo-protectants, reducing sugars, free amino acids, trehalose, proline, glycine betaine, and organic acids (fumaric acid, succinic acid, citric acid, and malic acid) (Khanna et al. 2019).

13.4.2 Host Plant to Plant Parasitic Nematodes

Pre-penetration or post-penetration resistances to plant parasitic nematodes are the two categories. Pre-penetration resistance describes a condition in which a nematode cannot enter the host plant because, for example, there are no metabolites required for host identification and presence repellent exudates released by host plant, or there is a physical barrier that the nematode cannot cross. While in post-penetration resistance the PPN inserts the host but is subsequently unable to sustain or reproduce because, for example, toxic metabolites are present, or it is unable to feed (Desmedt et al. 2020).

Volatile organic compound, DMDS (dimethyl disulfide), glucosinolate, myrosinase, pyrrolizidine alkaloids (PAs), and benzoxazinoids, such as 2, 4-dihydroxy-7- methoxy-2H-1,4-benzoxazin-3(4H)-one (DIMBOA), are secondary metabolites in various species of the Liliaceae, Boraginaceae, Fabaceae, Convolvulaceae, Asteraceae, Orchidaceae, and Apocynaceae and toxic to different life cycle stages of the plant parasitic nematodes *Meloidogyne incognita, M. hapla, Pratylenchus penetrans, and Heterodera schachtii* (Sikder and Vestergård 2020).

The autoimmune responses known as pathogen/microbes-associated molecular pattern (PAMP/MAMP)-triggered immunity (MTI), which depend on the sensing of conserved microbial or pathogenic signature molecules (M/PAMPs) by extracellular transmembrane receptors or pathogen recognition receptors (PRRs), are used by plants to protect against pathogen entry or plant parasitic nematodes. Thus, these responses further activate defense signaling cascades which alternately provide resistance against plant parasitic nematode attack (Fig. 13.3).

13.4.3 Plant Parasitic Nematodes to Host Plant

A broad class of obligatory phytopathogenic pathogens known as plant parasitic nematodes (PPNs) release chemicals termed effectors that are the causative agents in parasitic infection. Effectors have developed to affect many components of host metabolism, morphology, development, physiology, and immunology to make a host sensitive (Eves-van den Akker et al. 2021). They are generally characterized by nematode-derived compounds (typically, but not entirely proteins) released into the host plant. "Giant cells" cover the root-knot nematode (RKN) feeding location. These cells are created from a small number of vascular root cells that repeatedly divide their nuclei without dividing into separate cells. These cells multiply to form polynuclei and may be up to 200 to 300 times bigger than typical cells (Palomares-Rius et al. 2017; Mejias et al. 2019). Giant cells are enveloped with dividing cells, and because of their hypertrophy and hyperplasia, a gall is a model organ that is generated.

Three esophageal salivary glands produced most of these parasitic nematode effectors, which are subsequently delivered to plant cells via a needle stylet. Developmental factors control the esophageal glands' activity. The two subventral glands (SvG) release effectors that permit J2 movement and allow penetration in the root, whereas SvG, especially the dorsal gland, secrete proteins during parasitism (DG). Certain effectors are also synthesized in some other secretory organs, like chemosensory amphids, or are released directly through the PPN cuticle. Molecular conversation research has mostly concentrated on excreted proteinaceous effectors, even though other secreted substances, such phytohormones, have been found to encourage similar interactions (Nguyen et al. 2018; Vieira and Gleason 2019). These plant parasitic nematode secreted effector molecules create hindrance to plant pathogen resistance (Fig. 13.3).

13.5 Chemical Signaling Via Secondary Metabolites

Plant-soil organism interaction is driven primarily by chemical signaling that occurs near the roots. For example, *Ditylenchus* destructor was found to be drawn to crude root exudates from sweet potatoes in in vitro experiments (Xu et al. 2015). A variety of chemicals are released by plant roots, and these chemicals play a role in luring beneficial organisms and creating mutualistic interactions in the rhizosphere. These mixtures comprise polysaccharides, sugars, aromatic, aliphatic, amino acids, fatty acids, sterol, and phenolic acids; in addition, they may also contain secondary metabolites such as plant growth regulators and enzymes. Signals may be produced at a distance from the differentiating feeding site or after plant cell infusion of secretory chemicals. Here, procambial cells around the nematode's head transform into "giant cells" in response to signals that arrive from the nematode. The endoparasite relies on these enormous, multinucleate, metabolically active cells as a constant food supply (Huang 1985). To assess the direction of the PPN; the chemoreceptors in the anterior receptors; the amphids; and, in certain PPNs, the posterior receptors, the phasmids, simultaneously examine these signals (Curtis 2008; Rasmann et al. 2012).

13.5.1 Siderophore Production in Rhizosphere

Release of various allelochemicals like volatile compounds, for instance, toxins, antibiotics, degrading enzymes, and siderophores that elicit the defense system of plants (Kumar et al. 2017). Siderophore-mediated iron uptake becomes crucial to several disease-causing causal organisms including phytonematodes because abscission of this system greatly decreases the ability of a pathogen to colonize a host (Viljoen et al. 2019). Plant growth-promoting rhizobacteria (PGPR) has the potency to be used as biological control agents (Paul and Lade 2014; Viljoen et al. 2019). PGPR play a vital role to the host plant by producing siderophore phytohormones against many soil-borne pathogen including plant parasitic nematodes (Glick 2014; Borah et al. 2018). Siderophores are low molecular weight less than 10 KD iron-chelating compound, which is produced under iron-limited conditions by several bacteria, viz., *Azotobacter*, *Pseudomonas*, *Enterobacter*, *Bacillus*, *Serratia*, *Rhizobium*, *Azospirillum*, and *Enterobacter* (Glick et al. 1999; Loper and Henkels 1999; Ali and Vidhale 2013). Lewin (1984) determined that siderophore makes a complex with free iron and delivers it inside the cell through membrane receptor fragments, and these fragments are encrypted by five genes in the operon that remain off in sufficient iron availability. One or more than one siderophores are produced by many bacteria, which are used by different microorganism for iron and other metal accretion; the specific attribute of siderophore is to raise their utilization in clinical, environmental, and agricultural field.

Siderophores produced by bacteria have various biological impacts on both host and pathogen, simultaneously helping pathogens to occupy iron and disrupt the host tissues like mitochondrial degradation and causing upregulation of immune genes and mitophagy (Wilson et al. 2016). Production of siderophore is beneficial to plants by direct supply of iron and reducing competitiveness of pathogen in soil-borne disease suppression including root-knot disease (Tank et al. 2012). Species of *Pseudomonas* release a signaling molecule called SA molecule under limited iron conditions, which routed to SA-presenting siderophores (Mercado-Blanco and Bakker 2007). Siderophores have nematicidal action and suppress the activity of nematodes (Antil et al. 2021).

Using MALDI-IMS analytical technique insight into microbial interactions would be easily detectable. This technique identifies a specific organism responsible for producing a particular metabolite of interest within different species interactions

Table 13.1 Different siderophores produced by variable microbes

Siderophore	Structural variation	Source microbes
Hydroxamate Ferrichrome Ferribactin Gonobactin Nocobactin	Hydroxamate group [C(=O)N-(OH) R] supply two O_2 molecules, the formation of bidentate ligand with iron, a hexadentate octahedral complex with iron	*Ustilago sphaerogena, Pseudomonas fluorescence, Neisseria gonorrhea* and *N. meningitidis*
Phenolates/ Catecholate Enterochelin Agrobactin and parabactin	Enterochelin is a trimester of 2, 3-dihydroxybenzoylserine, each catecholate group provides two O_2 atoms for iron chelation so that a hexadentate octahedral complex is formed, wine colored complex is formed with ferric chlorite ($FeCl_3$) that absorbs 495 nm	*Escherichia coli, Klebsiella pneumonia* and *Salmonella typhimurium* *Agrobacterium tumefaciens* and *Paracoccus denitrificans*
Carboxylate/ complexones Rhizobactin Staphyloferrin A	DM4 and an amino poly (COOH) with ethylenediaminedicarboxyl and hydroxycarboxyl moieties DSM20459, consist of two citric acid and one D ornithine residues associated by two amide bonds	*Rhizobium meliloti* *Staphylococcus hyicus*

(Stasulli and Shank 2016). Moree et al. (2012) studied that *Aspergillus fumigatus* was delivering the PCA secreted from *Pseudomonas aeruginosa* into 1-HP, followed to transform on 1-MP and phenazine-1-sulphate. This generation of 1-HP elicited the production of two siderophores from the species *A. fumigatus*. This biotransformation event (lack of IMS) is responsible to *P. aeruginosa* directly promoting *A. fumigatus* to release siderophores instead of *A. fumigatus* functionally auto-eliciting this response (Moree et al. 2012). The macrolide AZM (antibiotic azithromycin) affects metabolite production in *P. aeruginosa* when exposed at range below the threshold inhibitory concentration; this AZM ceased the biosynthesis of specialized metabolite by enhancing quorum sensing (Tateda et al. 2001; Nalca et al. 2006). Phelan et al. (2014) demonstrated that a single gene (involved in phenazine biosynthesis) disruption leads to global metabolic alterations in *P. aeruginosa* metabolites generations. These changes also affected interspecific interactions; the gene phzF2 mutant promoted *A. fumigatus* to raise the synthesis of a siderophore as compared to co-cultured with wild-type *Pseudomonas aeruginosa*.

The variability found in the structure of siderophores (three important siderophores, i.e., hydroxamate, complexones, and carboxylate) from one species to another, based on their iron-binding moieties (Ali and Vidhale 2013) (Table 13.1).

13.5.2 Oxylipins

Oxylipins are involved in acyl-homoserine lactones (AHLs) priming. Bacteria perform quorum sensing; AHLs are auto inducers, which stimulated callose accumulation and deposition of phenolics, SA, and oxylipins in most of the plant species

(Schenk et al. 2014; Schikora et al. 2016). Deposition of oxylipins in distal cells stimulated closure of stamata, therefore altering plant resistance towards bacterial and other pathogen invasion (Schenk et al. 2014). Oxylipins play a crucial role in plant defense mechanism (Mhlongo et al. 2018). Synthesis of functional phyto-oxylipins is proceed either through LOX (lipoxygenesis) which place an oxygen atom at the C9 or C13 position over lipid chain or by the non-functional protein synthesis of structurally the same phytoprostanes (Sattler et al. 2006).

13.5.3 Flavonoids Produced in Response to Nematode

All terrestrial plants include a diverse group of secondary metabolites with a carbon basis called flavonoids. The description of flavonoids from numerous plant species ranges over 10,000 different varieties. Flavonoids are phenylpropanoids synthesized from the shikimate and acetate routes by a cytosolic multienzyme complex tethered to the endoplasmic reticulum. This fact is used to classify flavonoid subgroups. Flavonoids are diphenyl propane-based (C3-C6-C3) (Petrussa et al. 2013). Based on their structural characteristics, flavonoid subgroups can be divided into the chalcones, flavones, flavonols, flavandiols, anthocyanins, condensed tannins, aurones, isoflavonoids, and pterocarpans (Winkel-Shirley 2001, 2002; Hassan and Mathesius 2012). Multiple signaling pathways are activated when a plant detects the Nod factors, which leads to the infection of root hairs and the production of nodules. Xanthones, vanillin, and isovanillin, which are linked to flavonoids, can likewise trigger NodD gene expression, although in much higher amounts (Cooper 2007). ABC transporters and multidrug and toxic compound extrusion (MATE) transporters may transport flavonoids into the rhizosphere in aglycone and glycosidic forms (Sugiyama et al. 2007; Badri et al. 2008). Flavonoids including coumestrol, glyceollin (specific to soybeans), formononetin, medicarpin, and flavonols are often linked to PPN defense components (e.g., kaempferol and quercetin). According to some studies, it has been found that flavonoid glycosides like medicarpin glucoside malonate and formononetin glucoside malonate are probably involved in defense (Cook et al. 1995). Jasmonic acid, salicylic acid, ethylene, auxin, and ROS cross-talks can induce flavonoids biosynthesis when PPNs cause mechanical damage and injury during feeding and penetration (Goverse and Smant 2014; Holbein et al. 2016). Only plants with nematodes in the shoot were found to produce flavonoids in their roots, suggesting that systemic signals may be what trigger infected plants to produce flavonoids; however, these systemic signals are still unknown (Edwards et al. 1995).

Flavonoids that build up at PPN feeding sites may have an impact on nematode fertility and fecundity by reducing egg production or skewing the male-to-female ratio because more females are generated under conditions of ample nourishment and vice versa (e.g., *Heterodera* and *Meloidogyne* spp.) (Grundler et al. 1991). Jones et al. (2007) found that transparent testa (tt) mutants of *Arabidopsis*, including tt4/tt6, tt4/tt5, and tt6, which are lacking steps of the flavonoid pathway, were more prone to contract an infection. Yet a comparable investigation by Wuyts et al. (2006) using the *M. incognita*-infected *Arabidopsis* flavonoid mutants tt3, tt4, tt5, and tt7 found that the flavonoid pathway defects had no impact on the number of adult females, egg masses, eggs, or juveniles. Flavonoids may control polar auxin transport to increase auxin accumulation in nematode feeding sites. Auxin efflux transporters PIN (Pin-formed) and PGP are known to be blocked by certain flavonoids, which are also known to prevent cell-to-cell polar auxin transfer (P-Glycoprotein) (Peer et al. 2004, Peer and Murphy 2007). Moreover, certain flavonoids can modify the activity of the enzyme IAA, which in turn affects the quantity of auxin (indoleacetic acid oxidase) (Stenlid 1963). For cell division, cell differentiation, cell wall loosening, and the development of new vascular tissue, both types of feeding sites require local auxin accumulation and redistribution (Balasubramanian and Rangaswami 1962; Karczmarek et al. 2004; Ng et al. 2015). Auxin is redistributed in feeding sites and surrounding cells by PIN protein localization. To promote auxin transport into giant cells and syncytia, for instance, the expression of PIN2 and PIN7 was reduced. Additionally, transcriptome and proteomic studies in roots with root-knot and cyst nematode infections showed a link between the expression of flavonoid genes and proteins and auxin-inducible genes and proteins. For instance, Oliveira et al. (2014) found that cowpea roots infected with cyst nematode roots had upregulated levels of PIN2 transcripts and various flavonoids 4–6 days after *M. incognita* inoculation, while Ithal et al. (2007) found that cowpea roots infected with cyst nematode roots had upregulated levels of chalcone flavone isomerase and an auxin-induced protein (such as chalcone synthase, chalcone isomerase, and isoflavone reductase). The stimulation of CHS1 and CHS2 (chalcone synthase, the first enzyme in flavonoid production), which occurs in root-knot nematode galls, was found to be associated by an augmented auxin response spatially and temporally after 120 h of inoculation. Flavonoids can play a range of roles during plant-nematode interactions by acting as protective chemicals or signals that directly or indirectly change nematode fitness at different life stages.

The survival of nematode eggs, nematode fertility, and nematode attraction to host roots have all been found to be impacted by flavonoids, according to numerous studies. Most of these investigations, however, need to be validated in plants and use plant hosts that have clear flavonoid mutations. In general, it appears that some flavonoids are stimulated during plant-nematode interactions, particularly in feeding sites. Also, there is proof that certain interactions lead to increased amounts of flavonoids, which may function as phytoalexins, being accumulated by nematode-resistant plant genotypes. However, it has been demonstrated that the lack of flavonoids in host plants does not hinder the development of sedentary PPN feeding sites. Hence, it seems more plausible that flavonoids have defensive rather than developmental regulatory roles in the interactions between plants and nematodes. Future studies might focus on figuring out how flavonoids affect worm behavior and survival directly, as well as on developing host plants that contain more flavonoids that act as phytoalexins to promote nematode resistance.

13.5.4 Volatile Organic Compounds

Nematodes use chemosensory perception to understand their surroundings. Root exudate signals are commonly used by plant parasitic nematodes to choose their preferred host (Birds 2004). There are many chemical gradients around physiologically active roots, and it is possible that some of these chemicals serve as "long distance attractants," helping nematodes move towards root-occupied soil volumes as opposed to "short distance attractants," which may aid nematodes in moving to specific host roots (Perry 2005). The infectious J2 larvae of the root-knot nematodes *Meloidogyne incognita* and *M. graminicola* travel the longest distance to less suitable hosts yet take the most direct route. This implies that specific root metabolites function as both attractants and repellents, influencing the nematodes' movement patterns to reach their perfect host (Reynolds et al. 2011).

Attractants

Volatile compounds act as far-reaching cues that help infective root-knot nematode J2 larvae find suitable hosts in their natural habitat. Water-soluble compounds function more locally as signals for signaling in the root region (Curtis et al. 2009). For instance, *M. incognita* can be detected by using plant volatile organic molecules to determine the location of hosts (Kihika et al. 2017). Even so, we still know very little about the molecules that nematodes use to attract their hosts, but new research has revealed few hosts attracted attractants (Table 13.2). Five substances, including [2-isopropyl-3-methoxypyrazine, 2-(methoxy)-3-(1-methylpropyl) pyrazine, tridecane, and a- and b-cedrene], were found in the volatiles released from the roots of both tomato and spinach, while an additional three substances—-3-carene, sabinene, and methyl salicylate—were unique to tomato roots. In bioassays, the compounds 2-isopropyl-3-methoxypyrazine and tridecane attracted *M. incognita* J2 larvae to spinach roots, but methyl salicylate was more alluring to the J2s than these two substances, and subsequent experiments supported this finding, showing that methyl salicylate makes tomato roots more alluring to *M. incognita* than spinach roots (Murungi et al. 2018).

In a similar way, methyl salicylate, pinene, limonene, tridecane, and 2-methoxy3-(1-methylpropyl)-pyrazine were the root volatiles from *Capsicum annum* that had the most positive chemotactic effects on infective *M. incognita* J2 larvae (Kihika et al. 2017). Hence, according to two research (Kihika et al. 2017; Murungi et al. 2018), the most significant volatile attractant of *M. incognita* in the investigated solanaceous plants is methyl salicylate. In a test, salicylic acid attracted *M. incognita*, but *Radopholus similis* was drawn to dopamine (Wuyts et al. 2006). We know very little about the substances to which cyst nematodes are attracted. Potato cyst nematode *Globodera pallida* J2 larvae were attracted to unknown volatile compounds in potato root exudates (Farnier et al. 2012). In a bioassay, the compounds ethephon, methyl jasmonate, salicylic acid, indole acetic acid, mannitol, and salicylic acid all positively affected *G. pallida* J2s chemotaxis (Fleming et al. 2017). In in vitro nematode infection studies on *Arabidopsis* mutants, the cyst nematode *Heterodera schachtii* was less attracted to and less likely to invade them

Table 13.2 Impact of root exudates on nematode mobility

Root exudates	Target nematode	Action	References
2-isopropyl-3-methoxypyrazine, tridecane	*M. incognita*	Attractant	Murungi et al. (2018)
Zeatin	*M. incognita*	Attractant	Kirwa et al. (2018)
Dopamine	*Radopholus similis*	Attractant	Wuyts et al. (2006)
Salicylic acid	*M. incognita*	Attractant	Wuyts et al. (2006)
Methyl salicylate	*M. incognita*	Attractant	Kihika et al. (2017)
Palmitic acid and linoleic acid	*M. incognita*	Repellent	Dong et al. (2018)
Isoamyl alcohol, 1-butanol	*M. incognita*	Attractant	Shivakumara et al. (2018)
Small lipophilic molecules	*M. incognita*	Repellent	Dutta et al. (2012)
Small lipophilic molecules	*M. incognita*	Repellent	Dutta et al. (2012)
p-coumaric acid, caffeic acid	*M. incognita*	Repellent	Wuyts et al. (2006)
Protocatechuic acid	*Radopholus similis*	Repellent and nematicidal	Wuyts et al. (2006)
Unknown volatile metabolites in root exudates	*Globodera pallida*	Attractants	Farnier et al. (2012)
Trans-cinnamic acid	*M. incognita*	Repellent	Fleming et al. (2017)
Salicylic acid, methyl jasmonate	*G. pallida*	Attractants	Fleming et al. (2017)
Metabolites of ethylene pathway	*Heterodera glycines*	Ethylene (ET)-synthesis inhibitor and ET-insensitive mutations attractant to cyst nematode	Hu et al. (2017)
Erucin	*M. incognita*	Nematicidal	Aissani et al. (2015)

than the wild-type plant (Escudero Martinez et al. 2019). Some of the root metabolites and their action towards nematode are enlisted in Table 13.2.

Repellent

A critical initial step in creating more efficient control strategies may be identifying the chemicals that deter plant parasitic nematodes. The second-stage juveniles of three root-knot nematodes, *Meloidogyne hapla*, *Meloidogyne javanica*, and *Meloidogyne incognita*, were incredibly drawn to the root tips of both tomato plants and barrel clover (*Medicago truncatula*). Nonetheless, ethylene signaling-deficient mutant roots attracted more nematodes than the wild type (Čepulytė et al. 2018). Like this, *M. hapla* was attracted to roots of *Arabidopsis* whose ethylene synthesis was suppressed but not those of mutants whose ethylene production was increased. A mutant tomato with insensitive roots to ethylene also had more attractive roots (Fudali et al. 2013). These examples imply that root-knot nematodes are typically repelled by either ethylene or ethylene-responsive pathways.

The effect of ethylene on cyst nematodes is less pronounced. *Heterodera glycines* were attracted to and penetrated the roots of plants whose ethylene synthesis was suppressed than untreated soybean and *Arabidopsis* roots. Conversely, the wild-type roots of *Arabidopsis* accessions were less appealing to *H. glycines* than the ethylene-insensitive mutants (Hu et al. 2017). Roots of the ethylene-overproducing *A. thaliana* mutant were more susceptible to the beet cyst nematode (*Heterodera schachtii*), whereas the ethylene-insensitive mutant was less susceptible (Wubben et al. 2001). Similar to this, plant roots treated with ethylene were more appealing to the soybean cyst nematode and acquired infection much more quickly, leading to a higher infection rate (Kammerhofer et al. 2015). Future research should therefore seek to determine whether the repellence of root-knot nematodes is controlled by ethylene directly or by other substances in ethylene-responsive pathways. Yet, many particular substances have only been shown to repel a single nematode taxon in a single plant species. Examining several plant metabolites that successfully repelled plant parasitic nematodes in testing without plants may be helpful. For instance, root-knot, cyst, and stubby root nematodes exhibited negative chemotaxis in response to thymol produced from *Capsicum annum* (pepper) roots, either alone or in combination with other root volatiles of *C. annum* (Kihika et al. 2017). Certain flavonoids could deter plant parasitic nematodes as well; however, the effect seems to depend more on the species in question. For instance, the flavonoids kaempferol, quercetin, and myricetin were repulsive to *Radopholus similis* and *Meloidogyne incognita* but not to *Pratylenchus penetrans*. Other flavonoids including luteolin, daidzein, and genistein repelled *R. similis* while having no effect on *M. incognita* and *P. penetrans* (Wuyts et al. 2006). Some of the root metabolites and their action towards nematode are enlisted in Table 13.2.

13.6 Chemical Signaling Via Phytohormones

In nature, plants face a wide variety of threats, including microorganisms and insects that can restrict their development or even kill them. The interactions between plants and microbes, as well as the development and growth of plants, are profoundly affected by phytohormones. *Meloidogyne* spp. successfully infect plants by forming feeding cells, which they use to affect cell development and alter defensive responses (Gheysen and Mitchum 2011; Ji et al. 2013). Root-knot nematodes (RKN) cause the apparently observable growth of root galls by inducing the development of "giant cells" inside the tissues of the root, through which they take plant metabolites for nourishment (Mantelin et al. 2017). Interactions between plants and RKNs are known to include a wide variety of phytohormones, including auxin, salicylic acid, jasmonic acid, brassinosteroids, ethylene, gibberellic acid, and abscisic acid. Novel metabolic pathways may also be produced in the host plants because of nematodes altering the production of vital elements for their own nutrition (Hofmann et al. 2010).

13.6.1 Auxin Sensitivity and Signaling in Nematode Feeding Sites

The vascular tissues of plants are triggered and regulated by hormonal inductive stimuli. It is recognized that plant hormones regulate plant growth and development, with transport-dependent auxin gradients initiating the production of plant organs (Benkova et al. 2003). Young leaf-produced auxin is the key signal directing vascular differentiation. Its fundamental regulating mechanisms and polar and non-polar transport routes are elucidated. Plant growth regulators have been linked to *Meloidogyne* species-induced gall development. Balasubramanian and Rangaswami (1962) were the pioneers in identifying IAA-like compounds in *M. javanica*-infected root extracts. Bird (1962) found that tomato root extracts stimulated the development of wheat coleoptiles, demonstrating the existence of action of the auxin. Myuge and Viglierchio (1975) demonstrated that IAA increased root growth and galling in *M. incognita*-parasitized tomato plants. Several research demonstrated that root galls had a greater concentration of auxin than uninfected root tissues (VIglierchio and Yu 1968; Kochba and Samish 1972). In plants, four to eight founder cells finally transform into giant cells (GCs) by RKNs (Jones and Payne 1978). These cells can reach sizes of up to 1 mm in diameter after a rapid expansion. GCs maintain identity throughout their entire lifespan. Root galls are produced by both GCs and the surrounding tissue and may be observed with the naked eye. Intriguingly, the size of root galls does not strictly correlate with the size of the GC or the quantity of other tissues within the structure.

Auxin has an important function in root growth of plants, where it is primarily involved in the cell division as well as the formation and maintenance of root meristems (De Smet et al. 2010). RKN stimulates the production of large cells in the plant root, and it is known that auxin accumulates at these nematode feeding sites (Kyndt et al. 2016). This hormone is delivered from the apical meristem sites to the root tip by basipetal transport involving transporter proteins involved in influx and efflux. The AUXIN RESISTANT 1 (AUX1) and LIKE AUX1 (LAX) transmembrane protein families regulate auxin inflow, whereas the PIN family members are crucial for auxin efflux (Kyndt et al. 2016). The auxin transport system, which includes plant roots, is controlled by the spatial and subcellular localization of these proteins (Wisniewska et al. 2006). In both the root and shoot tissues of plants, elevated auxin levels are found near the areas where organ primordia are first formed (Tanaka et al. 2006) (Fig. 13.4). Hutangura et al. (1999) examined the expression of the auxin-responsive promoter (GH3) fused to the gusA reporter gene in white clover (*Trifolium repens* cv. Haifa) during the induction of root galls by *M. javanica* to determine if nematode infection alters auxin distribution in

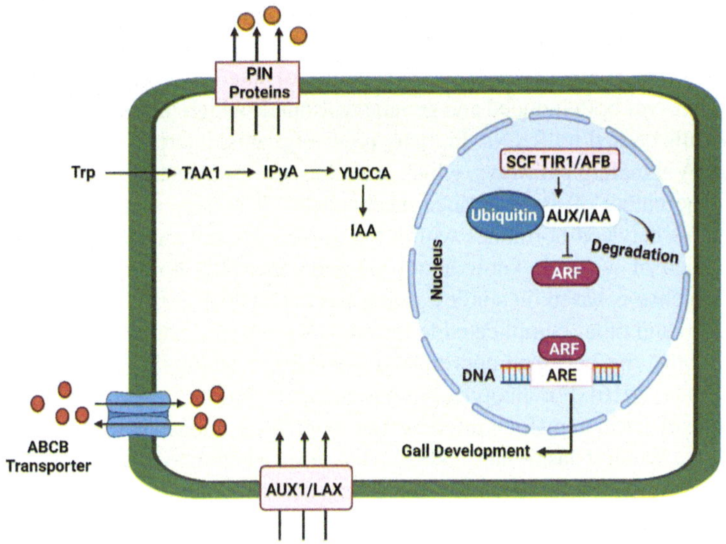

Fig. 13.4 A schematic presentation of biosynthesis, transport, and signaling pathway of auxin and role of auxin in the development of the galls. AUX/LAX—Influx Carrier; PIN Proteins-Efflux Carrier; ABCB Transporter—ATP-Binding Cassette B transporters; ARE—Auxin Responsive Element; ARF—Auxin Response Factors; Trp—Tryptophan; IAA—Indole Acetic Acid. (Modified by Zhang et al. 2022)

developing galls. Due to their ability to regulate auxin transport, flavonoids were investigated as a potential plant signal for mediating auxin localization shifts.

13.6.2 Jasmonic Acid (JA) and Salicylic Acid (SA)

The jasmonate family of chemicals is a known phytohormone that protects plants from nematodes, necrotrophic diseases, and a variety of abiotic stresses (Nahar et al. 2011). JA appears to have a significant role in all these activities, frequently associated with other phytohormones. Strigolactones (SLs) were first discovered as signaling molecules in the rhizosphere, but they have now been found to have a variety of roles throughout the plant (Cook et al. 1966; Akiyama et al. 2005; Umehara et al. 2008). The first identified SL, strigol, was obtained from root exudates of cotton and characterized as a seed germination stimulant for the root-parasitic plant *Striga lutea* (Cook et al. 1966).

The synthesis of phytohormones is coordinated across plants to activate defensive mechanisms. In order to prevent infection by RKNs that feed on living root tissues, called biotrophs, salicylic acid (SA)-dependent signaling is often induced in plants (Martinez-Medina et al. 2016). In contrast, signaling mediated by jasmonic acid (JA) is typically effective against necrotrophic pathogens and leaf-chewing insects that cause cellular damage in plants (Pieterse et al. 2009). Much research into plant signaling has focused on two phytohormones: salicylic (SA) and jasmonic (JA). This study applied the tomato (*Lycopersicon esculentum*) and the root-knot nematode (*M. incognita*) as a model system. The nature of interactions between JA and SA signals involves both induced and genetic resistance, which inhibits the development of root galls caused by RKNs. In many types of plants, interplay exists between the SA and JA signaling pathways, which often has an antagonistic effect (Pieterse et al. 2009). The nature of the resistance mechanisms elicited by a plant depends on the interactions between hormones within the immunological signaling network of the plant. Much of what we know about the ways in which hormone signals interact during defense is based on studies of leaf tissue (Lu et al. 2015). In the case of root nematodes and other complicated long-term parasitic associations, relatively little is known about the hormone-coordinated defensive reactions that occur (Martinez-Medina et al. 2016). For dicotyledons and monocotyledons, researchers have looked at the role of JA in nematode infection, although the evidence is few and sometimes conflicting. Methyl jasmonate (MeJA) has been shown to increase resistance to parasitic nematodes in previous studies on dicotyledonous plants, including the roots of spinach (*Spinacia oleracea*) and oat (*Avena sativa*) and the shoots of tomato. This may be due to an increase in the level of compounds that are toxic to nematodes, such as proteinase inhibitors, phytoecdysteroids (Soriano et al. 2004a, b; Cooper et al. 2005).

13.6.3 Strigolactones

Strigolactones (SLs), which operate as signaling molecules in the rhizosphere, are phytohormones that are secreted from roots. Initially discovered as signaling molecules in the rhizosphere, stigolactones (SLs) are plant hormones produced from carotenoid pigments (Cook et al. 1966; Akiyama et al. 2005; Umehara et al. 2008). The formation of carlactone is achieved through the sequential action of many enzymes, including the carotene isomerase (D27) and two carotenoid cleavage dioxygenases (CCD7 and CCD8) (Waters et al. 2017). The environment has a significant impact on SL biosynthesis because of their function as stress regulators in plants (Andreo-Jiménez et al. 2015).

SLs have an impact on the regulation of plant parasitic nematodes PPNs) in both positive and negative ways. Escudero Martinez et al. (2019) noticed that SLs influenced the nematode *H. schachtii*'s host attraction and root invasion in *Arabidopsis*. Lahari et al. (2019) also demonstrated that GR24 inhibited the typical accumulation of JA after nematode infection. Such results indicate that enhanced root-knot nematode susceptibility in rice requires SL signaling in rice; GR24 treatment restored the phosphate and nitrate deficiency-induced decline in lateral root density in WT, SL-biosynthesis mutants, but not in the SL-signaling mutant (Sun et al. 2014). SL signaling is crucial in shaping the root architecture of rice. The JA signaling pathway is widely established to be engaged in plant defense against PPNs.

In contrary to the previous studies favorable effect of SLs on PPNs infection, a negative effect has also been postulated. The authors noticed an additional drop in abscisic acid (ABA) levels, which controls nematode infection in a positive way (Nahar et al. 2012; Kammerhofer et al. 2015). Xu and co-workers claimed that ABA suppression was responsible for the increased resistance to PPNs mediated via SLs, rather than the JA route. Breeding plants with increased SLs production seems like a potential technique to limit this pest infection if SLs have a detrimental effect on PPN's efficiency and infestation. There is still little and unclear information available about the potential role of SLs in interactions between plants and PPNs.

13.7 Conclusion and Prospects

Biotrophic parasites known as plant parasitic nematodes, more specifically root-knot nematodes, have developed smart tactics to infest a variety of plant taxa. Here, the effective approaches considered for estimating metabolites include nuclear magnetic resonance, electrospray ionization, mass spectrometry imaging, chromatography, mass spectrometry, Fourier transform infrared spectroscopy, etc. As the host plant and nematodes interact, the chemical signaling is mediated by metabolites and phytohormones. They can perform a multitude of roles by behaving as defensive molecules or signals that both directly and indirectly alter nematode survival at various stages of development. Due to the complex nature of metabolites released by the microbes (mainly nematodes) and plants in the rhizosphere, identifying and probing them more precisely is challenging. To address these constraints, novel methods for examining soil root exudates are anticipated to be used in the upcoming decades.

References

Aissani N, Urgeghe PP, Oplos C, Saba M, Tocco G, Petretto GL, Eloh K, Menkissoglu-Spiroudi U, Ntalli N, Caboni P (2015) Nematicidal activity of the volatilome of Eruca sativa on *Meloidogyne incognita*. J Agric Food Chem 15:6120–6125

Akiyama K, Matsuzaki KI, Hayashi H (2005) Plant sesquiterpenes induce hyphal branching in arbuscular mycorrhizal fungi. Nature 435(7043):824–827

Ali SS, Vidhale NN (2013) Bacterial siderophore and their application: a review. Int J Curr Microbiol App Sci 2:303–312

Ali MA, Abbas A, Azeem F, Javed N, Bohlmann H (2015) Plant-nematode interactions: from genomics to metabolomics. Int J Agri Biol 1(6):17

Allwood JW, Goodacre R (2010) An introduction to liquid chromatography-mass spectrometry instrumentation applied in plant metabolomic analyses. Phytochem Anal 21(1):33–47

Allwood JW, Ellis DI, Goodacre R (2008) Metabolomic technologies and their application to the study of plants and plant-host interactions. Physiol Plant 132(2):117–135

Andreo-Jiménez B, Ruyter-Spira C, Bouwmeester H, López-Ráez JA (2015) Ecological relevance of strigolactones in nutrient uptake and other abiotic stresses, and in plant–microbe interactions below-ground. Plant Soil 394:1–19

Antil S, Kumar R, Pathak DV, Kumar A, Panwar A, Kumari A, Kumar V (2021) On the potential of Bacillus aryabhattai KMT-4 against *Meloidogyne javanica*. Egypt J Biol Pest Contl 31:1–9

Atolani O, Fabiyi OA (2020) Plant parasitic nematodes management through natural products: current progress and challenges. Management of phytonematodes: recent advances and future challenges, pp 297–315

Badri DV, Loyola-Vargas VM, Broeckling CD, De-la-Peña C, Jasinski M, Santelia D, Martinoia E, Sumner LW, Banta LM, Stermitz F, Vivanco JM (2008) Altered profile of secondary metabolites in the root exudates of Arabidopsis ATP-binding cassette transporter mutants. Plant Physiol 146:323–324

Balasubramanian M, Rangaswami G (1962) Presence of indole compounds in nematode galls. Nature 194:774–775

Bebber DP, Holmes T, Gurr SJ (2014) The global spread of crop pests and pathogens. Global Ecol Biogeog 23(12):1398–1407

Benkova E, Michniewicz M, Sauer M, Teichmann T, Seifertova D, Jurgens G, Friml J (2003) Local, efflux-dependent auxin gradients as a common module for plant organ formation. Cell 115(5): 591–602

Bharti SK, Roy R (2012) Quantitative ^{1}H NMR spectroscopy. TrAC. Trends Anal Chem 35:5–26

Bird AF (1962) The inducement of giant cells by *Meloidogyne javanica*. Nematology 8(1):1–10

Bird DM (2004) Signaling between nematodes and plants. Curr Opin Plant Biol 7:372–376

Borah B, Ahmed R, Hussain M, Phukon P, Wann SB, Sarmah DK, Bhau BS (2018) Suppression of root-knot disease in *Pogostemon cablin* caused by *Meloidogyne incognita* in a rhizobacteria mediated activation of phenylpropanoid pathway. Biol Control 119:43–50

Boughton BA, Thinagaran D, Sarabia D, Bacic A, Roessner U (2016) Mass spectrometry imaging for plant biology: a review. Phytochem Rev 15:445–488

Castro-Moretti FR, Gentzel IN, Mackey D, Alonso AP (2020) Metabolomics as an emerging tool for the study of plant-pathogen interactions. Meta 10(2):52

Čepulytė R, Danquah WB, Bruening G, Williamson VM (2018) Potent attractant for root-knot nematodes in exudates from seedling root tips of two host species. Sci Rep 8:10847

Chagas FO, Pessotti RC, Caraballo-Rodríguez AM, Pupo MT (2018) Chemical signaling involved in plant-microbe interactions. Chem Soc Rev 47(5):1652–1704

Cook CE, Whichard LP, Turner B, Wall ME, Egley GH (1966) Germination of witchweed (*Striga lutea* Lour.): isolation and properties of a potent stimulant. Science 154(3753):1189–1190

Cook R, Tiller SA, Mizen KA, Edwards R (1995) Isoflavonoid metabolism in resistant and susceptible cultivars of white clover infected with the stem nematode *Ditylenchus dipsaci*. J Plant Physiol 146:348–354

Cooper JE (2007) Early interactions between legumes and rhizobia: disclosing complexity in a molecular dialogue. J Appl Microbiol 103:1355–1365

Cooper WR, Jia L, Goggin L (2005) Effects of jasmonate-induced defenses on root-knot nematode infection of resistant and susceptible tomato cultivars. J Chem Ecol 31:1953–1967

Coulier L, Bas R, Jespersen S, Verheij E, van der Werf MJ, Hankemeier T (2006) Simultaneous quantitative analysis of metabolites using ion-pair liquid chromatography-electrospray ionization mass spectrometry. Anal Chem 78(18):6573–6582

Curtis RH (2008) Plant-nematode interactions: environmental signals detected by the nematode's chemosensory organs control changes in the surface cuticle and behaviour. Parasite 15(3): 310–316

Curtis RH, Robinson AF, Perry RN, Perry R, Moens M, Starr J (2009) Root-knot nematodes

De Koning JA, Hogenboom AC, Lacker T, Strohschein S, Albert K, Brinkman UT (1998) On-line trace enrichment in hyphenated liquid chromatography–nuclear magnetic resonance spectroscopy. J Chromatogr A 813(1):55–61

De Smet I, Lau S, VoB U, Vanneste S, Benjamins R, Rademacher EH, Beeckman T (2010) Bimodular auxin response controls organogenesis in Arabidopsis. Proc Natl Acad Sci 107(6): 2705–2710

Decraemer W, Hunt DJ (2006) Structure and classification. In: Plant nematology. CABI, Wallingford, pp 3–32

Denancé N, Sánchez-Vallet A, Goffner D, Molina A (2013) Disease resistance or growth: the role of plant hormones in balancing immune responses and fitness costs. Front Plant Sci 4:155

Desmedt W, Mangelinckx S, Kyndt T, Vanholme B (2020) A phytochemical perspective on plant defense against nematodes. Front Plant Sci 11:602079

Dong L, Li X, Huang C, Lu Q, Li B, Yao Y, Liu T, Zuo Y (2018) Reduced *Meloidogyne incognita* infection of tomato in the presence of castor and the involvement of fatty acids. Sci Horticul 237: 169–175

Dutta TK, Powers SJ, Gaur HS, Birkett M, Curtis RH (2012) Effect of small lipophilic molecules in tomato and rice root exudates on the behaviour of Meloidogyne incognita and *M. graminicola*. Nematology 14:309–320

Edwards R, Mizen T, Cook R (1995) Isoflavonoid conjugate accumulation in the roots of lucerne (Medicago sativa) seedlings following infection by the stem nematode (*Ditylenchus* dipsaci). Nematology 41:51–66

Ellis DI, Broadhurst D, Kell DB, Rowland JJ, Goodacre R (2002) Rapid and quantitative detection of the microbial spoilage of meat by Fourier transform infrared spectroscopy and machine learning. Appl Environ Microbiol 68(6):2822–2828

Escudero Martinez CM, Guarneri N, Overmars H, van Schaik C, Bouwmeester H, Ruyter-Spira C, Goverse A (2019) Distinct roles for strigolactones in cyst nematode parasitism of Arabidopsis roots. Eur J Plant Pathol 154:129–140

Eves-van den Akker S, Stojilković B, Gheysen G (2021) Recent applications of biotechnological approaches to elucidate the biology of plant-nematode interactions. Curr Opin Biotechnol 70: 122–130

Farnier K, Bengtsson M, Becher PG, Witzell J, Witzgall P, Manduríc S (2012) Novel bioassay demonstrates attraction of the white potato cyst nematode *Globodera pallida* (Stone) to non-volatile and volatile host plant cues. J Chem Ecol 38:795–801

Fiehn O (2001) Combining genomics, metabolome analysis, and biochemical modelling to understand metabolic networks. Comp Funct Genom 2:155–168

Fleming TR, Maule AG, Fleming CC (2017) Chemosensory responses of plant parasitic nematodes to selected phytochemicals reveal long-term habituation traits. J Nematol 49:462

Fudali SL, Wang C, Williamson VM (2013) Ethylene signaling pathway modulates attractiveness of host roots to the root-knot nematode *Meloidogyne hapla*. Mol Plant Microbe Inter 26:75–86

Fuller VL, Lilley CJ, Urwin PE (2008) Nematode resistance. New Phytol 180(1):27–44

Gheysen G, Mitchum MG (2011) How nematodes manipulate plant development pathways for infection. Curr Opin Plant Biol 14(4):415–421

Gika H, Virgiliou C, Theodoridis G, Plumb RS, Wilson ID (2019) Untargeted LC/MS-based metabolic phenotyping (metabonomics/metabolomics): the state of the art. J Chromatogr B Analyt Technol Biomed Life Sci 1117:136–147

Glick BR (2014) Bacteria with ACC deaminase can promote plant growth and help to feed the world. Microbiol Res 169:30–39

Glick BR, Patten CL, Holguin G, Penrose DM (1999) Biochemical and genetic mechanisms used by plant growth promoting bacteria. Imperial College Press, London

Goverse A, Smant G (2014) The activation and suppression of plant innate immunity by parasitic nematodes. Annu Rev Phytopathol 52:243–265

Grundler F, Betka M, Wyss U (1991) Influence of changes in the nurse cell system (syncytium) on sex determination and development of the cyst nematode Heterodera schachtii: total amounts of proteins and amino acids. Phytopathology 81(1):70–74

Hassan S, Mathesius U (2012) The role of flavonoids in root–rhizosphere signalling: opportunities and challenges for improving plant–microbe interactions. J Exp Bot 63(9):3429–3444

Heyman HM, Dubery IA (2016) The potential of mass spectrometry imaging in plant metabolomics: a review. Phytochem Rev 15:297–316

Hofmann J, El Ashry AE, Anwar S, Erban A, Kopka J, Grundler F (2010) Metabolic profiling reveals local and systemic responses of host plants to nematode parasitism. Plant J 62(6): 1058–1071

Holbein J, Grundler FM, Siddique S (2016) Plant basal resistance to nematodes: an update. J Exp Bot 67:2049–2061

Hu Y, You J, Li C, Williamson VM, Wang C (2017) Ethylene response pathway modulates attractiveness of plant roots to soybean cyst nematode *Heterodera glycines*. Sci Rep 7:1–3

Huang SC (1985) Formation, anatomy and physiology of giant cells induced by root-knot nematodes. In: Sasser JN, Carter CC (eds) An advanced treatise on *Meloidogyne*, Biology and control, vol I. North Carolina State University Graphics, Raleigh, pp 143–154

Hutangura P, Mathesius U, Jones MG, Rolfe BG (1999) Auxin induction is a trigger for root gall formation caused by root-knot nematodes in white clover and is associated with the activation of the flavonoid pathway. Func Plant Biol 26:221–231

Ithal N, Recknor J, Nettleton D, Hearne L, Maier T, Baum TJ, Mitchum MG (2007) Parallel genome-wide expression profiling of host and pathogen during soybean cyst nematode infection of soybean. Mol Plant-Microbe Inter 20:293–305

Jandera P (2011) Stationary and mobile phases in hydrophilic interaction chromatography: a review. Anal Chim Acta 692(1–2):1–25

Jeckel AM, Beran F, Züst T, Younkin G, Petschenka G, Pokharel P, Dreisbach D, Ganal-Vonarburg SC, Robert CAM (2022) Metabolization and sequestration of plant specialized metabolites in insect herbivores: current and emerging approaches. Front Physiol 13:1001032

Ji H, Gheysen G, Denil S, Lindsey K, Topping JF, Nahar K, Haegeman A, De Vos WH, Trooskens G, Van Criekinge W (2013) Transcriptional analysis through RNA sequencing of giant cells induced by *Meloidogyne graminicola* in rice roots. J Exp Bot 64:3885–3898

Jones MG, Payne HL (1978) Early stages of nematode-induced giant-cell formation in roots of *Impatiens balsamina*. J Nematol 10(1):70

Jones JT, Furlanetto C, Phillips MS (2007) The role of flavonoids produced in response to cyst nematode infection of Arabidopsis thaliana. Nematology 9(5):671-7

Jones JT, Haegeman A, Danchin EG, Gaur HS, Helder J, Jones MG, Kikuchi T, Manzanilla-López R, Palomares-Rius JE, Wesemael WM, Perry RN (2013) Top 10 plant-parasitic nematodes in molecular plant pathology. Mol Plant Pathol 14(9):946–961

Kammerhofer N, Radakovic Z, Regis JM, Dobrev P, Vankova R, Grundler FM, Siddique S, Hofmann J, Wieczorek K (2015) Role of stress-related hormones in plant defence during early infection of the cyst nematode *Heterodera schachtii* in Arabidopsis. New Phytol 207: 778–789

Kang WS, Chen LJ, Wang YY, Zhu XF, Liu XY, Fan HY, Duan YX (2020) Bacillus simplex treatment promotes soybean defence against soybean cyst nematodes: a metabolomics study using GC-MS. PLoS One 15(8):e0237194

Karczmarek A, Overmars H, Helder J, Goverse A (2004) Feeding cell development by cyst and root-knot nematodes involves a similar early, local and transient activation of a specific auxin-inducible promoter element. Mol Plant Pathol 5:343–36P

Kaur M, Sodhi H S (2022) Implication of microbial signals: plant communication. In Plant-microbe interactions (pp 41–57). CRC Press

Khanna K, Sharma A, Ohri P, Bhardwaj R, Abd Allah EF, Hashem A, Ahmad P (2019) Impact of plant growth promoting Rhizobacteria in the orchestration of *Lycopersicon esculentum* Mill. Resistance to plant parasitic nematodes: a metabolomic approach to evaluate defense responses under field conditions. Biomol Ther 9(11):676

Khanna K, Kohli SK, Ohri P, Bhardwaj R (2021) Plants-nematodes-microbes crosstalk within soil: a trade-off among friends or foes. Microbiol Res 248:126755

Kihika R, Murungi LK, Coyne D, Ng'ang'a M, Hassanali A, Teal PE, Torto B (2017) Parasitic nematode Meloidogyne incognita interactions with different *Capsicum annum* cultivars reveal the chemical constituents modulating root herbivory. Sci Rep 7:2903

King BV (2003) Sputter depth profiling. In: O'Connor DJ, Sexton BA, Smart RSC (eds) Surface analysis methods in materials science. Springer series in surface sciences, vol 23. Springer, Berlin, pp 107–125

Kirwa HK, Murungi LK, Beck JJ, Torto B (2018) Elicitation of differential responses in the root-knot nematode Meloidogyne incognita to tomato root exudate cytokinin, flavonoids, and alkaloids. J Agric Food Chem 66:11291–11300

Knights HE, Jorrin B, Haskett TL, Poole PS (2021) Deciphering bacterial mechanisms of root colonization. Environ Microbiol Rep 13(4):428–444

Kochba J, Samish RM (1972) Levels of endogenous cytokinins and auxin in roots of nematode resistant and susceptible peach rootstocks. J Am Soc Hortic Sci 97:115–119

Kumar D, Thakur NS, Gunaga RP (2017) Allelopathic influence of leaf aqueous extract and leaf litter of Indian lilac (*Melia azedarach* L.) on germination, growth, biomass and grain yield of green gram (*Vigna radiata* L.) and black chickpea (Cicer arietinum L.). Int J Curr Microbiol Appl Sci 6:2669–2683

Kyndt T, Goverse A, Haegeman A, Warmerdam S, Wanjau C, Jahani M, Gheysen G (2016) Redirection of auxin flow in *Arabidopsis thaliana* roots after infection by root-knot nematodes. J Exp Bot 67(15):4559–4570

Lahari Z, Ullah C, Kyndt T, Gershenzon J, Gheysen G (2019) Strigolactones enhance root-knot nematode (Meloidogyne graminicola) infection in rice by antagonizing the jasmonate pathway. New Phytol 224:454–465

Lewin R (1984) How microorganisms transport iron: in the midst of plenty, microorganisms are often in danger of iron-starvation; the mechanism by which they transport iron has now been elucidated. Science 225:401–402

Li J, Wang C, Liang W, Liu S (2021) Rhizosphere microbiome: the emerging barrier in plant-pathogen interactions. Front Microbiol 12:772420

Liu X, Locasale JW (2017) Metabolomics: a primer. Trends Biochem Sci 42(4):274–284

Liu R, Bao ZX, Zhao PJ, Li GH (2021) Advances in the study of metabolomics and metabolites in some species interactions. Molecules 26(11):3311

Loper JE, Henkels MD (1999) Utilization of heterologous siderophores enhances levels of iron available to Pseudomonas putida in the rhizosphere. Appl Environ Microbiol 65(12):5357–5363

Lu Q, Zhang W, Gao J, Lu M, Zhang L, Li J (2015) Simultaneous determination of plant hormones in peach based on dispersive liquid–liquid microextraction coupled with liquid chromatography–ion trap mass spectrometry. J Chromatogr B 992:8–13

Mantelin S, Bellafiore S, Kyndt T (2017) Meloidogyne graminicola: a major threat to rice agriculture. Molecular Plant Pathology 18(1):3

Martinez-Medina A, Pozo MJ, Cammue BP, Vos CM (2016) Belowground defence strategies in plants: the plant–*Trichoderma* dialogue. Belowground defence strategies in plants, 301–327

Mejias J, Truong NM, Abad P, Favery B, Quentin M (2019) Plant proteins and processes targeted by parasitic nematode effectors. Front Plant Sci 10:970

Mercado-Blanco J, Bakker PA (2007) Interactions between plants and beneficial *Pseudomonas* spp.: exploiting bacterial traits for crop protection. Antonie Van Leeuwenhoek 92:367–389

Mhlongo MI, Piater LA, Madala NE, Labuschagne N, Dubery IA (2018) The chemistry of plant-microbe interactions in the rhizosphere and the potential for metabolomics to reveal signaling related to defense priming and induced systemic resistance. Front Plant Sci 9:112

Miura D, Fujimura Y, Wariishi H (2012) In situ metabolomic mass spectrometry imaging: recent advances and difficulties. J Proteome 75(16):5052–5060

Moree WJ, Phelan VV, Wu CH, Bandeira N, Cornett DS, Duggan BM, Dorrestein PC (2012) Interkingdom metabolic transformations captured by microbial imaging mass spectrometry. Proc Natl Acad Sci 109:13811–13816

Murungi LK, Kirwa H, Coyne D, Teal PE, Beck JJ, Torto B (2018) Identification of key root volatiles signaling preference of tomato over spinach by the root knot nematode Meloidogyne incognita. J Agric Food Chem 66:7328–7336

Myuge SG, Viglierchio DR (1975) Influence of growth promoters and inhibitors on tomato plants infected with *Meloidogyne incognita* and *M. hapla*. Nematology 21:476–477

Nahar K, Kyndt T, De Vleesschauwer D, Höfte M, Gheysen G (2011) The jasmonate pathway is a key player in systemically induced defense against root knot nematodes in rice. Plant Physiol 157(1):305–316

Nahar K, Kyndt T, Nzogela YB, Gheysen G (2012) Abscisic acid interacts antagonistically with classical defense pathways in rice–migratory nematode interaction. New Phytol 196:901–913

Nalca Y, Jansch L, Bredenbruch F, Geffers R, Haussler BJ (2006) Quorum-sensing antagonistic activities of azithromycin in *Pseudomonas aeruginosa* PAO1: a global approach. Antimicrob Agents Chemother 50:1680–1688

Ng JL, Perrine-Walker F, Wasson AP, Mathesius UT (2015) The control of auxin transport in parasitic and symbiotic root–microbe interactions. Plan Theory 4:606–643

Nguyen CN, Perfus-Barbeoch L, Quentin M, Zhao J, Magliano M, Marteu N, Da Rocha M, Nottet N, Abad P, Favery B (2018) A root-knot nematode small glycine and cysteine-rich secreted effector, MiSGCR1, is involved in plant parasitism. New Phytol 217(2):687–699

Nicol JM, Turner SJ, Coyne DL, Nijs LD, Hockland S, Maafi ZT (2011) Current nematode threats to world agriculture. Genomics and molecular genetics of plant-nematode interactions 21–43

Nihorimbere V, Ongena M, Smargiassi M, Thonart P (2011) Beneficial effect of the rhizosphere microbial community for plant growth and health. Biotechnol Agron Soc Environ 15(2)

Oliveira JT, Araujo-Filho JH, Grangeiro TB, Gondim DM, Segalin J, Pinto PM, Carlini CR, Silva FD, Lobo MD, Costa JH, Vasconcelos IM (2014) Enhanced synthesis of antioxidant enzymes, defense proteins and leghemoglobin in rhizobium-free cowpea roots after challenging with *Meloydogine incognita*. Proteomes 2:527–549

Palomares-Rius JE, Escobar C, Cabrera J, Vovlas A, Castillo P (2017) Anatomical alterations in plant tissues induced by plant parasitic nematodes. Front Plant Sci 8:1987

Paul D, Lade H (2014) Plant-growth-promoting rhizobacteria to improve crop growth in saline soils: a review. Agron Sustain Dev 34:737–752

Peer WA, Murphy AS (2007) Flavonoids and auxin transport: modulators or regulators? Trends Plant Sci 12:556–563

Peer WA, Bandyopadhyay A, Blakeslee JJ, Makam SN, Chen RJ, Masson PH, Murphy AS (2004) Variation in expression and protein localization of the PIN family of auxin efflux facilitator proteins in flavonoid mutants with altered auxin transport in Arabidopsis thaliana, Plant Cell. 16: 1898–1911

Perry RN (2005) An evaluation of types of attractants enabling plant parasitic nematodes to locate plant roots. Russian J Nematol 13:83

Petrussa E, Braidot E, Zancani M, Peresson C, Bertolini A, Patui S, Vianello A (2013) Plant flavonoids—biosynthesis, transport and involvement in stress responses. Int J Mol Sci 14: 14950–14973

Phelan VV, Moree WJ, Aguilar J, Cornett DS, Koumoutsi A, Noble SM, Dorrestein PC (2014) Impact of a transposon insertion in phzF2 on the specialized metabolite production and interkingdom interactions of *Pseudomonas aeruginosa*. J Bacteriol 196:1683–1693

Pieterse CM, Leon-Reyes A, Van der Ent S, Van Wees SC (2009) Networking by small-molecule hormones in plant immunity. Nat Chem Biol 5(5):308–316

Pott DM, Osorio S, Vallarino JG (2019) From central to specialized metabolism: an overview of some secondary compounds derived from the primary metabolism for their role in conferring nutritional and organoleptic characteristics to fruit. Front Plant 10:835

Rasmann S, Ali JG, Helder J, van der Putten WH (2012) Ecology and evolution of soil nematode chemotaxis. J Chem Ecol 38:615–628

Ren JL, Zhang AH, Kong L, Wang XJ (2018) Advances in mass spectrometry-based metabolomics for investigation of metabolites. RSC Adv 8(40):22335–22350

Reynolds AM, Dutta TK, Curtis RH, Powers SJ, Gaur HS, Kerry BR (2011) Chemotaxis can take plant parasitic nematodes to the source of a chemo-attractant via the shortest possible routes. J R Soc Interface 8:568–577

Rolli E, Marasco R, Vigani G, Ettoumi B, Mapelli F, Deangelis ML, Gandolfi C, Casati E, Previtali F, Gerbino R, Pierotti Cei F, Borin S, Sorlini C, Zocchi G, Daffonchio D (2015)

Improved plant resistance to drought is promoted by the root-associated microbiome as a water stress-dependent trait. Environ Microbiol 17(2):316–331

Ryan D, Robards K (2006) Metabolomics: the greatest omics of them all? Anal Chem 78:7954–7958

Sasse J, Martinoia E, Northen T (2018) Feed your friends: do plant exudates shape the root microbiome? Trends Plant Sci 23(1):25–41

Sattler SE, Mene-Saffrane L, Farmer EE, Krischke M, Mueller MJ, DellaPenna D (2006) Nonenzymatic lipid peroxidation reprograms gene expression and activates defense markers in Arabidopsis tocopherol-deficient mutants. Plant Cell 18:3706–3720

Schenk ST, Hernández-Reyes C, Samans B, Stein E, Neumann C, Schikora M, Schikora A (2014) N-acyl-homoserine lactone primes plants for cell wall reinforcement and induces resistance to bacterial pathogens via the salicylic acid/oxylipin pathway. Plant Cell 26:2708–2723

Schikora A, Schenk ST, Hartmann A (2016) Beneficial effects of bacteria-plant communication based on quorum sensing molecules of the N-acyl homoserine lactone group. Plant Mol Biol 90: 605–612

Shafi A, Zahoor I, Habib H (2021) Omics technologies to unravel plant-microbe interactions. In: Pirzadah TB, Malik B, Hakeem KR (eds) Plant-microbe dynamics: recent advances for sustainable agriculture. CRC Press, Boca Raton, pp 201–220

Shivakumara TN, Dutta TK, Rao U (2018) A novel in vitro chemotaxis bioassay to assess the response of towards various test compounds. J Nematol 50:487–494

Sikder MM, Vestergård M (2020) Impacts of root metabolites on soil nematodes. Front Plant Sci 10:1792

Singh S, Singh B, Singh AP (2015) Nematodes: a threat to sustainability of agriculture. Procedia Environ Sci 29:215–216

Soriano IR, Asenstorfer RE, Schmidt O, Riley IT (2004a) Inducible flavone in oats (*Avena sativa*) is a novel defense against plant parasitic nematodes. Phytopathology 94(11):1207–1214

Soriano IR, Riley IT, Potter MJ, Bowers WS (2004b) Phytoecdysteroids: a novel defense against plant parasitic nematodes. J Chem Ecol 30:1885–1899

Stasulli NM, Shank EA (2016) Profiling the metabolic signals involved in chemical communication between microbes using imaging mass spectrometry. FEMS Microbiol Rev 40:807–813

Stenlid G (1963) The effects of flavonoid compounds on oxidative phosphorylation and on the enzymatic destruction of indoleacetic acid. Physiol Plantar 16:110–120

Stevie FA, Wilson RG, Simons DS, Current MI, Zalm PC (1994) Review of secondary ion mass spectrometry characterization of contamination associated with ion implantation. J Vac Sci Technol B 12(4):2263–2279

Sugiyama A, Shitan N, Yazaki K (2007) Involvement of a soybean ATP-binding cassette-type transporter in the secretion of genistein, a signal flavonoid in legume-rhizobium symbiosis. Plant Physiol 144:2000–2008

Sun H, Tao J, Liu S, Huang S, Chen S, Xie X, Yoneyama K, Zhang Y, Xu G (2014) Strigolactones are involved in phosphate-and nitrate-deficiency-induced root development and auxin transport in rice. J Exp Bot 65(22):6735–6746

Syed Ab Rahman SF, Singh E, Pieterse CMJ, Schenk PM (2018) Emerging microbial biocontrol strategies for plant pathogens. Plant Sci 267:102–111

Tanaka H, Dhonukshe P, Brewer PB, Friml J (2006) Spatiotemporal asymmetric auxin distribution: a means to coordinate plant development. Cell Mol Life Sci 63:2738–2754

Tank N, Rajendran N, Patel B, Saraf M (2012) Evaluation and biochemical characterization of a distinctive pyoverdin from a *Pseudomonas* isolated from chickpea rhizosphere. Brazilian J Microbiol 43:639–648

Tateda K, Comte R, Pechere JC, Köhler T, Yamaguchi K, Van Delden C (2001) Azithromycin inhibits quorum sensing in *Pseudomonas aeruginosa*. Antimicrob Agents Chemother 45:1930–1933

Tsugawa H (2018) Advances in computational metabolomics and databases deepen the understanding of metabolisms. Curr Opin Biotechnol 54:10–17

Uhrig RG, Labandera AM, Moorhead GB (2013) Arabidopsis PPP family of serine/threonine protein phosphatases: many targets but few engines. Trends Plant Sci 18(9):505–513

Umehara M, Hanada A, Yoshida S, Akiyama K, Arite T, Takeda-Kamiya N, Yamaguchi S (2008) Inhibition of shoot branching by new terpenoid plant hormones. Nature 455(7210):195–200

van Dam NM, Bouwmeester HJ (2016) Metabolomics in the rhizosphere: tapping into belowground chemical communication. Trends Plant Sci 21(3):256–265

Vieira P, Gleason C (2019) Plant-parasitic nematode effectors - insights into their diversity and new tools for their identification. Curr Opin Plant Biol 50:37–43

Viljoen JJ, Labuschagne N, Fourie H, Sikora RA (2019) Biological control of the root-knot nematode *Meloidogyne incognita* on tomatoes and carrots by plant growth-promoting rhizobacteria. Trop Plant Pathol 44:284–291

Vlglierchio DR, Yu PK (1968) Plant growth substances and plant parasitic nematodes. II. Host influence on auxin content. Exp Parasitol 23:88–95

Waters MT, Gutjahr C, Bennett T, Nelson DC (2017) Strigolactone signaling and evolution. Annu Rev Plant Biol 68:291–322

Weckwerth W (2003) Metabolomics in systems biology. Ann Rev Plant Biol 54(1):669–689

Weckwerth W (2011) Unpredictability of metabolism—the key role of metabolomics science in combination with next-generation genome sequencing. Anal Bioanal Chem 400:1967–1978

Wilson BR, Bogdan AR, Miyazawa M, Hashimoto K, Tsuji Y (2016) Siderophores in iron metabolism: from mechanism to therapy potential. Trends Mol Med 22:1077–1090

Winkel-Shirley B (2001) Flavonoid biosynthesis. A colorful model for genetics, biochemistry, cell biology, and biotechnology. Plant Physiol 126(2):485–493

Winkel-Shirley B (2002) Biosynthesis of flavonoids and effects of stress. Current Opinion in Plant Biology 5(3):218–223

Wisniewska J, Xu J, Seifertová D, Brewer PB, Ruzicka K, Blilou I, Friml J (2006) Polar PIN localization directs auxin flow in plants. Science 312:883–883

Wubben MJ, Su H, Rodermel SR, Baum TJ (2001) Susceptibility to the sugar beet cyst nematode is modulated by ethylene signal transduction in *Arabidopsis thaliana*. Mol Plant-Microbe Inter 14: 1206–1212

Wuyts N, Swennen R, De Waele D (2006) Effects of plant phenylpropanoid pathway products and selected terpenoids and alkaloids on the behaviour of the plant-parasitic nematodes *Radopholus similis*, *Pratylenchus penetrans* and *Meloidogyne incognita*. Nematology 8:89–101

Xu Z, Zhao YQ, Yang DJ, Sun HJ, Zhang CL, Xie YP (2015) Attractant and repellent effects of sweet potato root exudates on the potato rot nematode, Ditylenchus destructor. Nematology 17(1):117–124

Yin C, Casa Vargas JM, Schlatter DC, Hagerty CH, Hulbert SH, Paulitz TC (2021) Rhizosphere community selection reveals bacteria associated with reduced root disease. Microbiome 9(1):86

Zeng C, Lin H, Liu Z, Liu Z (2020) Metabolomics analysis of Camellia sinensis with respect to harvesting time. Food Res Int 128:108814

Zinovieva SV (2014) Co-adaptation mechanisms in plant-nematode systems. Parazitologiia 48(2): 110–130

Springer Nature

GPSR Compliance

The European Union's (EU) General Product Safety Regulation (GPSR) is a set of rules that requires consumer products to be safe and our obligations to ensure this.

If you have any concerns about our products, you can contact us on ProductSafety@springernature.com

In case Publisher is established outside the EU, the EU authorized representative is:

Springer Nature Customer Service Center GmbH
Europaplatz 3
69115 Heidelberg, Germany

The manufacturer's authorised representative in the EU is Springer Nature Customer Service Centre GmbH, Europaplatz 3, 69115 Heidelberg, Germany. If you have any concerns regarding our products, please contact ProductSafety@springernature.com

Printed and bound by CPI Group (UK) Ltd, Croydon, CR0 4YY
01/10/2026
02225900-0012